Lecture Notes in Electrical

C000256173

Volume 399

About this Series

"Lecture Notes in Electrical Engineering (LNEE)" is a book series which reports the latest research and developments in Electrical Engineering, namely:

- Communication, Networks, and Information Theory
- Computer Engineering
- Signal, Image, Speech and Information Processing
- Circuits and Systems
- Bioengineering

LNEE publishes authored monographs and contributed volumes which present cutting edge research information as well as new perspectives on classical fields, while maintaining Springer's high standards of academic excellence. Also considered for publication are lecture materials, proceedings, and other related materials of exceptionally high quality and interest. The subject matter should be original and timely, reporting the latest research and developments in all areas of electrical engineering.

The audience for the books in LNEE consists of advanced level students, researchers, and industry professionals working at the forefront of their fields. Much like Springer's other Lecture Notes series, LNEE will be distributed through Springer's print and electronic publishing channels.

More information about this series at http://www.springer.com/series/7818

Canjun Yang · G.S. Virk · Huayong Yang
Editors

Wearable Sensors and Robots

Proceedings of International Conference
on Wearable Sensors and Robots 2015

Editors
Canjun Yang
Department of Mechanical Engineering
Zhejiang University
Hangzhou, Zhejiang
China

Huayong Yang
Department of Mechanical Engineering
Zhejiang University
Hangzhou, Zhejiang
China

G.S. Virk
University of Gävle
Gävle
Sweden

ISSN 1876-1100   ISSN 1876-1119   (electronic)
Lecture Notes in Electrical Engineering
ISBN 978-981-10-9607-5   ISBN 978-981-10-2404-7   (eBook)
DOI 10.1007/978-981-10-2404-7

Jointly published with Zhejiang University Press, Hangzhou, China

Printed on acid-free paper

This Springer imprint is published by Springer Nature
The registered company is Springer Science+Business Media Singapore Pte Ltd.

Preface

The International Conference on Wearable Sensors and Robots (ICWSR 2015) held during October 16–18, 2015 in Hangzhou, China. ICWSR 2015 was sponsored by Zhejiang University, and co-sponsored by the National Natural Science Fund of China (NSFC), and International Organisation for Standardisation's working group on personal care robot safety (ISO/TC184/SC2/WG7).

With rapid progress in mechatronics and robotics, wearable sensing and robotic technologies have been widely studied for various applications including exoskeleton robots for rehabilitation, exoskeleton robots for supporting the daily lives of elderly people, wearable medical devices for monitoring vital signs, etc. However, some key technology challenges need to be addressed for achieving better research results, more effective application demonstrators and realistic commercialization. The conference brought together academics, researchers, engineers, and students worldwide to focus on and discuss the state of the art of the technology and to present the latest results on the various aspects of wearable sensors and robots.

The conference received 61 papers from experts and researchers in China and all over the world. 46 papers were reviewed and accepted, including 20 invited papers and 26 general papers. Meanwhile, the conference received 11 keynote speech abstracts from international professors and researchers. The proceedings consist of detailed papers on wearable sensors, design of sensors and actuators, advanced control systems, wearable robots, visual recognition applications, clinical applications, rehabilitation robotics, biological signal based robotics, intelligent manufacturing and industry robots, and research progress from keynote speakers. In addition, readers will obtain the latest information on medical device regulation and international standardization, wearable robots for training and support of human gait, design of exoskeleton for elderly persons, ergonomics design considerations driving innovation in assistive robotics, and analysis of human–machine interaction.

It is our desire that the proceedings of the International Conference on Wearable Sensors and Robots (ICWSR 2015) will provide an opportunity to share the perspectives of academic researchers and practical engineers on wearable sensors and robot research and development.

Hangzhou, China   Prof. Canjun Yang
Gävle, Sweden   Prof. G.S. Virk
October 2015   Program Chair, General Chair of ICWSR 2015

Contents

Part I
Wearable Sensors

The Design of E Glove Hand Function Evaluation Device Based on Fusion of Vision and Touch

Jing Guo, Cui-lian Zhao, Yu Li, Lin-hui Luo and Kun-feng Zhang

Abstract This paper presents an E glove hand function evaluation device based on visual and haptic fusion, and uses the Principal Component Analysis (PCA) algorithm to establish hand sensor distribution model. The PCA analysis chart shows that three sensors distributed on the thumb, forefinger, and middle finger could effectively estimate the grasp motions. Moreover, threshold values for all category models can be selected by the way of adaptive pressure threshold integrating visual aid. At last, five subjects dressed E glove judging the grasp motions under different combinations of sensors. The results show that: the classification accuracy rate depended on the pressure and visual sensor fusion method reached 94 %; the identification rate of the adaptive pressure threshold method to judge the grasp motions can be increased 1.6–1.7 times than only using single camera vision sensor or pressure sensor. Next step, the E glove hand function evaluation device will be further improved such as function of active control to the collected data will be added.

Keywords E glove · PCA · Pressure sensing · Visual and touch fusion

1 Introduction

Hands are primarily responsible for the sophisticated activities and work in motor function, its degree of flexibility and movement accuracy are closely related to human activities of daily living, quality of life, and social activities. In medical rehabilitation, stroke patient hand is with motor dysfunction, and Parkinson's disease patients with hand tremors and slow movement will lead to the hand motor

---

J. Guo (✉) · C. Zhao · Y. Li · L. Luo · K. Zhang
Shanghai Key Laboratory of Intelligent Manufacturing and Robotics,
School of Mechatronic Engineering and Automation, Shanghai University,
Shanghai 200072, China
e-mail: andy_guojing421@163.com

© Zhejiang University Press and Springer Science+Business Media Singapore 2017
C. Yang et al. (eds.), *Wearable Sensors and Robots*, Lecture Notes in Electrical
Engineering 399, DOI 10.1007/978-981-10-2404-7_1

dysfunction. i; Hand dysfunction becomes a difficult problem in rehabilitation training and medical evaluation at present (Gabriele et al. 2009).

So far, there is no unified standard for motor dysfunction in international evaluation; each method owns its emphasis, and there have not yet been a more perfect and accurate evaluation method (Meng et al. 2013; Zampieri et al. 2011). Action Research Arm Test (ARAT) is one of the commonly used test evaluation methods of hand movement function (Lyle et al. 1981). Compared with other commonly used evaluation methods, ARAT pays more attention to comprehensive hand function in daily life, and classify and quantify the type and size of grasping object. In recent years, domestic and foreign scholars have verified the reliability and validity of ARAT by using clinical case (Weng et al. 2007; Yozbatiran et al. 2008). In view of the complex hand diverse sports demand, a lot of researches have been committed to design better wearable devices and data processing methods at home and abroad. Nathan et al. (2009) designed a wearable data glove applied in auxiliary rehabilitation training in patients; angle sensors associated with the hand acquire grasp-aperture prediction model to calculate the distance between thumb and forefinger point, then the hand grasping state was defined with the distance; although the data glove device has high accuracy and stability, the data glove device must be equipped with the Activities of Daily Living Exercise Robot (ADLER) system which is huge and thus with great limitations. Jakel et al. (2010a, b, 2012) and Palm et al. (2010, 2014) and Skoglund et al. (2010) designed a wearable data glove applied in controlling mechanical arm, which utilizes pressure data of the pressure sensors, position, velocity, and acceleration data of marks on the glove to define the hand grasping state, and then control manipulator grasping; while the data glove device can reach a high level in stability and accuracy and real-time performance, 5–6 sensors and 6–8 sets of marks or even more on glove device make information processing complex and data glove with mechanical auxiliary device reduces the flexibility and practicability; besides, sole grasp pressure threshold cannot adapt to grasp kinds of project. Liang et al. (2013) and Han et al. (2012) designed an electronic nose detection device, by acquiring reasonable information through reasonably designing the number and distribution of sensors to reduce the information redundancy; but there are many sensors, and it need to reduce the number of sensors. E glove in this paper is a glove mounted with three sensors and cloth glove without mechanism. Differences from traditional data glove used to acquire sensing information, E glove can implement active control function through dealing with the data collected in the future.

Wearable data glove device applied in hand function evaluation possesses complex structure and low flexibility and practicability, aiming at motivating patients' active movement function, wearable pressure sensing data glove device based on fusion of vision data and touch data is developed in this paper. Pressure data from sensors and marks data from the glove are acquired for evaluating the hand grasping state. Combined with the actual grasp function, sensors, and marks on this wearable pressure sensing data glove device can accomplish tracking and evaluation at the cost of lesser sensors and marks; moreover, visual feedback technology is applied to distinguish pressure threshold of various grasping models, hence it is adaptive to grasp different projects; finally, the tracking accuracy of the different sensors combination are compared.

2 Glove Pressure Sensors Distribution Design Based on Principal Component Analysis

2.1 ARAT Grasping Motion Hand Partition Experiment

The distribution of the pressure sensors is associated with the contact region between hand and object model. In this paper, grasping motion is based on the theory of ARAT. ARAT consists of four subtests: grasp, grip, pinch, and gross motor and grasping objects of ARAT all are geometric objects with standard size. Grasping motion is one of the basic movements of hand, according to different functions grasps are divided into power grasps and precision grasps (Cutkosky et al. 1989), the two parts include all dimensions of grasping objects involved in ARAT and their corresponding grasping movements in Fig. 1.

The hand area can be divided into 0–18 areas in Fig. 2.

According to grasping motion similarity, grasp motion is divided into five classifications numbered as No. 1–No. 5. Beforehand, ARAT models are painted with blue pigment and five subjects without any illness or injury are selected and familiar with the experiment process. Let each participant grasps color model as ARAT method introduced, recording contact area number of hand after each object grasped. Table 1 shows one participant's full ARAT test contact area. Count the contact frequency for each numbered 0–18 contact area, result shows in Fig. 3.

As seen in Fig. 3, the contact area corresponding to former five highest contact frequency descending order is 12, 0, 3, 6, 9.

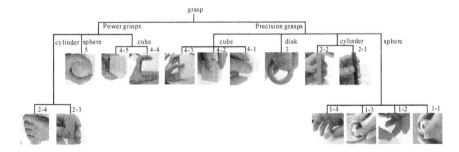

Fig. 1 Grasp classification

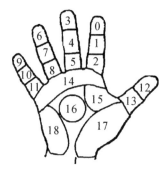

Fig. 2 Hand partition (Meng et al. 2010)

Table 1 Contact area

Grasp number	No. 1	No. 2	No. 3	No. 4	No. 5
Before grasp					
After grasp					

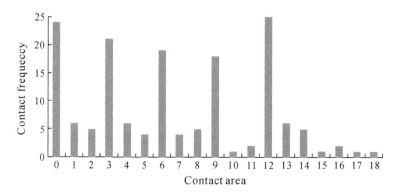

Fig. 3 Contact frequency and contact area

2.2 Sensor Layout Design Based on Principal Component Analysis

A. Principal Component Analysis

Principal Component Analysis (PCA) is a kind of data compression and feature information extraction technology and it converts a set of possibly correlated variables into a set of values of linearly uncorrelated variables thus reduced data redundancy; so, the data is processed in a low-dimensional feature space and meanwhile keep most of the original data information (Li et al. 2011).

Assume a set of p data which composes a vector X: X_1, X_2,..., X_P, for each X_j ($i = 1, 2,..., p$) corresponding to a coefficient a variable. Reassembled a new set of unrelated number denoted as composite indicator Fm replaces original indicators. The principal component model is expression in Eq. (1)

$$\begin{cases} F_1 = a_{11}X_1 + a_{12}X_2 + \ldots + a_{1p}X_p \\ F_2 = a_{21}X_1 + a_{22}X_2 + \ldots + a_{2p}X_p \\ \quad \ldots\ldots \\ F_m = a_{m1}X_1 + a_{m2}X_2 + \ldots + a_{mp}X_p \end{cases} \tag{1}$$

(1) is denoted as $F = AX$

Where F_i is the ith principal component, $i = 1, 2, \wedge, m$; Coefficient matrix A_{ij} row vectors as unit eigenvector corresponding m eigenvalues $\lambda_1, \lambda_2, ..., \lambda_m$.

B. Determine the number of sensors

It is known from experiment of Sect. 2.1 part that the contact area number with descending contact frequency is 12, 0, 3, 6, 9. Put the five-dimension pressure data in formula (1), PCA is used to reduce data dimension. Figure 4 illustrates five sensor data of grasping motion. Calculate five eigenvalues from five sensors data according to PCA algorithm model. The results are shown in Table 2.

In Table 2, principal components numbered 1–5 are the thumb, index finger, middle finger, ring finger, and little finger in turn. Table 2 demonstrates that the first three principal component's total contribution rates are 99.775 %, almost representing all the variable information. In Fig. 5, when the number of factor exceeds 3, the decrease extent of eigenvalue is very little, thus it is enough to reflect the original variable information, which implies that close last two sensors does not affect the recognition effect of grasping judgment, so the number of sensors is three, distributing in the thumb, forefinger, and middle finger fingertip position.

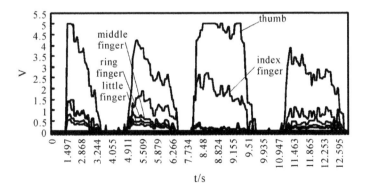

Fig. 4 Grasping action pressure distribution

Table 2 The contribution of each component of principal component analysis

Principal component number	Eigenvalue	Contribution rate (%)
1	3.9041	77.7523
2	1.0211	20.3368
3	0.0847	1.6868
4	0.0106	0.2111
5	0.0007	0.0139

Fig. 5 Eigenvalue factor graph

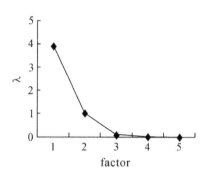

3 Adaptive Pressure Threshold Acquirement Method Based on Visual Feedback

3.1 Object Detection Based on Camshift Feedback Codebook

In this paper, moving target detection algorithm-based Camshift feedback codebook model is used in visual tracking. Camshift algorithm transforms the input image into a probability distribution by target color histogram, and then calculates the moments of the target area in the transformed probability distribution. In order to achieve continuous tracking, the continuous iterative method is utilized to calculate the target rectangular window position and size, and regards the expanded rectangle window as an image processing area for the next frame. Camshift target tracking steps are as follows:

1. Initialize track objects rectangular area;
2. Extract H component images from HSV color space of each frame, and calculate the gravity position of the window;
3. Move the center of the rectangular window to the gravity position and update the rectangular window;
4. Return the rectangular window position and size of targets.

If the moving distance is greater than the convergent minimum moving distance or the number of iterations is less than the maximum number of iterations, repeat the third and fourth step until it astringes.

Camshift algorithm is mainly for tracking and recognition by identifying the color of HSV. HSV color model is a model for the user perception, focusing on color representation, including color, depth, light and shade, which can be transformed from the RGB values. In the RGB values table, the best colors can be identified are red, green, and blue. So, ARAT models are blue and marks on the wearable pressure sensor data glove are red and green. To enhance the tracking performance, marks are designed as toroidal, and red and green marks are distributed at the thumb and index fingertips.

The rectangle upper-left vertex coordinates width w_{a1} and height h_{a1} can be obtained by Camshift algorithm. The minimum of w_{a1} and h_{a1} labeled L is taken as the model of classification recognition. The computational formula is

$$L = \mathbf{min}(w_{a1}, h_{a1}) \tag{2}$$

3.2 Adaptive Pressure Threshold Acquirement

The mass of ARAT models and grasp way will affect the contact pressure threshold, so the models will be divided into three categories on the basis of model mass. Test the contact pressure threshold value for each type ARAT model. The categories are shown in Table 3 and the pressure value test is shown in Fig. 6.

The model mass has a direct relationship with the grasp pressure. The greater the mass of the model is, the larger the grasp pressure threshold. The type I grasp pressure threshold test experiments are shown in Fig. 6. A weighted fusion algorithm (Song et al. 2013) is used to normalize three fingers the pressure values. The results show that the type I contact pressure threshold is $f_1 = 0.8$. According to this algorithm, the type II contact pressure threshold is $f_2 = 0.5$, and the type III contact pressure threshold is $f_3 = 0.1$.

Table 3 ARAT object classification

TypeID	Shape	3D size (mm)
Type I	Cube	97 × 97 × 97
Type II	Cube	74 × 74 × 74
	Sphere	70 (diameter)
Type III	Cube	50 × 50 × 50
	Cube	25 × 25 × 25
	Cuboid	100 × 25 × 10
	Cylinder	200 (high) × 20 (diameter)
	Cylinder	200 (high) × 10 (diameter)
	Disk	5 (high) × 35 (diameter)
	Sphere	10 (diameter)

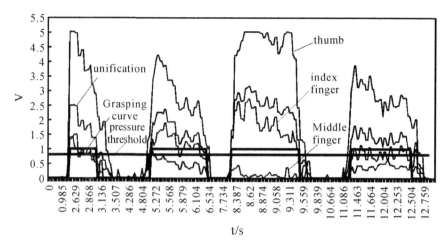

Fig. 6 Grasp pressure threshold test

The three-dimensional size of the models is directly related to the tracking rectangle. Before grasp, the model is in stationary state. According to the model rectangle achieved by target detection and the formula (2), the model belongs to which category it can be estimated. From Table 3, the rectangle threshold of the type I model is $L = 97 \pm 2$, and the type II model is $L = 70 \pm 2$. The rest of L is for the type III model. In practical application, the unit should be changed into pixel, and the value L is related to the camera installation height. Adaptive threshold for pressure calculation are as follows:

I. **Camshift**(w_{a1}, h_{a1});
CvRect(0, 0, image.cols, image.rows)
w_{a1} = image.cols
h_{a1} = image.rows
II. $L = \mathbf{min}(w_{a1}, h_{a1})$
III. **if**($L \geq L_1$) **then** $f = f_1$

if($L \geq L_2$) **then** $f = f_2$

$f = f_3$

4   Design and Testing of Wearable Sensor Data Glove

4.1   Design of Wearable Sensor Data Glove

Wearable pressure sensor data glove is shown in Fig. 7. It includes the data acquisition part and control module encapsulated within the back side of the glove and marks encapsulated fingertip position of external part of glove and pressure

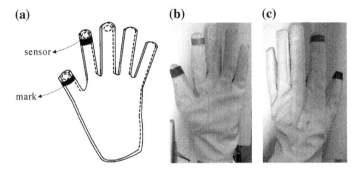

Fig. 7 Layout chart of wearable pressure sensing data glove (**a**), positive of wearable pressure sensing data glove (**b**), and back of wearable pressure sensing data glove (**c**)

sensors encapsulated in the palm side of inside the glove. Pressure sensor locates in the point where hand contacts with grasping object when hand is grasping an object.

4.2 The Analysis of the Combination of Different Sensors for Grasp Accuracy

In order to analyze the influence of the combination of different sensors for grasp accuracy, create three devices such as: single camera device, single sensors device, and combination of single camera device and single sensors device. Five subjects (No. 1–No. 5) use the above three devices to grasp the same objects. The developed software will record the visual data and the pressure data separately based on the above devices.

As shown in Fig. 8 are the grasp experiments on the condition of vision and touch fusion. Before grasp, minimum length of object rectangular is obtained through the camera, and judging object gripping pressure threshold according to rectangular threshold of object. Before the subject's hand touches object, grasp pressure is less than the grasping threshold, grasp count is 0. After grasp, hand touches object and when grasp pressure is more than the grasping threshold, at

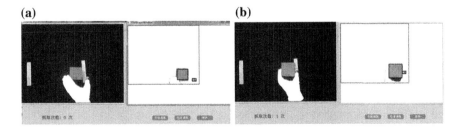

Fig. 8 Judgment number of grasp, before grasp (**a**), after grasp (**b**)

Table 4 Grasping contrast on the condition of three

	Camera	Sensor	Camera and sensor
No. 1	12	14	20
No. 2	11	10	19
No. 3	10	11	18
No. 4	13	12	18
No. 5	9	10	19
Success rate (%)	55	57	94

present grasp successful count is 1. Grasp successful counts of 5 subjects on the condition of three situations are shown in Table 4.

From the experimental result, we can know that occlusion issue is very serious on the condition of single camera device, so the success rate is lowest at 55 %; it cannot automatically adjust grasp pressure threshold without vision data when using only one grasp pressure threshold to judge all objects, so success rate is 57 %; Finally, pressure threshold cooperate with vision data to judge grasp, it can automatically adjust grasp, with a high success rate of 94 %.

5 Conclusion

An E glove hand function evaluation device based on vision and touch integration was presented in this paper, and hand sensor distribution model was established through the principal component analysis algorithm. The PCA analysis chart shows that three sensors distributed on the thumb, forefinger, and middle finger can effectively estimate the grasp motion. Moreover, it can accurately select the grasp threshold values for each category model by the adaptive pressure threshold method. Finally, five subjects dressed E glove sensors judge the grasp motion under the condition of homogeneous and heterogeneous sensors, and the highest accuracy rate of classification depended on heterogeneous sensors fusion reaches 94 %. The results show that: the identification rate of using the adaptive pressure threshold as well as vision fusion method of this simple device to judge the grasp process is better. Since the E glove hand function evaluation device is still in the laboratory stage, there are many issues worthy of further study.

Acknowledgments The authors wish to thank Dr. Zhi-jian Fan for providing guide.

References

Cutkosky MR et al (1989) On grasp choice, grasp models, and the design of hands for manufacturing tasks. IEEE Trans J Robot Autom 5(3):269–279
Gabriele W, Renate S (2009) Work loss following stroke. J Disabil Rehabil 31(18):1487–1493

Han J, Qiu D, Song S (2012) Electronic nose to detect flavor substances fermented using plant lactobacillus in grass carp. J Food Sci 33(10):208–211 (in Chinese)

Liang Wei, Zhang L, Wang H et al (2013) Wine classification detection method based on the technology of electronic nose. J Sci Technol Eng 13(4):930–934 (in Chinese)

Lyle RC et al (1981) A performance test for assessment of upper limb function in physical rehabilitation treatment and research. J Int J Rehabil Res 4(4):483–492

Li N et al. (2011) Study on the forecast and prevention of waterqnrush from the lower-group coal floor in Zhaoguan coal mine. Shandong University of Science and Technology, Qingdao (in Chinese)

Jakel R, Schmidt-Rohr SR, Losch M, Dillmann R (2010) Representation and constrained planning of manipulation strategies in the context of programming by demonstration. In: Proceedings of the IEEE international conference robotics and automation (ICRA), pp 162–169

Jakel R, Schmidt-Rohr SR, Xue Z, Losch M, Dillmann R (2010) Learning of probabilistic grasping strategies using programming by demonstration. In: Proceedings of the IEEE international conference on robotics and automation, ICRA

Jäkel R, Schmidt-Rohr SR, Rühl SW, Kasper A, Xue Z, Dillmann R (2012) Learning of planning models for dexterous manipulation based on human demonstrations. Int J Soc Robot 4:437–448

Meng H et al (2010) Handle pressure distribution test system based on LabVIEW. Zhejiang University of Technology, Hangzhou (in Chinese)

Meng D, Xu G et al (2013) The research progress of upper limb function assessment tools in patients with hemiplegia. J Chin J Rehabil Theor Pract 38(7):1032–1035 (in Chinese)

Nathan DE, Johnson MJ, McGuire JR (2009) Design and validation of low-cost assistive glove for hand assessment and therapy during activity of daily living-focused robotic stroke therapy. J Rehabil Res Dev 46(5):587–602

Palm R, Iliev B (2014) Programming by demonstration and adaptation of robot skills by Fuzzy time modeling. J Appl Mech Mater:648–656

Palm R, Iliev B, Kadmiry B (2010) Grasp recognition by Fuzzy modeling and hidden Markov models. In: Programming-by-demonstration of robot motions

Skoglund A, Iliev B, Palm B (2010) Programming by demonstration of reaching motions a next state planner approach. J Robot Auton Syst 58(5):607–621

Song C et al (2013) Study on motion control and stably grab of for arm-hand system. Hunan University, Changsha (in Chinese)

Weng C, Wang J, Wang G et al (2007) Reliability of the action research arm test in stroke patients. J China Rehabil Theor Pract 13(9):868–869 (in Chinese)

Yozbatiran N, Der-Yeghiaian L, Cramer SC (2008) A standardized approach to performing the action research arm test. J Neurorehabil Neural Repair 22(1):78–90

Zampieri C, Salarian A, Carlson-Kuhta P et al (2011) Assessing mobility at home in people with early Parkinson's disease using an instrumented timed Up and Go test. J Parkinsonism Relat Disord 17(4):277–280

An Emotion Recognition System Based on Physiological Signals Obtained by Wearable Sensors

Cheng He, Yun-jin Yao and Xue-song Ye

Abstract Automatic emotion recognition is a major topic in the area of human–robot interaction. This paper presents an emotion recognition system based on physiological signals. Emotion induction experiments which induced joy, sadness, anger, and pleasure were conducted on 11 subjects. The subjects' electrocardiogram (ECG) and respiration (RSP) signals were recorded simultaneously by a physiological monitoring device based on wearable sensors. Compared to the non-wearable physiological monitoring devices often used in other emotion recognition systems, the wearable physiological monitoring device does not restrict the subjects' movement. From the acquired physiological signals, one hundred and forty-five signal features were extracted. A feature selection method based on genetic algorithm was developed to minimize errors resulting from useless signal features as well as reduce computation complexity. To recognize emotions from the selected physiological signal features, a support vector machine (SVM) method was applied, which achieved a recognition accuracy of 81.82, 63.64, 54.55, and 30.00 % for joy, sadness, anger, and pleasure, respectively. The results showed that it is feasible to recognize emotions from physiological signals.

Keywords Emotion recognition · Physiological signals · Wearable sensors · Genetic algorithm · Support vector machine

1 Introduction

Automatic emotion recognition is a major topic in the area of human–robot interaction. People express emotions through facial expressions, tone of voice, body postures, and gestures which are accompanied with physiological changes. Facial expressions, tone of voice, body postures, and gestures are controlled by the

C. He · Y. Yao · X. Ye (✉)
Department of Biomedical Engineering and Instrument Science,
Zhejiang University, 310027 Hangzhou, China
e-mail: yexuesong@zju.edu.cn

© Zhejiang University Press and Springer Science+Business Media Singapore 2017   15
C. Yang et al. (eds.), *Wearable Sensors and Robots*, Lecture Notes in Electrical
Engineering 399, DOI 10.1007/978-981-10-2404-7_2

somatic nervous system while physiological signals, such as electroencephalogram (EEG), heart rate (HR), electrocardiogram (ECG), respiration (RSP), blood pressure (BP), electromyogram (EMG), skin conductance (SC), blood volume pulse (BVP), and skin temperature (ST) are mainly controlled by the autonomous nervous system. That means facial expressions, tone of voice, body postures, and gestures can be suppressed or masked intentionally while physiological signals can hardly be masked. Using physiological signals to recognize emotions is also helpful to those people who suffer from physical or mental illness thus exhibit problems with facial expressions, tone of voice, body postures or gestures.

Researches have shown a strong correlation between emotions and physiological signals. However, whether it is reliable to recognize emotions from physiological signals is still problematic. Numerous researches were investigating the problem (Picard et al. 2001; Lisetti and Nasoz 2004; Kim and André 2008; Rattanyu et al. 2010; Verma and Tiwary 2014).

This paper presents an emotion recognition system based on physiological signals obtained by wearable sensors. Some common emotion models and emotion induction methods are described briefly. The data collection procedure during which a physiological monitoring device based on wearable sensors was used is introduced. The strategy for feature extraction from the acquired physiological signals and the feature selection method based on genetic algorithm are illustrated. The support vector machine (SVM) method which was used to classify the physiological features into four kinds of emotions is demonstrated. The experiment implementation procedure is presented as well. Finally, the results of the experiments are discussed, which contribute to a conclusion.

2   Method

2.1   Emotion

In discrete emotion theory, all humans are thought to have an innate set of basic emotions that are cross-culturally recognizable (Ekman and Friesen 1971). In dimensional emotion theory, however, emotions are defined according to multiple dimensions (Schlosberg 1954). Although it is problematic which emotions are basic in discrete emotion theory (Gendron and Barrett 2009) and in which dimensions emotions should be defined in dimensional theory (Rubin and Talarico 2009), it's no doubt that joy, sadness, anger, and pleasure are four different common emotions in humans. Those four emotions were chosen as the classification categories in our study.

To obtain the physiological signals associated with the specific emotions, an effective emotion induction procedure is of significance. Numerous emotion or mood induction procedures (MIPs) have been reported including presenting subjects with emotional stimuli (pictures, film clips, etc.), and letting subjects play games (van't Wout et al. 2010) or interact with human confederate (Kučera and Haviger 2012). Several picture, audio, or video databases for emotion induction have also been created (Biehl et al. 1997; Bai et al. 2005; Bradley and Lang 2008).

In our study, we did not use the emotion induction materials from those databases above because those materials did not induce the expected emotions effectively in our experiments. Instead, we selected several contagious video clips which performed better in our emotion induction experiments.

2.2 Physiological Signals Processing

2.2.1 Data Collection

Several kinds of physiological signals including ECG and RSP signals have been revealed to be correlated with emotions. To collect ECG and RSP signals, a physiological monitoring device based on wearable sensors which monitors multiple physiological signals simultaneously in real time (Zhou et al. 2015) was used. The ECG signals were sampled at 250 Hz and the RSP signals were sampled at 10 Hz. The schematic representation of a normal ECG waveform is shown in Fig. 1 and the ECG and RSP waveforms obtained by the physiological monitoring device are shown in Figs. 2 and 3, respectively.

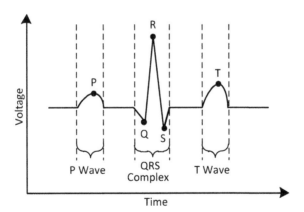

Fig. 1 Schematic representation of a normal electrocardiogram (ECG) waveform. An ECG waveform consists of a P wave, a QRS complex and a T wave. The QRS complex usually has much larger amplitude than the P wave and the T wave. P is the peak of a P wave. Q is the start of a QRS complex. R is the peak of a QRS complex. S is the end of a QRS complex. T is the peak of a T wave

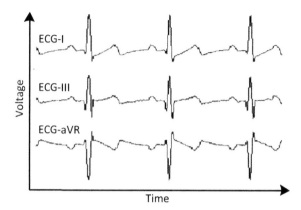

Fig. 2 Electrocardiogram (ECG) signals obtained by the physiological monitoring device. ECG-I is the voltage between the left arm electrode and right arm electrode. ECG-III is the voltage between the left leg electrode and the right leg electrode. ECG-aVR is the voltage between the right arm electrode and the combination of the left arm electrode and the left leg electrode

Fig. 3 Respiration (RSP) signals obtained by the physiological monitoring device

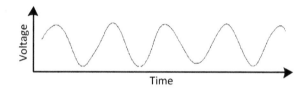

2.2.2 Feature Extraction

After the P-waves, the QRS complexes, and the T waves of the ECG signals were determined, a total of 78 ECG signal features were extracted as follows:

1. The mean value, median value, standard variance, minimum value, maximum value, and value range of R–R, P–P, Q–Q, S–S, T–T, P–Q, Q–S, and S–T time intervals;
2. The mean value, median value, standard variance, minimum value, maximum value, and value range of the amplitudes of P waves, QRS complexes, and T waves divided by the mean value of the corresponding ECG waveforms;
3. The mean value, median value, standard variance, minimum value, maximum value, and value range of HRD (the histogram distribution of R-R time intervals);
4. HR50 (the number of pairs of adjacent R-R time intervals differing by more than 50 ms divided by the total number of R-R time intervals);
5. HRDV (sum of HRD divided by the maximum value of HRD)
6. Each spectrum power of ECG signals in four frequency band (0–0.2 Hz, 0.2–0.4 Hz, 0.4–0.6 Hz, and 0.6–0.8 Hz).

Before RSP features were extracted, a low-pass filter was applied to the raw RSP signals. After that, a total of 67 RSP signal features were extracted as follows:

1. The mean value, median value, standard variance, minimum value, maximum value, value range, and peak ratio (the number of peaks divided by the length of data) of the following signals:

   (a) RSP waves, RSP peak–peak intervals, and RSP peak amplitudes;
   (b) The first difference of RSP waves, RSP peak-peak intervals, and RSP peak amplitudes
   (c) The second difference of RSP waves, RSP peak-peak intervals, and RSP peak amplitudes

2. Each spectrum power of RSP signals in four frequency band (0–0.1 Hz, 0.1–0.2 Hz, 0.3–0.3 Hz, and 0.3–0.4 Hz).

Considering the seventy-eight ECG signal features and the sixty-seven RSP signal features, a total of one hundred and forty-five features were extracted.

2.2.3 Feature Selection

More features usually provide more information about the original signals, but also lead to an increase in computational complexity. Besides, the random noise in those signal features which make little contribution to identify different emotions might leads to overfitting in supervised machine learning such as SVM. Therefore, an effective feature selection method to select only a key subset of measured features to create a classification model is needed. Emotion recognition can be looked as a pattern recognition issue. For a pattern recognition issue, the selection criterion usually involves the minimization of a specific measure of predictive error for models which fit to different subsets. A common method is sequential feature selection (SFS) (Cover and Van Campenhout 1977), which adds features from a candidate subset while evaluating the criterion. Another novel method is using genetic algorithm (Deb et al. 2002) to select features, which will be described here.

The genetic algorithm (GA) is a method based on natural selection which drives biological evolution. The GA repeatedly modifies a population of individual solutions. At each step, the GA selects individuals at random from the current population to be parents and uses them to produce the children for the next generation. There are some rules like crossover at each step to create the next generation from the current population. At each step, the individual selection is random, but the survival opportunity of each individual is not equal. The individuals who have higher survival opportunity are more likely to be selected and keep evolving till the optimization goal is reached. In our study, the survival opportunity was evaluated by the emotion recognition error.

Through the GA algorithm described above, fourteen features were selected from the original one hundred and forty-five features.

2.3 Emotion Recognition

To recognize emotions from the key features selected by GA, a modified support vector machine (SVM) method was used. An SVM classifies data by finding the optimal hyperplane that separates all data points of one class from those of another class (Cortes and Vapnik 1995). The optimal hyperplane for an SVM means the one with the maximum margin between the two classes. A margin is the maximal width of two slabs parallel to the hyperplane that have no interior data points. A larger margin assures the hyperplane is more likely to classify new data correctly. The data points that are on the boundary of the slab are called support vectors. The complexity of the classifier is characterized by the number of support vectors rather than the dimensionality of the transformed hyperspace. An example of SVM is shown in Fig. 4.

Sometimes the data might not allow for a separating hyperplane. As shown in Fig. 5, the outliers caused by error such as artifact during data collection make it difficult to find a proper separating hyperplane. Even if a separating hyperplane is found, the margin is small. In that case, a soft margin method is proposed which chooses a hyperplane that splits the examples as cleanly as possible while still maximizing the distance to the nearest cleanly split (Cortes and Vapnik 1995).

Some binary classification problems do not have an effective linear separating hyperplane, so-called nonlinear classification, as shown in Fig. 6a. In this case, the initial hyperspace S is transformed to a higher dimensional hyperspace S', as shown in Fig. 6b. In the higher dimensional hyperspace S', there is a linear hyperplane to successfully separate the two classes. Usually, the analysis formula of the

Fig. 4 Linear Support vector machine. The optimal linear hyperplane separates all samples into two classes with a maximum margin

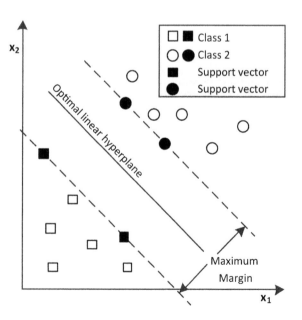

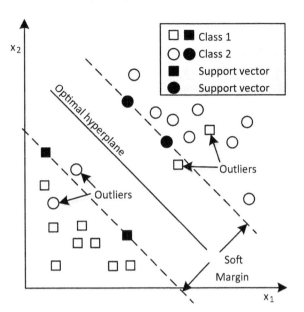

Fig. 5 Support vector machine with a soft margin. The soft margin allows for some mislabeled samples to maximize the margin

transformation is difficult to get. However, It is found that all the calculations for hyperplane classification use nothing more than dot products, then a nonlinear kernel function in linear hyperspace S is developed to replace the dot products in the higher dimensional hyperspace S'(Boser et al. 1992). Some common kernels are listed here: polynomial function, Gaussian radial basis function (RBF) and multi-layer perceptron (neural network, NN).

Support vector machines (SVMs) are originally designed for binary classification. But there have been some extensions for multiclass classification (Hsu and Lin 2002). One of the multiclass classification approaches using SVM is building binary classifiers which distinguished between every pair of classes, so-called one-against-one (Knerr et al. 1990). For a one-against-one approach, classification is done by a max-wins voting strategy that every classifier assigns the instance to one of the two classes, then the vote for the assigned class is increased by one vote, and finally, the class with the most votes determines the instance classification.

In our multi-emotion recognition system, we applied linear and nonlinear SVMs with a soft margin as the classifiers and one-against-one method as the multiclass classification approach.

2.4 Experiment Implementation

First, we used Beck Depression Inventory-II (BDI-II) (Beck et al. 1996) and Toronto Alexithymia Scale (TAS) to select 11 subjects who had no depression (BDI-II score < 4) and were capable of expressing emotions clearly (TAS < 60)

Fig. 6 Transform a nonlinear SVM to a linear SVM. Kernels are used to fit the maximum margin nonlinear hyperplane in a transformed linear hyperplane without knowing the analysis formula the transformation function

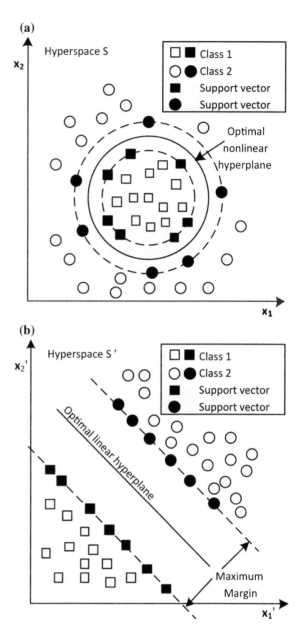

from the experiment volunteers. This study was approved by the Institutional Review Board of Zhejiang University. Informed written consent was obtained from all experiment volunteers. Then, we prepared a quiet multimedia experiment room equipped with an air conditioner, a computer, a 17 inch LCD screen, and a pair of stereo headphones. In each experiment, one subject wearing the physiological sensors sat alone in the multimedia room with air temperature setting to 25 °C and watched the video clips or listened to the music we had prepared. Before and after each experiment, the subject was asked to fill out a questionnaire about his or her experience and emotion. The signals of those subjects who did not report expected emotion were labeled invalid and discarded. For the valid signals, the time slot when the subjects were most likely in an expected emotion state was determined.

3 Results and Discussion

The SVM classifiers were tested with leave-one-out cross validation. Leave-one-out cross validation involves using one observation as the validation set and the remaining observations as the training set. This is repeated on all ways to cut the original sample on a validation set of one observations and a training set. The linear SVM and the nonlinear SVMs with polynomial function, Gaussian radial basis function, and multilayer perceptron as the kernel function were all tested. The results are shown in Table 1.

As shown in Table 1, the linear SVM classifier achieved the highest recognition accuracy for four emotions in total while the nonlinear SVM classifier with a polynomial function as the kernel achieved the lowest recognition accuracy for four emotions in total. The linear SVM classifier performed better than the nonlinear SVM classifiers in total. That is probably because the number of physiological features was reduced from one hundred and forty-five to fourteen during the feature selection procedure and the linear SVM was able to provide a relatively good classifier.

As to the recognition accuracy for each emotion, the linear SVM and nonlinear SVMs with different functions as the kernel performed differently. However, each SVM classifier achieved higher recognition accuracy for joy than for the other three emotions. There might be two reasons for that. One is that joy causes greater physiological changes than the other three emotions. The other one is that the induction for joy was more effective than the other three emotions in the emotion induction experiments. In addition, although the physiological data from the subjects who did not report expected emotions were labeled as invalid and discarded,

Table 1 Emotion recognition accuracy using the linear SVM and the nonlinear SVMs with different functions as the kernel. RBF is Gaussian radial basis function. Poly is polynomial function. NN is multilayer perceptron (neural network) function

	Linear (%)	RBF (%)	Poly (%)	NN (%)
Joy	81.82	63.64	45.45	90.91
Sadness	63.64	18.18	18.18	63.64
Anger	54.55	18.18	27.27	36.36
Pleasure	30.00	20.00	20.00	10.00
Total	58.14	30.23	27.91	51.16

there exists the possibility that the subjects reported their emotions inaccurately. Another possible reason is that pleasure is close to joy and the SVM classifier failed to distinguish them.

To improve the performance of the presented emotion recognition system, the following methods could be taken into account in further work. Precisely designed emotion induction experiments could be conducted on more subjects. Some other supervised machine learning algorithms could be developed. And some other physiological signals like EMG signals might be obtained together with ECG and RSP signals, but it should be noted that the acquisition process of physiological signals should not make the subjects feel uncomfortable.

4 Conclusion

As physiological signals cannot be masked intentionally, recognizing emotions from physiological signals has advantages over from facial expressions, tone of voice, body postures, and gestures. Based on a combination of a feature selection method and a support vector machine method, it is feasible to recognize emotions from physiological signals.

Acknowledgments This work is supported by National Science and Technology Major Project of the Ministry of Science and Technology of China (No. 2013ZX03005008). And the authors would like to thank Congcong Zhou, Chunlong Tu, Jian Tian, Jingjie Feng, and Yun Gao for their physiological monitoring device and advice.

References

Bai L, Ma H, Huang YX (2005) The development of native chinese affective picture system–a pretest in 46 college students. Chin Ment Health J 19(11):719–722

Beck AT, Steer RA, Brown GK (1996) Manual for the beck depression inventory-II. Technical report. Psychological Corporation, San Antonio, Texas

Biehl M, Matsumoto D, Ekman P et al (1997) Matsumoto and Ekman's Japanese and Caucasian facial expressions of emotion (JACFEE): reliability data and cross-national differences. J Nonverbal Behav 21(1):3–21. doi:10.1023/A:1024902500935

Boser BE, Guyon IM, Vapnik VN (1992) A training algorithm for optimal margin classifiers. In: Proceedings of the fifth annual workshop on Computational learning theory, pp 144–152

Bradley MM, Lang PJ (2008) International affective picture system (IAPS): affective ratings of pictures and instruction manual. Technical report, The Center for Research in Psychophysiology, University of Florida, Gainesville, Florida

Cortes C, Vapnik V (1995) Support-vector networks. Mach Learn 20(3):273–297. doi:10.1007/BF00994018

Cover TM, Van Campenhout JM (1977) On the possible orderings in the measurement selection problem. IEEE Trans Syst Man Cybern 7(9):657–661. doi:10.1109/TSMC.1977.4309803

Deb K, Pratap A, Agarwal S et al (2002) A fast and elitist multiobjective genetic algorithm: NSGA-II. IEEE Trans Evol Comput 6(2):182–197. doi:10.1109/4235.996017

Ekman P, Friesen WV (1971) Constants across cultures in the face and emotion. J Pers Soc Psychol 17(2):124. doi:10.1037/h0030377

Gendron M, Barrett LF (2009) Reconstructing the past: a century of ideas about emotion in psychology. Emot Rev 1(4):316–339. doi:10.1177/1754073909338877

Hsu CW, Lin CJ (2002) A comparison of methods for multiclass support vector machines. IEEE Trans Neural Networks 13(2):415–425. doi:10.1109/72.991427

Kim J, André E (2008) Emotion recognition based on physiological changes in music listening. IEEE Trans Pattern Anal Mach Intell 30(12):2067–2083. doi:10.1109/TPAMI.2008.26

Knerr S, Personnaz L, Dreyfus G (1990) Single-layer learning revisited: a stepwise procedure for building and training a neural network. Neurocomputing. Springer, Heidelberg, pp 41–50. doi:10.1007/978-3-642-76153-9_5

Kučera D, Haviger J (2012) Using mood induction procedures in psychological research. Procedia Soc Behav Sci 69:31–40. doi:10.1016/j.sbspro.2012.11.380

Lisetti CL, Nasoz F (2004) Using noninvasive wearable computers to recognize human emotions from physiological signals. EURASIP J Appl Sig Process 2004(11):1672–1687. doi:10.1155/S1110865704406192

Picard RW, Vyzas E, Healey J (2001) Toward machine emotional intelligence: analysis of affective physiological state. IEEE Trans Pattern Anal Mach Intell 23(10):1175–1191. doi:10.1109/34.954607

Rattanyu K, Ohkura M, Mizukawa M (2010) Emotion monitoring from physiological signals for service robots in the living space. In: 2010 International conference on control automation and systems. Gyeonggi-do, pp 580–583

Rubin DC, Talarico JM (2009) A comparison of dimensional models of emotion: evidence from emotions, prototypical events, autobiographical memories, and words. Memory 17(8):802–808

Schlosberg H (1954) Three dimensions of emotion. Psychol Rev 61(2):81. doi:10.1037/h0054570

van't Wout M, Chang LJ, Sanfey AG (2010) The influence of emotion regulation on social interactive decision-making. Emotion 10(6):815–821. doi:10.1037/a0020069

Verma GK, Tiwary US (2014) Multimodal fusion framework: a multiresolution approach for emotion classification and recognition from physiological signals. Neuroimage 102(P1):162–172. doi:10.1016/j.neuroimage.2013.11.007

Zhou CC, Tu CL, Tian J et al (2015) A low power miniaturized monitoring system of six human physiological parameters based on wearable body sensor network. Sens Rev 35(2):210–218. doi:10.1108/SR-08-2014-687

Integrated Application Research About the Necklace Type Wearable Health Sensing System Under the Internet of Things

Jian-jun Yu and Xing-bin Chen

Abstract The Wearable Health Sensing System (WHSS) is to integrate the various sensors on the necklace, which is convenient for the real-time detection of human life and health, meanwhile to use the data processing ability of portable handheld terminal and coordination with the application of APP interaction to make the health assessment, diet care information pushing, and the emergency medical treatment true. This paper puts forward a scheme that integrating the multiple health sensors on the necklace, which includes the movement frequency, heart rate, sweat, skin tissue fluid, ECG, blood oxygen saturation, skin moisture, body fat, etc., objecting to setup a dynamic and collaborative monitoring physiological data under the daily life and the complicated working environment by means of a lower physiological and psychological load. Three modules are designed for WHSS: sensor integration module, power supply and power module, and signal integration and network transmission module. Corresponding sensors are chosen for WHSS according to various application requirements. Existing disturbances affect the accuracy and reliability of the judgment about the health feature points, accordingly the anti-interference processing is carried out. WHSS has only integrated multi-sensor without setting powerful data processing unit, so the captured signal is extended to handheld terminals to display and control in accordance with the processing ability of formalities terminal. The function model of data fusion system is setup, the level of the data fusion system is analyzed, the data fusion algorithm is realized, and the Multi-sensor system identification technology is studied. Due to the automatic comprehensive analysis capabilities of the multi-sensor system identification technology, it could be more effective and accurate to reflect the

J. Yu
Department of Industrial Engineering, South China University of Technology, Guangzhou 510640, China

X. Chen (✉)
School of Mechanical and Automotive Engineering,
South China University of Technology, Guangzhou 510640, China
e-mail: scutba@126.com

© Zhejiang University Press and Springer Science+Business Media Singapore 2017
C. Yang et al. (eds.), *Wearable Sensors and Robots*, Lecture Notes in Electrical
Engineering 399, DOI 10.1007/978-981-10-2404-7_3

physical status of the object than any single sensor in the system. Under the environment of the Internet of things, the fusion of multiple data display control from WHSS, can accurately obtain the health information, and sets up the health system security network to a certain extent.

Keywords Wearable health sensing system · Data fusion processing · Multi-sensor system identification · Internet of things

1   Introduction

The wearable sensor technology is the portable testing equipment technology which can wear on people directly or incorporate into clothing and accessories. In the early 1960s, MIT had put forward using the technology of multimedia, sensor and wireless communication to inset in people's clothes (Liang 2014).

With the development of mobile communication technology, the wearable devices, such as the Google glasses, apple bracelet, smart watch, make the combination of "wearable technology" and mobile Internet come true. At the same time, the technology about the digital production and the precision manufacturing can further promote the development of the lightweight materials and the electronic chip integration, thus provides a lifestyle of the mobile intelligent (Zhang et al. 2008).

After in-depth researching and analyzing on the current situation of the development of wearable devices, this paper puts forward a scheme that integrating the multiple health sensors on the necklace, which includes the movement frequency, heart rate, sweat, skin tissue fluid, ECG, blood oxygen saturation, skin moisture, body fat, etc., objecting to setup a dynamic and collaborative monitoring physiological data under the daily life and the complicated working environment by means of a lower physiological and psychological load. Therefore, the scheme can provide daily nurse such as recommending diet control, exercise index, and related health care advice by use of the Internet of things processing environment, even under the extreme health can offer emergency assistance such as stroke and high blood pressure.

2   Hardware System Design

In this paper, the design of WHSS mainly includes three modules: sensor integration module, power supply and power module, and signal integration and network transmission module.

1. Sensor integration module

① Motion sensors, such as accelerometer, gyroscope, magnetometer, pressure sensor (calculated of altitude by measuring the atmospheric pressure), etc.

The user can know the running steps, swimming laps, cycling distance, energy consumption, and sleep time through the motion sensor to measure, record, and analysis of the movements of the human anywhere at any time.

② Biosensor: blood glucose sensor, blood pressure sensor, ECG sensors, EMG sensors, body temperature sensors, brain wave sensor, etc.

With the aid of wearable technology, the doctor can improve the level of diagnosis and the family also can be better communicated with the patient. For example, using the wearable medical devices with blood pressure sensor, the device can cooperate with professional medical organizations to tracking and monitoring the body data of tens of millions users for a long term, then to analyze the extracting medical diagnosis model, to predict and shape the healthy development of the user, finally, to provide users with an individual special personal medical and health management solutions.

③ Environmental sensors: temperature and humidity sensor, gas sensor, pH sensor, UV sensor, ambient light sensor, particle sensor, baroceptor, microphone, and thin magnetic navigation sensor, etc.

People often are part of some threatening environment for health, such as air/water pollution, noise/light pollution, electromagnetic radiation, extreme climate, complex road traffic system, etc. Most frightening of all, many times we walk in such an environment, but know nothing, such as PM2.5 pollution causing all kinds of chronic diseases.

2. Power supply and power modules

In order to ensure normal operation of the sensor integration on the necklace and the signal transceiver module, the WHSS needs to be provided with a convenient stable miniaturization of electricity power supply and energy storage module independently. Based on the further study of the knowledge of the micro vibration generator, the mechanical watches, the pocket temperature difference battery, and the charging principle, this paper improves and uses them in WHSS.

① Principle of Micro vibration generator

The vibration generator is a device, which can turn vibrate mechanical energy into electrical energy by using the electromagnetic induction principle. The scientists at the University of Michigan have developed a small generator, which can produce enough electricity to provide energy for watches, pacemakers, and wireless sensor devices through the random vibration of the cars moving, factories activity, and human movement.

According to the authors, the micro generator can produce 0.5 mW of power (500 amounts) by using body's unique amplitude, which is enough to drive a watch that only needs between 1 and 10 amount of energy, also enough to drive a pacemaker that needs 10–50 amount of energy. What's more, this generator has a long service life and can be used for very expensive to replace the depleted battery, and somewhere need to provide power for the sensor by means of attachment.

② Principle of Mechanical watch

The mechanical watch usually can be divided into two kinds: hand-winder and automatic-winder (Shi 2007). The power source of two mechanical watches is powered by movement within the clockwork to drive gear to promote the clock, but through a different way of power source. Hand-winder mechanical watch relies on power from manual twist spring, whose thickness is thinner than the automatic-winder and the weight is lighter. Relatively, the automatic-winder watch is to drive the clockwork spring by using the power from the rotate swinging of the automatic movement.

Due to the pursuit of convenient type, this paper will refer to the energy principle of automatic-winder watch and follow the law of conservation of energy that convert the potential energy producing from the rotor swing to the kinetic energy of the WHSS internal energy module through a complete set of automatic transmission system and then add to the original system. The rotor swings under the action of gravity and different azimuth transform which is loaded in the front or back of the watch movement.

③ Principle of Pocket thermoelectric cell

The pocket thermoelectric cell uses the temperature difference to access the electricity by the semiconductor. The fundamental principle of the thermoelectric power generation is the basic principle of "Seebeck" effect that is a closed loop formed with two different kinds of metal connecting, which can produce a tiny voltage if the temperature of the two points is different.

The thermoelectric power generation can only produce about 200 mV very low voltage because of the limited difference about the temperature between the body and the surrounding environment. Some researchers from Germany's Fraunhofer Institute have developed a new circuit system, which can be induced to 200 mV voltage so that the temperature difference generator can generate the electricity by the body's temperature (Tong 2014). Similarly, the new circuit system can be induced to the minimum temperature of 0.5 °C, in this way, it can produce enough energy to a series of portable devices through our temperature and ambient temperature.

④ Principle of Wireless charging

The wireless charger adopts the latest wireless charging technology rather than not the traditional charging connecting the charger wire to the recharge terminal equipment, on which a variety of devices can use a charging base (Wei 2014). The wireless charging technology adopts the principle of the electromagnetic induction and the related technology of AC induction that means sending and receiving communication signal to induce to charge by using the corresponding coils between the sending and receiving terminal. Therefore, the inductive coupling technology will become a bridge that crosses from the charging base and equipment through the magnetic field between the coils, as shown in Fig. 1.

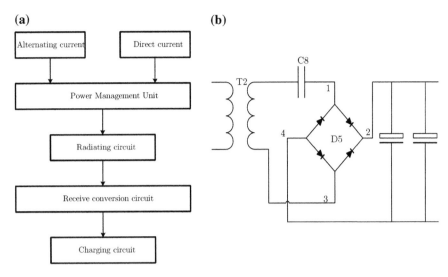

Fig. 1 **a** Wireless charging diagram. **b** Receiving circuit diagram

3. Signal integration and network transmission module

It is one of the key technical problems for this paper to design the WHSS successfully, whether to send and receive the nondestructive transmission signal steadily when the signal processing largely depends on the portable terminals, considering the premise of the limited carrying capacity of the wearable carrier, so it can safely and efficiently process the sensing signal by selecting the correct signal transmission technology. Nowadays, the major wireless transmission mode includes electromagnetic wave, infrared light, and sound waves, of which the most useful is radio frequency electromagnetic wave transmission mode. What's more, it's widely used in the medical health care systems including ZigBee, Bluetooth, and WiFi (Chen et al. 2011). As shown in Table 1, it is necessary to compare the difference of the above three kinds of wireless communication technology to choose the right means of transport.

As shown in Table above, ZigBee, Bluetooth, and WiFi can work at 2.4 GHz ISM frequency band, what the difference between them is

① From the point of power, the power consumption of the ZigBee is extremely low, supporting sleep mode, and short time about the node access and activation. The common Bluetooth can work sustainably several weeks, but the Bluetooth 4.0 technology issuing in 2011 supports low power consumption, which can be reduced to one over ten of the normal one (Sandhya and Devi 2012). But WiFi is the maximum power consumption that only can insist on for hours.

② From the aspects of network topology, ZigBee support three kinds of topology structure, which is convenient to network expansion that can accommodate 256

Table 1 The technical manual of ZigBee, Bluetooth, and WiFi

Technical manual	ZigBee	Bluetooth	WiFi
Frequency range	2.4 GHz/915 MHz/868 MHz	2.4 GHz	2.4 GHz/5.8 GHz
Transmission rate	250 Kbps/40 Kbps/20 Kbps	1 Mbps	11 Mbps/54 Mbps
Distance communication	10–75 m	5–10 m	100 m
The net time	30 ms	3 s	5 s
Connect time	15 ms	>3 s	3 s
Power dissipation	Sending power 1 mW	Working condition 30 mA	250–350 mA
Networking number	$\leq 65,535$	≤ 256	Big capacity
Networking structure	StarLAN, Smartlink, Mesh type	Ad hoc, Scatternet	Large-scale network

nodes in a single network to be suitable for large-scale sensor network, but the Bluetooth technology is suitable only for small sensor network.

The human sensor network of the WHSS system includes the ECG breathing, temperature, and blood pressure sensor node, but no need of a large network. The ECG acquisition frequency is 200 Hz, if is 7 channel, whose transmission of data is about 2 KB per second. The respiratory wave sampling frequency is common 50 Hz, and whose transmission data is 0.1 KB per second. The temperature and blood pressure acquisition frequency is low, whose transmission data can be neglected. The traditional Bluetooth transmission rate is about 120 KB/s that can satisfy the above requirements. The WHSS usage scenarios are family health care or mobile health monitoring, whose sensor node will be not far away from the family gateway or the mobile gateway. What's more, the Bluetooth technology has become one of the standard configurations of the smartphone, whose corresponding application development framework also has been very mature. Above all, comparing with other two kinds of wireless communication technology, Bluetooth technology in power consumption, development of network, transmission distance, and development difficulty degree is more in-line with the requirements of this system.

2.1 The Realization of the Hardware Carrier

The WHSS is integrated multi-sensor skillfully on the necklace, both fashion and wear characteristics, of which the sensors are arranged orderly in the necklace according to the test factors, as shown in Fig. 2.

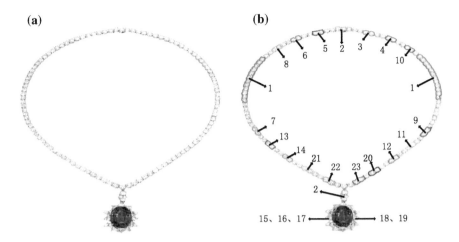

Fig. 2 a Fashional designing WHSS. **b** Sketch of different sensors layout on the WHSS

(1) Power supply and power modules. (2) Signal integration and network transmission module, (3) Ambient light sensor. (4) UV sensor. (5) Temperature and humidity sensor. (6) Particle sensor. (7) Baroceptor. (8) pH sensor. (9) Thin magnetic navigation sensor. (10) Gas sensor. (11) Gyroscope. (12) Accelerometer. (13) Pressure sensor. (14) Magnetometer. (15) ECG sensors. (16) EMG sensors. (17) Body temperature sensor. (18) Blood pressure sensor. (19) Blood glucose sensor.

The arrangement of each sensor according to their functions and categories that should be fully considered the position of necklace contacting with human body and the detecting target. Biosensor, for example, such as blood glucose sensor, blood pressure sensor, ECG sensors, EMG sensors, body temperature sensors, etc., is needed to detect the signal of the pulse and myoelectricity with closing to the human body chest, to detect the signal of the blood glucose, and body temperature with contacting the sweat. Why the arrangement of the power supply and power modules should give full consideration to the principle of the micro vibration generator, mechanical watches, pocket thermoelectric cell, and wireless charging is, on the one hand, to fully induce the vibration produced by the body movement, on the other hand, to fully contact and induce the temperature difference between atmosphere and body. Signal integration and network transmission module not only need to consider the convenience of connecting with various sensors and the miniaturization of arrangement, but also need to consider the stability of the terminal signal transceiver, most need to consider to minimize the impact from the sending and receiving radiation on human body. Environmental sensors should be exposed to the space direct contacting with human body and the environment to detect all kinds of signal of the natural environment.

2.2 Sensor Selection

Most sensors in this paper output analog voltage and have the advantages of high sensitivity, strong antijamming capability, large overload capacity, stable and reliable performance, at the same time, whose power supply voltage are 5–6 V DC voltage, $-50 \sim +300$ mmHg pressure range, 2000 µV/mmHg sensitivity, 1 * $10 - 4/C$ temperature sensitivity coefficient and 0.5 % accuracy, 0.5 % repeatability, 0.5 % hysteresis, and 100 times overload. In addition, most of them are low power consumption, usually only a few millivolts. The following will take the temperature sensor as an example to detail.

2.3 The Sensor Signal Acquisition Process

In the process of the human body temperature measurement, the temperature sensor is applied to a normal working voltage V_{in} at first, as shown in Fig. 3, a certain type of temperature sensor RT is put on the bias circuit, which is shunted in a fixed R2 resistor in parallel, while the offset R1 is a fixed resistor too. When temperature changes, the resistance of temperature test chip changes, the sampling voltage V_{out} also changes in A/D recognition and conversion. In the integrated A/D recognition and conversion circuit, we need to identify about the changed inputting voltage and the voltage of the corresponding temperature, and then output the digital signal of the changed temperature, which shows the temperature in the APP through to the decode identification of handset.

Fig. 3 Bias circuit about a certain type of temperature sensor RT

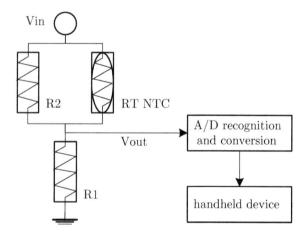

2.4 Sensor Signal Preprocessing

Healthy life signals are random, including pulse is not exceptional, at the same time, because WHSS has noninvasive contact detection and is exposed to the outside world, which contains a large amount of disturbance such as noise, temperature, skin's obstacle, hence it is needed to preprocess the signal to reduce the system load and improve the quality of detection.

The pulse signal extracted from the bosom is inevitable with a lot of random disturbance, of which the influence can use moving average method to eliminate. Moving average method is a kind of flexible filtering method, it can be used as a low-pass filter to eliminate the noise existed in various waveforms, can also be used as a high-pass filter to eliminate the baseline drift of high frequency signal. The algorithm requires the use of a smooth coefficient in process to specify the number of waveform data or sampling points. Moving average technique shows the real trend that is used to eliminate the abnormal points in a piece of data. It is equivalent in essence to a finite impulse response FIR filter.

Assuming the average points as m, it can be concluded by the type that the average point $x(n)$

$$x(n) = \frac{1}{s} \sum_{n}^{n-(m-1)} y(n) \begin{cases} s = m \ldots \text{Low pass} \\ s = -m \ldots \text{High pass} \end{cases}$$

In this type, x is average, n is the location of the data points, and y is the actual values in n. Here the choice of m is associated with two factors, one is signal sampling period, the other is the peak-to-peak value of pulse signal. The relationship between them should be satisfied

$$m = \text{VPP}/\text{SP}$$

VPP: peak-to-peak value, SP: sampling period.

Usually the pulse signal range is 30–300 bpm that is between 0.5 and 5 Hz, the peak-to-peak value range is 2–0.2 s. The sampling rate is 360 Hz, and the sampling period is 0.003 s. So the number of average points should be between 66 points to 666 points. But the average points should achieve the maximum in theory to ensure that the filter effect reaches the ideal state in all sorts of cases. This paper uses the method of low-pass filtering, the effect is as shown in Fig. 4.

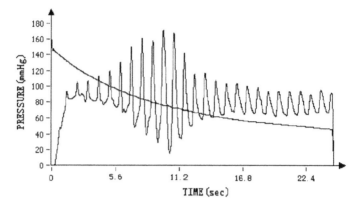

Fig. 4 The waveform of flatted average filter

3   WHSS Implementation

WHSS is needed to choose corresponding sensors for various application requirements, but it is easy to cause mutual interference between each other because of the portable integrated. These existing disturbances affect the accuracy and reliability of the judgment about the health feature points, so it must be carried out the anti-interference processing. However, WHSS is only integrated multi-sensor without setting powerful data processing unit, of which the captured signal should be extended to handheld terminals to display and control in accordance with the processing ability of formalities terminal.

3.1   Multi-sensor Anti-interference Design

Sensor circuit is usually used to measure the weak signal, which has a high sensitivity but also very easy to receive some random noise or interference signal from outside or inside, if the size of these noise and interference can be compared with the useful signal, the useful signals may be submerged in the output of the circuit, or if the useful signal is difficult to distinguish from the noise components, the sensor circuit will hinder the useful signal measurement. So the antijamming is the key to success in the sensor circuit design.

The method of filtering and threshold interruption from the sensors themselves can reduce the interference of low frequency and small amplitude, but will still be the possibility of error count, especially the counting. There is necessary to take the software antijamming filter method to further filter out the unwanted signals. First of all, the millivolt level output signal of all the sensors must be feed into the multiple core shielded wire to overcome the interference of electrostatic induction, then adding the terminal matching resistor on a transmission line and adding filter on the signaling pathways to weaken the high frequency noise, finally, using the filtering software in data processing to remove peak noise.

Filter circuit is mainly composed of two parts, one part is low-pass filter, the main peak of the pulse signal frequency is about 1 Hz, even the strong component is 20 Hz, the cutoff ceiling frequency of the low-pass filter is about 100 Hz with resistance R about 10 k and 104 (about 0.1 uF) ceramics capacitors here, it carries on the preliminary filter. The other part is 50 Hz trap, in this system, we adopt SC0073 pulse sensor to collect the pulse signal; however in the signal acquisition, processing, and analysis, it will be added the 50 Hz frequency interference, although affecting our signal processing, so we need to choose the filter with 50 Hz to filter out some certain frequency information in the input signal.

$$f_0 = 50\,\text{Hz}, \quad f_0 = \frac{1}{2\pi RC}, \quad \text{set } C = 4.7 * 10^5 \,\text{pf}$$

Then

$$R = \frac{1}{2\pi \times 50 \times 4.7 \times 10^{-7}} \Omega = 67.7\,\text{k}\Omega$$

This system chooses double T network that is formed by two letters; T parallel by the passive RC about low-pass and high-pass network, which is used to prevent a signal in the frequency. Such as inhibition of 50 Hz power frequency signal, the center frequency of the filter with resistor must be selected as 50 Hz and the voltage magnification of the center frequency is 0.

$$f_0 = 50\,\text{Hz}, \quad f_0 = \frac{1}{2\pi RC},$$

set $C = 4.7 * 10^5$ pf, obtained by

$$R = \frac{1}{2\pi \times 50 \times 4.7 \times 10^{-7}} \Omega = 67.7\,\text{k}\Omega$$

3.2 Transmission Extension of the Portable Sensing System Signal

RFCOMM is a simple transfer protocol, its purpose is how to ensure a complete communication path in application between two different devices and maintain a communication between them (Guo 2007). RFCOMM protocol layer provides multiplex data link between the application protocol and L2CAP protocol layer, at most up to 60 road communications connections between two devices.

RFCOMM regulates the concrete way of control signals and data exchanging between Bluetooth devices. RFCOMM protocol is based on L2CAP logic channel, the key factors of time needed for message request is closely related to the L2CAP protocol, including the link protocol maintenance time and the efficiency of transmission medium. RFCOMM is only for the connection between the inter-connected equipment directly, or between equipment and network access devices, but the premise is that the equipment on both telecommunication terminals must be compatible with the RFCOMM protocol.

3.3 Display and Control Response in the Handheld Terminal

With the rapid development of smartphones and the widespread popularity of 4G networks in recent years, the smartphone hardware configuration already has a very high increase, coupled with mobile included sensors, such as 3D acceleration sensor, light sensor, GPS, and BeiDou Navigation Satellite System can offer more features for WHSS.

First of all, APP should be hung when the Activity is not operation in the front desk. When the Activity is to suspend or resume, handset IOS or Android will trigger the event handler, so when the application is not visible, the UI updates and network searching should be put on hold.

The application is likely to be frequently switched between background and foreground because of multitasking of mobile devices. It is very important to switch rapidly and efficiently when needed to do like this. The uncertain process man-agement mechanism of IOS or Android means that if the application is in the background, it may be possible to finish for releasing resources. It should not be visible for the user, but the program state can be ensured to save and put the update in the queue that the users do not notice the differences between restarting and replying the procedures. When switching back to the application, the user can see coherence smoothly of the UI and program status they saw last time. But the application cannot grab attention or interrupt the user's current activity absolutely. When the application is not at the front desk operation, it is required to inform or remind users to pay attention to the application with using the Notification and Toast. There are various ways for mobile phones to notify the user, for example, when received a phone call, it will ring; when unread messages, mobile LED will flicker; when a new voice mail, a small mail icon will appear in the status bar. Through the Notification mechanism, WHSS can use all these technologies and more available technology to display and control response in the handheld terminal.

The application may be one of the multiple using programs at any time, so it is important easy to use the provided UI. Do not force the user to understand or learn it again at the time of each loading application. It should be easy, simple, and smart to use. It is very important to keep the Activity and ensure fast response by using a worker thread and the background of service when the connection may lead to delay in the trough and unreliable network, more importantly, it must be timely to avoid them interfere with other applications for quick response.

If they do not comply with the display control response, especially in health care applications, users will feel difficulty or confusion in the application. The result is that the user may stop using the application or lead to the change of vital signs for an unpleasant experience, which violates the original intention of designing. If the users stop using the application, the medical system cannot determine or monitor of vital signs, even miss some important events for the diagnosis of patients, thus cannot provide the corresponding treatment.

4   Multiple Data Fusion Processing

Sensor data fusion is a technology that various of sensors information acquisition, expresses, and comprehensive processing of its inner link and optimization, it is processed and synthesized from the angle of multiple information, then get all kinds of inner link and law of information, as to eliminate the useless and false information, keep the correct and useful components, finally realizes the optimization of information (Stiller et al. 2011).

4.1   The Function Model of Data Fusion System

The general function model of data fusion system, as shown in Fig. 5, mainly includes four levels of processing.

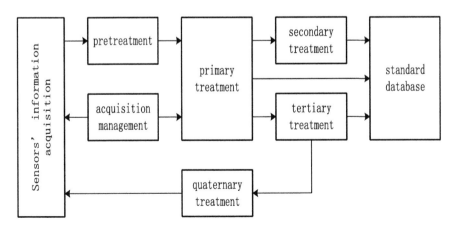

Fig. 5 The general function model of data fusion system

First level processing content is the data fusion of the primary treatment for information, including data registration and data association. Data registration is to make all the sensors with the same space and time by calibration technology. Data association is to associate each point trace produced by sensor with the existent standard point marked in the database, at the same time, to predict target state to achieve the purpose of continuous tracking target.

Second level processing content is attribute information fusion. To merge of attribute data from multiple sensors, in order to get the joint estimation of target identity.

Third level processing content is situational extraction and threat assessment. The situational extraction refers to the representation obtained by integration of unified situation from the incomplete data. The threat assessment is to assess the threat level of the integrated environment system.

Fourth level processing content is to optimize finally the data fusion results, such as optimizing the utilization of resources, optimization of sensor management in order to improve the system fusion effect.

4.2 The Level of the Data Fusion System

In multi-sensor data fusion, it must be fused gradually aimed at the different characteristics and the data acquisition method about the specific problems because of the diversification of the data for the different working principle and characteristics of different sensors. According to the degree of abstraction of data processing, data fusion is classified into three levels: the data level, feature level, and decision level.

1. The data level belongs to the lowest level of data fusion, as shown in Fig. 6. It is direct to fusion process for the raw data of same level sensors, after that, to carry

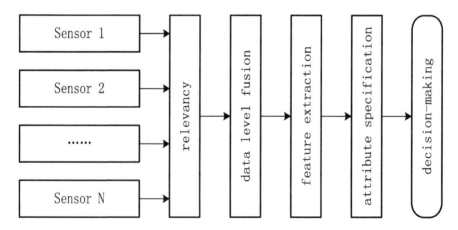

Fig. 6 The data level fusion

on the feature extraction and situational estimation if not information loss of single processing.

2. The feature level data fusion is to observe the target with each sensor and extract the representative characteristic information, thus to produce a single eigen-vector to estimate the identity fusion. The feature information extracted by this method should be a fully labeled amount or statistic of the data information, which must be related processing, as shown in Fig. 7.

The main advantage of the feature level data fusion can be compressed obser-vation information, which is easy to extract feature vector for real-time processing. The level of data fusion can also be divided into two types of the target feature fusion and the target state fusion, in which the target feature fusion must process the characteristics before fusion and classify the feature vector into combination. The target state fusion should complete the data matching and then realize the related parameters and state estimation. This method is more applied in the field of heterogeneous sensor target tracking.

3. The decision level belongs to the high level fusion, as shown in Fig. 8, the fusion results provide the basis for decisions directly. Multiple heterogeneous sensors are observed of the same goal, finishing independently extracting the feature information, and then through the relevant processing judgment, deci-sion level fusion to obtain the joint inference results finally.

Decision level data fusion is not high requirement for the bandwidth because of flexibility and good real-time performance; it can still give a final decision even one or several sensors is failure, so it has strong anti-interference ability and good fault tolerance.

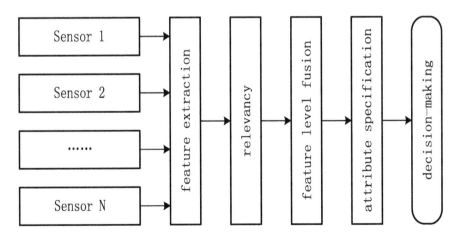

Fig. 7 The feature level fusion

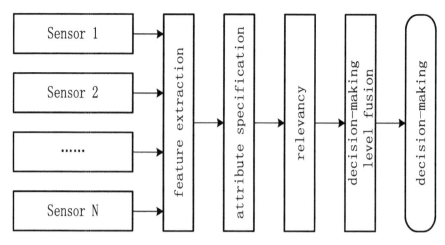

Fig. 8 The decision level fusion

4.3 Realization of Data Fusion Algorithm

At present, there are many data fusion methods that the commonly used classic methods include Kalman filter algorithm, weighted average method, the Bayes fusion estimation, D–S evidential reasoning method, fuzzy logic method, method of production rules, as well as the method of neural network, etc. (Wang 2013). In the multiple methods of data fusion, there is not the most optimal algorithm, each algorithm itself has advantage and disadvantage, and corresponds to the respective application fields, so the selection of fusion methods should accord to the practical problems on the advantage and consider the different methods of adaptive hybrid algorithm to solve the problem.

This paper introduces the filtering algorithm to improve the accuracy of calculation, reduces calculation error when the larger error appearing in the calculation results because of the error of the sensor itself, and the influences of the environment.

$\{X(t_k)\}$ is estimated state vector sequence of linear dynamic systems with random noise. Assuming its dynamic differential equation or state equation as follows:

$$X_k = \Phi_{k,k-1} X_{k-1} + \Gamma_{k-1} w_{k-1}$$

Here $X_k = X(t_k)$ is n-dimension state vector. $\Phi_{k,k-1}$ is $n \times n$ order one step state transition matrix.

For any positive integer k

$$X_k = \Phi_{k,0} X_0 + \left(\Phi_{k,k} \Gamma_{k-1} w_{k-1} + \Phi_{k,k-1} \Gamma_{k-2} w_{k-2} + \Phi_{k,k-2} \Gamma_{k-3} w_{k-3} + \cdots + \Phi_{k,1} \Gamma_0 w_0 \right)$$
$$= \Phi_{k,0} X_0 + \sum_{i=0}^{k-1} \Phi_{k,i+1} \Gamma_i w_i$$

The type is the linear relation that the system in the moment of state X_k at vector t_k, the initial state X_0, and the dynamic noise vector $w_0, w_1, \cdots w_{k-1}$ in the moment from t_0 to t_{k-1}.

5 Multi-sensor System Identification Technology

Due to the automatic comprehensive analysis capabilities of the multi-sensor system identification technology, it could be more effective and accurate to reflect the physical status of the object than any single sensor in the system.

5.1 Target Identification in Data Fusion of Attribute Level

The fusion center should identify the target of attribute level if each sensor in the multi-sensor system just reported the target attribute information to fusion center but could not identify itself. That is the fusion center fused the information from the sensors or subsystem and then assessed the state of the target. Generally, the diagnosis results of the target were based on the basic state which is identified by the attribute level.

Supposing that n sensors are made an effective measure on the characteristic of healthy, the j sensor output data is X_j, and $j = 1, 2, \cdots, n$

1. Direct average method

Supposing that n sensors are making a measure, so the output data fusion results of n sensors are

$$\overline{X} = \sum_{j=1}^{n} X_j/n$$

2. Method of weighted mean

Supposing the weight is $W_j, \sum_{j=1}^{n} W_j = 1$, so the output data fusion results of n sensors are

$$\overline{X} = WX = [W_1, W_2, \cdots, W_n][X_1, X_2, \cdots, X_n]^T$$

X_j follows the normal distribution $N\left(\mu_j, \sigma_j^2\right)$, by multivariate statistical theory, $\overline{X}$ follows the normal distribution $N\left(\sum_{j=1}^{n} W_j\mu_j, \sum_{j=1}^{n} W_j^2\sigma_j^2\right)$, Indicates that the expectations obtained after output fusion x is the weighted average of various sensors, the precision is

$$\sigma^{\overline{X}} = \sqrt{\sum_{j=1}^{n} W_j^2 \sigma_j^2}$$

It can be calculated the characteristic parameters after fusion of multi-sensor by using the method of Lagrange multiplier

$$\overline{X} = WX = \sum_{j=1}^{n} W_j X_j = \frac{\sum_{j=1}^{n} \frac{X_j}{\sigma_j^2}}{\sum_{j=1}^{n} \frac{1}{\sigma_j^2}}$$

Here, the optimal weight of Lagrange is $W_j = \frac{1}{\sigma_j^2 \sum_{j=1}^{n} \frac{X_j}{\sigma_j^2}}$

There is different importance in different parameters when the characteristic parameters involved in calculation, which can give the big weighting parameters to the relatively bigger one and give the small importance of parameter to the relatively small weight.

5.2   Target Recognition of the Decision Level Data Fusion

Each sensor and subsystem in the multi-sensor system identifies the data level when observed the information, and then reports the concluded estimation of the health state to the fusion terminal, in the end, makes the target recognition based on decision level fusion. That means the basic health state estimation information reported from each sensor or subsystem should be fused, then to make a diagnostic decision on the basis of the fusion results.

In the target recognition of the decision level fusion, the results of the combination of evidence are obtained by using of D–S evidence theory calculation formula, which can make health estimate with the judgment about the fusion results after according to the rules.

Setting m_1 and m_2 are two independent basic probability assignment in the same framework of 2^U, the focal elements, respectively, are $A_1, A_2, \cdots, A_k$ and $B_1, B_2, \cdots, B_k$, and setting

$$K = \sum_{\substack{i,j \\ A_i \cap B_j = \Phi}} m_1(A_i) m_2(B_j) \prec 1$$

The combination probability assignment function $m : 2^U \rightarrow [0, 1]$ for all nonempty sets A of basic probability assignment is

$$m(C) = \begin{cases} \dfrac{\displaystyle\sum_{\substack{i,j \\ A_i \cap B_j = \Phi}} m_1(A_i)m_2(B_j) \prec 1}{1-k} & \forall C \subset U, C \neq \Phi \\ 0 & C = \Phi \end{cases}$$

On the type, if $K \neq 1$, m is sure of a basic probability assignment; If $K = 1$, regarding m_1, m_2 as contradiction, but cannot combine the basic probability assignment. So the combination rules of the multiple attributes are as follows:

$$\begin{cases} m(\Phi) = 0 \\ m(A) = \dfrac{\sum_{\cap A_i = A} \prod_{i=1}^{n} m_i(A_i)}{1 - \sum_{\cap A_i = \Phi} \prod_{i=1}^{n} m_i(A_i)} = \dfrac{\sum_{\cap A_i = A} \prod_{i=1}^{n} m_i(A_i)}{\sum_{\cap A_i \neq \Phi} \prod_{i=1}^{n} m_i(A_i)} \quad \forall A \subset U, A \neq \Phi \end{cases}$$

Decision level fusion recognition is to fuse the preliminary identification results provided from each sensor or subsystem, especially using the evidence combination rule of D–S theory, to get the results of multiple sensor information fusion, which eliminates the uncertainty in a certain extent, then completes the health diagnosis after the decision basis on the fusion result.

5.3 Internet of Things Data Fusion About the Life and Health Monitoring

The current Medical Internet of Things data fusion is mainly concentrated in the level of application and network.

1. The application level of data fusion technology

The application level of data fusion based on the query mode, which used the relative mature distributed data base to applied to the data collection of Medical Internet of Things. This technology similar to the fusion technique of wireless sensor network who using interface protocols of SQL.

2. The network level of data fusion technology

The network level of data fusion could be seen as a special routing algorithm due to it combined with routing technology. The fusion algorithm focused on establishing the path which was the necessary way for data transmission and fusion from source node to sink node. Then the data collected by sensor nodes could along with this path to transmission and fusion per the special intermediate node. In the process of data fusion in wireless sensor network, previous node could treatment the collected data to simplify and compress. The most effective data will be fused to less number of packets and be sent to the next node.

The wireless sensor network of Medical Internet of Things focuses on the whole health state of detection area rather than emphasizes on the data of certain sensor.

For emergency monitoring, for example, it is necessary to get the whole information including the physical health status, environmental conditions, and so on. In this processing, not only the specific monitoring data were reported, but the original data transform to the state of health were more important.

6 Conclusions

In this research, the portable health sensing system containing multi-sensor integrated was introduced to as the study object. The main research contents including heterogeneous information unification, incomplete information null value interpolation and simplification, and distributed data base fusion processing. This system combined with the D–S evidence theory to manage the information conflict and improve the information fusion effectively, finally to realize the efficiency of the information monitoring and the results real-time feedback. Facing the massive data, the accuracy and effectiveness of information was very important, especially for application of the Medical Internet of Things. The data fusion technology, which introduced in this research, can successfully be reducing the consumption of network resources and it is a promising novel technology.

Acknowledgements Project supported by the National Natural Science Foundation of China (No. 71071059, 71301054), training plan of Guangdong province outstanding young teachers in higher education institutions (Yq2013009), open project of laboratory of innovation method and decision management system of Guangdong province (2011A060901001-03B).

References

Chen M, Gonzalez S, Vasilakos A et al (2011) Body area networks: a survey. J Mobile Netw Appl 16(2): 171–193. Available from http://www.cnki.net/

Guo L (2007) Research and implementation of bluetooth RFCOMM protocol. MS Thesis, XiDian University, Xi'an, China (in Chinese)

Liang W (2014) Research on wearable sensor network based human activity recognition technologies. MS Thesis, Nanjing University Nanjing, China (in Chinese)

Sandhya S, Devi KAS (2012) Analysis of bluetooth threats and v4.0 security features. In: Proceedings of the international conference on computing, communication and applications (ICCCA), pp 1–4. Available from http://www.cnki.net/

Shi ZG (2007) Design of intelligent safe-guard system based on embedded system. MS Thesis, Hunan University, Hunan, China (in Chinese)

Stiller C, Leon FP, Kruse M (2011) Information fusion for automotive applications—an overvise. J Inf Fusion 12(4):244–252. Available from http://www.cnki.net/

Tong L (2014) Study on human thermoregulation model based on local heat exchange in low pressure environment. MS Thesis, Qingdao Technological University, Qingdao, China (in Chinese)

Wang R (2013) Research on the algorithm of dissimilar sensors data fusion in wearable sensor network. MS Thesis, Nanjing University of Posts and Telecommunications, Nanjing, China (in Chinese)

Wei H (2014) The design and implementation of the wireless charging system based on electromagnetic induction. MS Thesis, Soochow University, Suzhou, China (in Chinese)

Zhang ZB, Yu M, Zhao X et al (2008) Wearable concurrent monitoring system for physiological parameters. J Space Med Medical Eng 21(1): 66–69. Available from http://www.cnki.net/

A Multi-scale Flexible Tactile-Pressure Sensor

Xiao-zhou Lü

Abstract The Bionic Electronics industry often requires that flexible artificial skin should have human skin-like multi-sensory functions. However, present sensors are difficult to achieve tactile and pressure sensory functions simultaneously due to the trade-off between measurement range and measurement precision. In order to deal with this problem, this paper presents a flexible tactile-pressure sensor based on bionic structure. The bionic structure combines several sensors to simulate different types of pressure stimulation receptors of the human skin. The upper sensors have high measurement accuracy to achieve the tactile sensory function. The lower sensor has a wide measurement range to achieve pressure sensory function. These two measurement ranges are able to realize multi-scale stress measurement of external pressure stimulation. The sensor, in turn, is able to perform the tactile and pressure sensory functions simultaneously.

Keywords Flexible tactile-pressure sensor · Multi-scale measurement · Bionic structure

1 Introduction

With the development of Bionic, Biomechanics and Biomedical Engineering, tactile or pressure sensors, especially those applied in wearable artificial skins or robots, have become a hot topic in Artificial Robot, Orthopedic Surgery, Prosthesis, rehabilitation and the footwear industry (Silvera-Tawil et al. 2015). For example, in artificial robots, tactile sensors are applied in the manipulator to mimic human tactile function to complete the grip action (Yousef et al. 2011). In orthopedic surgery, pressure sensors are used to measure the distribution of plantar stress (Zhang et al. 2010). In artificial prostheses, pressure sensors are applied for the

X. Lü (✉)
School of Aerospace Science and Technology,
Xidian University, Xi'an 710071, China
e-mail: xzlu@xidian.edu.cn

© Zhejiang University Press and Springer Science+Business Media Singapore 2017 49
C. Yang et al. (eds.), *Wearable Sensors and Robots*, Lecture Notes in Electrical
Engineering 399, DOI 10.1007/978-981-10-2404-7_4

measurement of interfacial stress between the amputation and the prosthetic limb (Sundara-Rajan et al. 2012). In rehabilitation medicine, tactile sensors are expected to create artificial skin to provide rehabilitation training for stroke patients (Lenzi et al. 2011). In the footwear industry, pressure sensors can be used to measure pressure distribution to aid engineers during the design of more comfortable shoes (Shu et al. 2010). These requirements and applications demonstrate the important role tactile and pressure sensors have come to play a major role in many fields.

Many approaches have been presented for the development of tactile or pressure sensors (Tiwana et al. 2012). These sensors are designed to mimic human skin sensory functions. According to the type of sensory function, these sensors can be divided into the following types: (1) Tactile sensors (De Maria et al. 2012), which are used to measure light pressure and have a smaller measurement range; (2) Pressure sensors (Lipomi et al. 2011), used to measure intense pressure and have a wider measurement range; and (3) Multi-sensory sensors, used to mimic multiple sensory functions such as tactile-heat and tactile-slipping functions.

For human skin, the tactile and pressure sensory functions are stimulated through the application of external pressure. In essence, they have the same function, which is called tactile-pressure sensory function. However, the tactile sensors presented above usually have a high sensitivity and can mimic tactile functions to measure light pressure, but do not have a wide measurement range to mimic pressure functions. On the other hand, the pressure sensors presented above usually have a wide measurement and can mimic pressure functions to measure intense pressure but do not have high sensitivity. There are few sensors that mimic the tactile and pressure function simultaneously. The reason is that the ranges required for measuring light pressure and intense pressure differs by several orders of magnitude. Therefore, there usually is a trade-off between measurement range and measurement precision.

To deal with the problem presented above, this paper presents a flexible tactile-pressure sensor based on a bionic structure. The bionic structure is based on plate capacitance and super elastic materials, and couples several sensors to simulate different types of pressure stimulation receptors found in human skin. The upper sensors have a high measurement sensitivity to achieve the tactile function, while the lower sensor has a wide measurement range to achieve pressure function. These two measurement ranges are able to realize multi-scale stress measurement of external pressure stimulation. The sensor, in turn, is able to perform the tactile and pressure functions simultaneously. The principle, design, structure, and fabrication of the flexible tactile-pressure sensor is also presented in this work.

2 Principle of the Tactile and Pressure Sensory Function of Human Skin

The human skin is a complex sensing system, which utilizes many different types of sensory functions to achieve sensing of external stimulation. The human skin sensory functions that react to external pressure stimulation are known as touch and

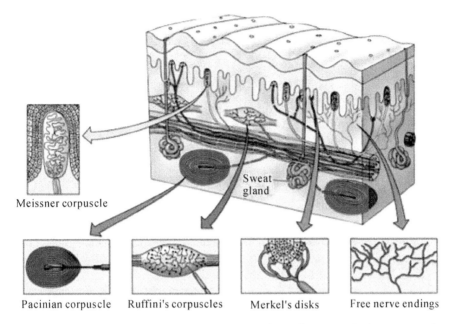

Fig. 1 The touch and pressure receptor can be divided into four types: Meissner corpuscle, Pacinian corpuscle, Merkel's disk, and Ruffini's corpuscles

pressure. The sensations of touch and pressure are produced by receptors in the skin. Touch and pressure receptors can be divided into four types: Meissner corpuscle, Pacinian corpuscle, Merkel's disk, and Ruffini's corpuscles (Abraira and Ginty 2013). The Meissner corpuscle and Merkel's disk have high sensitivity and can measure light pressure, while the Pacinian corpuscle and Ruffini's corpuscles have wide measurement range and can measure intense pressure. These four receptors allow the human skin to have tactile and pressure sensory functions (McGlone and Reilly 2010) (as shown in Fig. 1).

3 Bionic Structure of the Flexible Tactile-Pressure Sensor

The flexible sensor presented in this work consists of three layers, which mimic the three layer structure of the human skin. These three layers are: the upper tactile electrodes, which mimics the Meissner corpuscle and Merkel disk to achieve the tactile sensory function; the middle driving electrode; and the lower pressure electrode, which mimics the Pacinian corpuscle and Ruffini endings to achieve pressure sensory function (as shown in Fig. 2).

The upper layer is 10 mm × 10 mm in size and 1 mm thick. It couples four upper electrodes in the lower side. It also contains four bumps on its upper side. Each bump corresponds to an electrode on the other side. Every upper electrode

Fig. 2 The flexible sensor
consists of three layers, which
mimic the three layer structure
of the human skin

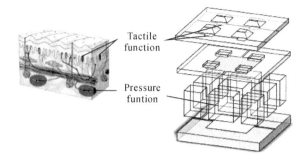

Tactile
function

Pressure
funtion

functions as a capacitor plate in conjunction with the middle driving electrodes. The aim of the bump is to detect light pressure applied on the surface of the sensor.

The middle layer is 10 mm × 10 mm in size and 1 mm thick. It couples one driving electrode in the lower side. It also contains empty space to allow each of the upper layer electrodes to move, thus increasing sensitivity by allowing a wider range of capacitance values to be measured.

The lower layer is 10 mm × 10 mm in size and 4 mm thick. It couples one lower electrode in the upper side. The lower electrode forms a capacitor along with the driving electrode. It also contains super elastic material to provide sensitivity for lower capacitances.

Therefore, the bionic structure is composed of the upper bumps, upper electrodes, upper space, driving electrode, the super elastic material, and the lower electrodes. This bionic structure forms the flexible tactile-pressure sensor. The sensor can be assembled in a sensor array and can be used for the creation of artificial skin (as shown in Fig. 3).

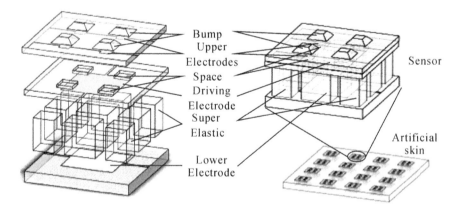

Fig. 3 The flexible tactile-pressure sensor is composed of a bionic structure, and can be used to create artificial skin

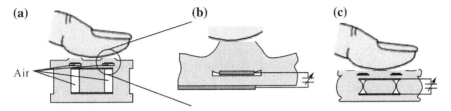

Fig. 4 The upper capacitances have a high sensitivity (**a**), and can measure light pressure (**b**), while the lower capacitance has a wide measurement range, and can measure intense pressure (**c**)

4 Working Principle of the Flexible Tactile-Pressure Sensor

If light pressure is applied on the surface of the flexible tactile-pressure sensor, the upper bumps are subject to the pressure and the upper layer is compressed, reducing the distance between the upper electrodes and the driving electrode. The upper capacitances, in turn, are changed. Measurement of light pressure can be associated with the measurement of the upper capacitances, thus achieving imitation of the tactile function (as shown in Fig. 4a). Since the upper electrodes have a small size, measurement of the upper capacitances will have high sensitivity. The upper space also can increase the sensitivity of the upper capacitances (as shown in Fig. 4b).

If intense pressure is applied on the surface of the flexible tactile-pressure sensor, the volume of the upper space does not vary substantially, while the lower super elastic material is compressed. The distance between the driving electrode and lower electrode is changed, which means that the lower capacitance is changed. Intense pressure can be measured according to the variation of the lower capacitance. Since the lower electrode has a large size, the lower capacitance has a wide measurement range, and the imitation of the pressure function of the human skin is achieved (as shown in Fig. 4c).

5 Conclusions

In this work, we present a flexible tactile-pressure sensor based on a bionic structure. The bionic structure couples several sensors to simulate different types of pressure stimulation receptors of the human skin. The upper sensors have high measurement accuracy to achieve the tactile sensory function. The lower sensor has a wide measurement range to achieve pressure sensory function. These two measurement ranges are able to realize multi-scale stress measurement of external pressure stimulation. The sensor, in turn, is able to perform the tactile and pressure sensory functions simultaneously.

References

Abraira VE, Ginty DD (2013) The sensory neurons of touch. Neuron 79(4):618–639

De Maria G, Natale C, Pirozzi S (2012) Force/tactile sensor for robotic applications. Sensor Actuat A Phys 175:60–72

Lenzi T, Vitiello N, De Rossi SMM, Persichetti A, Giovacchini F, Roccella S, Vecchi F, Carrozza MC (2011) Measuring human-robot interaction on wearable robots: a distributed approach. Mechatronics 21(6):1123–1131

Lipomi DJ, Vosgueritchian M, Tee BCK, Hellstrom SL, Lee JA, Fox CH, Bao ZN (2011) Skin-like pressure and strain sensors based on transparent elastic films of carbon nanotubes. Nat Nanotechnol 6(12):788–792

McGlone F, Reilly D (2010) The cutaneous sensory system. Neurosci Biobehav R 34(2):148–159

Shu L, Hua T, Wang YY, Li QA, Feng DD, Tao XM (2010) In-shoe plantar pressure measurement and analysis system based on fabric pressure sensing array. IEEE T Inf Technol B 14(3): 767–775

Silvera-Tawil D, Rye D, Velonaki M (2015) Artificial skin and tactile sensing for socially interactive robots: a review. Robot Auton Syst 63:230–243

Sundara-Rajan K, Bestick A, Rowe GI, Klute GK, Ledoux WR, Wang HC, Mamishev AV (2012) An interfacial stress sensor for biomechanical applications. Meas Sci Technol 23(8)

Tiwana MI, Redmond SJ, Lovell NH (2012) A review of tactile sensing technologies with applications in biomedical engineering. Sensor Actuat A Phys 179:17–31

Yousef H, Boukallel M, Althoefer K (2011) Tactile sensing for dexterous in-hand manipulation in robotics—a review. Sensor Actuat A Phys 167(2):171–187

Zhang X, Zhao Y, Xu Y (2010) Design and development of a novel MEMS force sensor for plantar pressure measurement. In: Book design and development of a novel MEMS force sensor for plantar pressure measurement. IEEE Computer Society, pp 1–5

Design of a Wearable Thermoelectric Generator for Harvesting Human Body Energy

Haiyan Liu, Yancheng Wang, Deqing Mei, Yaoguang Shi and Zichen Chen

Abstract This paper presents the design and fabrication of a wearable thermo-electric generator (TEG) with high power density for harvesting the human body heat energy. The proposed TEG was fabricated using a flexible printed circuit board (FPCB) as the substrate. The P-type and N-type thermoelectric blocks were made of Bi_2Te_3-based thermoelectric material and welded on the FPCB, and they were surrounded by the soft PDMS material. The prototyped TEG consisted of 18 thermocouples, which was connected by FPCB and silver paste over an area of 42×30 mm^2. The fabricated TEG could generate a voltage of 48 mV for a temperature difference of 12 K. Then, the TEG was mounted onto the human wrist skin to harvest the human body heat energy. Results showed that the measured output power was 130.6 nW at ambient temperature of 25 °C. Thus, the developed TEG has the potential for human body heat energy harvesting and utilized for the development of wearable self-powered mobile devices.

Keywords Wearable device · Thermoelectric generator · Human body heat · High power density

1 Introduction

Wearable electronic devices have been widely utilized in mobile electronics, consumer electronics, etc. To supply the auxiliary power, micro-batteries for these devices have been developed (Sammoura et al. 2003). However, the power gen-

H. Liu · Y. Wang (✉) · D. Mei · Y. Shi · Z. Chen
Key Laboratory of Advanced Manufacturing Technology of Zhejiang Province,
College of Mechanical Engineering, Zhejiang University, Hangzhou 310027, China
e-mail: yanchwang@zju.edu.cn

Y. Wang · D. Mei · Z. Chen
State Key Laboratory of Fluid Power and Mechatronic Systems,
College of Mechanical Engineering, Zhejiang University,
Hangzhou 310027, China

© Zhejiang University Press and Springer Science+Business Media Singapore 2017   55
C. Yang et al. (eds.), *Wearable Sensors and Robots*, Lecture Notes in Electrical
Engineering 399, DOI 10.1007/978-981-10-2404-7_5

erated by micro-batteries will not permanently last, which limits the application of the micro-batteries in wearable electronic devices. Also, the changes of the battery will lead to a lot of integration problems for wearable electronic devices. A flexible thermoelectric generator (TEG), worn on the human skin, can harvest the human body wasted energies for supplying the powers for wearable electronic devices, and have been widely studied (Yang et al. 2014).

The selection of the thermoelectric materials is a critical factor to affect the TEGs' power generation. Recently, conducting polymers, such as polyaniline and polypyrrole, have been utilized as the thermoelectric materials (Park et al. 2013). The polymers exhibit high flexibility and have been demonstrated as the potential for use on human skin, thus have been used for the development of wearable TEGs, but the thermoelectric performance is not high enough (Kim et al. 2013). In this study, we utilized the Bi_2Te_3-based powder as the thermoelectric material for the design of a novel flexible wearable TEG. To realize high power density generation and integration on the human skin, the structural design and fabrication of the wearable TEG will be studied and become one goal of this research.

In the past two decades, several wearable thermoelectric devices (TEDs) have been developed. Settaluri et al. (2012) reported a wearable TEG for human body energy harvesting. The design of the heat transfer system had been conducted. Lee et al. (2013) presented a flexible hybrid cell for simultaneously harvesting thermal and mechanical energies from human body. Kim et al. (2014a, b) developed a wearable TEG based on the glass fabric for harvesting the human body energy. Leonov (2013) fabricated a completely hidden TEG and extensively studied in different conditions. From the above research, the fabrication processes for the wearable TEGs are complicated and time-consuming (Itoigawa et al. 2005; Weber et al. 2006). In this paper, a hot-pressing process was utilized to fabricate the thermopiles, and they were welded onto a flexible printed circuit board (FPCB) to generate the flexible and wearable TEG. The fabrication method and process will be investigated.

In this study, we present a flexible wearable TEG for harvesting human body heat energy. The structural design and fabrication methods for the thermoelectric module and TEG were first described. Then, experimental characterization of the performance of the TEG was carried out. This is followed by the results and discussion.

2 Design

2.1 Design of the Wearable TEG

Figure 1a shows the concept view of the proposed wearable TEG, which can be easily worn on the human wrist. The TEG has 18 thermocouples with P-type and N-type thermoelectric blocks and they were welded onto a flexible PCB. To protect these blocks, a thin-layer of PDMS material (Glatz et al. 2006) was pasted around these blocks. As shown in Fig. 1b, the top surface of each thermocouple was

Fig. 1 Schematic diagram of the designed wearable TE device: **a** TEG wears on the wrist, **b** schematic view of the wearable TE device on the skin, **c** electron flow of diagram of the wearable TE device

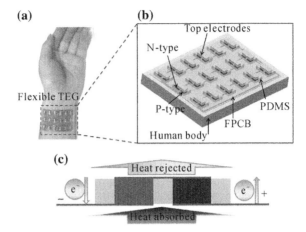

connected by the silver paste conductor. To prevent electrical contact between the TEG and skin, a polyimide film was designed on the top of silver paste due to its superior thermal and electrical resistance.

Geometrical details of the prototype TEG are shown in Fig. 2. The TEG has 18 integrated thermocouples and has an area of 42×30 mm^2. Both P-type and N-type thermoelectric blocks have the same square shape with dimensions of 2 mm $\times$ 2 mm $\times$ 1.5 mm, and were welded onto the FPCB. The distance between the P-type and N-type elements in one thermocouple is 1 mm, as shown in Fig. 2b. The protective PDMS film has a thickness of 1.5 mm, the P-type and N-type junctions were surrounded by the PDMS. As in B-B' cross-sectional view shown in Fig. 2c, the distance between the adjacent thermocouples is 5.5 mm (Suemori et al. 2013).

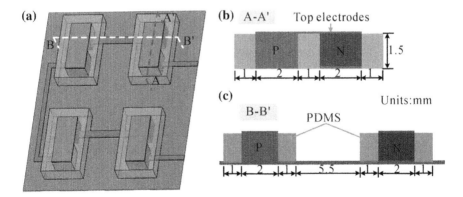

Fig. 2 **a** Schematic view of four P–N thermocouples, **b** dimensions of each P–N junction pair, **c** dimensions of the adjacent P–N junction pairs in each row

2.2 Modeling of the Wearable TEG

For the proposed TEG, the number of the thermocouple is n, the relative Seebeck coefficient of the used thermoelectric material is α ($=\alpha_p - \alpha_n$), where α_p and α_n represent the Seebeck coefficients of the N-type and P-type thermoelectric blocks. The output voltage (V) could be generated as

$$V = n\alpha\Delta T_G \tag{1}$$

where $\alpha = \alpha_p - \alpha_n$, ΔT_G is the temperature difference between hot and cold junctions.

For the designed TEG, the analytical loading voltage (V_{AL}) is

$$V_{AL} = V\frac{R_L}{R_L + R_G}, \tag{2}$$

where R_G and R_L are the internal electrical resistance and external loading, respectively.

Then, the output power (P) for the designed TEG can be calculated as

$$P = V^2\frac{R_L}{(R_L + R_G)^2} \tag{3}$$

When R_G equals R_L, the maximum power (P_{max}) can be obtained as

$$P_{max} = \frac{V^2}{4R_G} = \frac{n^2\alpha^2}{4R_G}\Delta T_G^2 \tag{4}$$

As in Eqs. (1) and (4), the output voltage (V) and output power (P) can be increased by increasing the number of TE thermocouples, or reducing the internal electrical resistance, etc.

The power density (E) defined as the generated output power versus the surface area of the TEG. Thus, E can be calculated as

$$E = \frac{P}{A} = \frac{P}{a \times b}, \tag{5}$$

where a and b are the length and width of the TEG, respectively.

3 Fabrication

3.1 Thermoelectric Materials and Fabrication

To harvest the heat at low temperature, the Bi_2Te_3-based thermoelectric powder (Photoelectric Inc.) with the particle size of 10–20 μm (1200 mesh) was selected. The process to fabricate the thermocouple blocks was shown in Fig. 3. First, the powder (Fig. 3a) was put into a hollow steel die and fabricated by hot-pressing process to make the thermoelectric block samples, as shown in Fig. 3b. During the hot processing, the pressure and temperature were kept at 60 MPa and 550 °C and lasted for 30 min. Then, the block samples were cooled at room temperature. By using the hot-pressing process, the mechanical properties and electrical conductivity of the thermoelectric block sample could be enhanced due to the finer grain size of the thermoelectric material (Mix et al. 2015). After hot processing, the block samples have a cylindrical shape with the diameter of 12.7 mm and thickness of 4 mm, as shown in Fig. 3c. The measured relative density of the P-type and N-type samples are about 92.1 and 94.8 %, respectively. Finally, these bulk samples were cut into a rectangular shape of 2 mm × 2 mm × 1.5 mm by wire-cutting process, as in Fig. 3d.

3.2 Fabrication of the Wearable TEG

After fabrication of the P-type and N-type thermoelectric blocks, we have developed a method to manufacture the wearable TEG. The detailed fabrication process

Fig. 3 The process of fabricating the rectangular thermoelectric block samples: **a** thermoelectric powder; **b** hot-pressing; **c** cylindrical block sample after pressing; **d** rectangular block sample after wire-cutting process

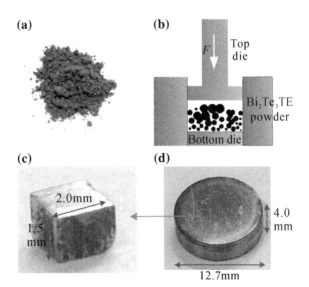

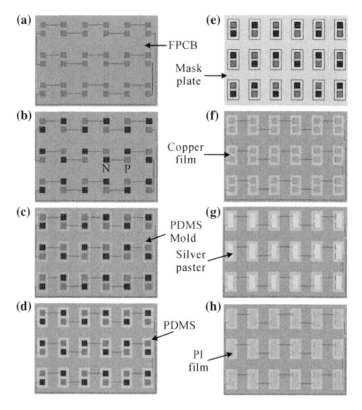

Fig. 4 Fabrication process of the TEG: **a** FPCB with bottom electrodes fixed on the glass wafer; **b** fixing the TE legs on the FPCB by soldering process; **c** a PDMS mold placed around the TE legs; **d** generating a PDMS film; **e** placing the mask plane; **f** deposition of the copper film; **g** screen-printing the silver paste film; **h** covering the PI film

to manufacture the TEG was shown in Fig. 4. To increase the flexibility, the flexible circuit printed board (FPCB) was selected as the bottom electrical connection layer of the wearable TEG. The fabricated FPCB for the device was shown in Fig. 4a. Then, the TE rectangular bulks, FPCB, and DI wafer were cleaned by acetone (>99.7 %) and alcohol (>99.5 %), respectively. By using the soldering process, the TE bulks were bonded onto the FPCB at the corresponding right position, as in Fig. 4b. In Fig. 4c, an open mold made of PDMS was placed around the TE bulks. Then, the PDMS compositions were poured into the mold and cured at 80 °C for 180 min to generate a thin film around the TE bulks (Fig. 4d). It can be used to fill the gaps between the nearby P-type and N-type TE bulks and provide a surface plane for the top electrical connection. In Fig. 4e, a mask was placed on the top surface of the PDMS. A thin copper film was deposited on the top surface of TE bulks by chemical deposition method, as in Fig. 4f. The thickness of the copper thin film is about 20 µm. The contact resistance between the copper film and TE bulks was measured with a low value, which means that the TE bulks are firmly

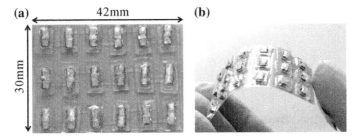

Fig. 5 The fabricated TEG **a** in top view, and **b** bending to a curved shape

connected with the copper film. It is well known that the adhesion between the copper film and PDMS layer will resulted in the exfoliation of the copper film (Kim et al. 2014a, b). Here, we fabricated a screen-printed silver film to form the interconnection between the TE bulks. The silver film was annealed at 120 °C for 10 min to enhance the hardness of silver paste, as shown in Fig. 4g. Finally, a polyimide film was cut into a rectangular shape to cover the silver paste to prevent the physical and electrical contact of the TEGs, and may also protect the silver paste from damaging.

Figure 5a shows the photograph of the fabricated TEG, it has 18 P–N thermo-couples with the overall dimensions of 42 mm × 30 mm. The total internal resistance of these thermocouples is about 60 Ω. This TEG has a remarkable flexibility and can be integrated into a curved surface, as in Fig. 5b. After several times of bending, there have little changes of the electrical properties of the TEG. That means the fabricated TEG can withstand repeated bending, but the fabrication process for the designed TEG still needs to be improved.

4 Experimental Characterization

The performance of the designed TEG was tested and characterized by using the experimental setup, as shown in Fig. 6 (Kim et al. 2014a, b). The TEG with bottom surface was placed on a heating platform, the other side was exposed to the air. To reduce the thermal contact resistance, a thin thermal paste was applied to the contact surface between the TEG and the heating platform. A FPCB transfer board was designed for the connection of the TEG, it can measure the voltage, current, and resistance values of the output signals. The internal resistance was measured with the value of 60 Ω.

Then, a thermal infrared (IR) camera (T2 Large infrared thermal imager) was utilized to measure the temperature difference of the wearable TEG (Francioso et al. 2013). Figure 7 shows the measured temperature difference of the TEG at $T_h = 120$ °C. Results show that the TEG can generate a relatively uniform tem-perature difference.

Fig. 6 Experimental setup to characterize the performance of the TEG

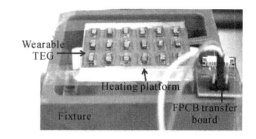

Fig. 7 Thermal camera image of the wearable TEG

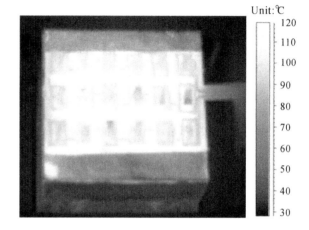

5 Results and Discussion

5.1 TEG Characterization

Figure 8a shows the measured and analytical model calculated output voltages of the TEG versus temperature difference. The temperature difference between the top and bottom surfaces of the TEG was measured by the IR camera. As shown in Fig. 8a, increasing the temperature difference from 2 to 12 K, the output voltage is increased correspondingly. At $\Delta T = 12$ K temperature difference, the measured output voltage is 48 mV. Results showed that there has a linear relationship between the measured output voltage and temperature difference. Because the TEG has 18 thermocouples, the output voltage generated by the developed TEG is higher than that of some reported wearable TEGs. To predict the output voltage, it can be calculated by Eq. 1. Here, $n = 18$, $\alpha_p = 120$ μV/K, $\alpha_N = -110$ μV/K, thus $\alpha = 230$ μV/K. So, the analytical calculated results of output voltage for the TEG are showed in Fig. 8a. It has the same trend as the measured results. There has no significant change (less than 5 %) between the experimental measured and analytical calculated output voltages. It demonstrated that the model can predict the output voltage of the developed TEG.

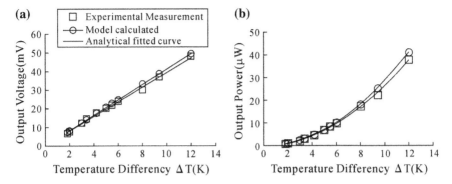

Fig. 8 Experimental measured and analytical calculated output voltage (**a**) and power (**b**) versus temperature differences for the developed TEG

Figure 8b shows the measured and analytical predicted output power of the developed TEG. The predicted output power was calculated by using Eq. 3. As the increasing of the temperature difference, the measured and predicted output power values were also increased. Generally, the same trend was observed both in the experimental and model predicted output power. At $\Delta T = 12$ K, the measured and model predicted output powers are about 37.8 and 41.1 μW, respectively. The maximum discrepancy between the measured and model calculated output power is about 8.73 % and demonstrated that the output power of the TEG can be predicted.

The output voltage and power were also measured as increasing of the loading resistance and shown in Fig. 9a, b, respectively. The model calculated output voltage and power were also shown in Fig. 9. In Fig. 9a, the output voltage was increased as the increasing of the loading resistance at $\Delta T = 11$ K temperature difference. The highest discrepancy between the experimental measured and model calculated output power was found at the loading resistance of 900 Ω with the value of 16.7 %. It was contributed by the wire resistance and the contact resistance during the experiment. The effects of these factors we had omitted during the model calculation. For the generated output power (Fig. 9b), it first increased as the increasing of the loading resistance, and then reached a peak value of 8.3 μW at loading resistance of 60 Ω, and then decreased as the increasing of loading resistance. For the maximum output power, the voltage is about 22.4 mV and current is 0.37 mA.

For the developed TEG, power density is a critical parameter to characterize the heat harvesting performance. It can be calculated by using Eq. 5. For the developed TEG, at $\Delta T = 11$ K, the maximum power density equals to 0.67 μW/cm^2 was observed at matched load resistance. For the designed TEG, approximately 90 % of the surface area contributed little effect on the thermoelectric conversion due to these areas only consists of the substrate and electrode layers.

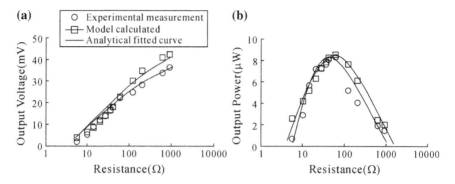

Fig. 9 Experimental measured and analytical calculated output voltage (**a**) and power (**b**) versus loading resistance at $\Delta T = 11$ K for the developed TEG

5.2   TEG Test on the Human Body

For further demonstration of the use of the wearable TEG on the human body, we applied the TEG to human wrist skin as a body heat energy harvester, as shown in Fig. 10. The TEG generates an open-circuit output voltage of 2.8 mV and an output power of 130.6 nW at an air temperature of 25 °C. The output power per cm^2 can be calculated to be a power density of 10.4 $nWcm^{-2}$. The human body surface is about 1.5 m^2, given these results, the proposed wearable TEG could obtain 1.56 μW from the total skin area of a typical adult. The measured power density is enough to active sub-microwatt wearable devices such as a temperature sensor or CMOS image sensor. This verifies that the developed TEG is useful in generating electricity by body heat and providing power autonomy to portable microwatt electronic devices.

Fig. 10 Photographs of the wearable TE device wearing on the wrist. (The authors declare that they have no conflict of interest. Research involved human participant compliance with ethical standards.)

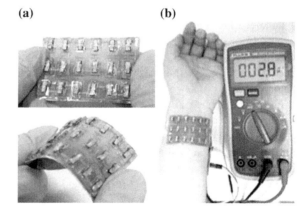

6 Conclusions

A novel wearable TEG with high flexibility and output power density has been developed and utilized for human body heat energy harvesting. The structural design, working principle, and fabrication method for the proposed TEG have been presented. Experimental characterization showed that the developed wearable TEG can produce 8.3 μW output power (about 0.37 mA and 22.4 mV) at the temperature difference of 11 K. When the TEG was worn on the wrist, it could generate 130.6 nW at ambient temperature of 25 °C.

The developed TEG can be easily applied to human wrist to harvest the body heat energy. It could be used to activate sub-microwatt wearable electronic devices, such as a temperature sensor or LED lights. Future work will be focused on optimal structural design of the TEG with efficient energy harvesting abilities for wearing on the human body.

Acknowledgments The author would like to acknowledge the supports from the National Basic Research Program (973) of China (2011CB013300), National Natural Science Foundation of China (51275466), Science Fund for Creative Research Groups of National Natural Science Foundation of China (51221004), and the Fundamental Research Funds for the Central Universities (2015QNA4004).

References

Francioso L, De Pascali C, Bartali R et al (2013) PDMS/Kapton interface plasma treatment effects on the polymeric package for a wearable thermoelectric generator. ACS Appl Mater Interfaces 5(14):6586–6590. doi:10.1021/am401222p

Glatz W, Muntwyler S, Hierold C (2006) Optimization and fabrication of thick flexible polymer based micro thermoelectric generator. Sens Actuators A 132:337–345. doi:10.1016/j.sna.2006.04.024

Itoigawa K, Ueno H, Shiozaki M et al (2005) Fabrication of flexible thermopile generator. J Micromech Microeng 15:S233–S238. doi:10.1088/0960-1317/15/9/S10

Kim G-H, Shao L, Zhang K et al (2013) Engineered doping of organic semiconductors for enhanced thermoelectric efficiency. Nat Mater 12(8):719–723. doi:10.1038/NMAT3635

Kim M-K, Kim M-S, Lee S et al (2014a) Wearable thermoelectric generator for harvesting human body heat energy. Smart Mater Struct 23:105002–105007. doi:10.1088/0964-1726/23/10/105002

Kim SJ, We JH, Cho BJ (2014b) A wearable thermoelectric generator fabricated on a glass fabric. Energy Environ Sci 7:1959–1965. doi:10.1039/c4ee00242c

Lee S, Bae S-H, Lin L et al (2013) Flexible hybrid cell for simultaneously harvesting thermal and mechanical energies. Nano Energy 2:817–825. doi:10.1016/j.nanoen.2013.02.004

Leonov V (2013) Thermoelectric energy harvesting of human body heat for wearable sensors. IEEE Sens J 13(6):2284–2291. doi:10.1109/JSEN.2013.2252526

Mix T, Müller K-H, Schultz L et al (2015) Formation and magnetic properties of the L10 phase in bulk, powder and hot compacted Mn–Ga alloys. J Magn Magn Mater 391:89–95. doi:10.1016/j.jmmm.2015.04.097

Park T, Park C, Kim B et al (2013) Flexible PEDOT electrodes with large thermoelectric power factors to generate electricity by the touch of fingertips. Energy Environ Sci 6:788–792. doi:10.1039/c3ee23729j

Sammoura F, Lee KB, Lin L (2003) Water-activated disposable and long shelf life microbatteries. Sens Actuators A 111:79–86. doi:10.1016/j.sna.2003.10.020

Settaluri KT, Lo H, Ram RJ (2012) Thin thermoelectric generator system for body energy harvesting. J Electron Mater 41(6):984–988. doi:10.1007/s11664-011-1834-3

Suemori K, Hoshino S, Kamata T (2013) Flexible and lightweight thermoelectric generators composed of carbon nanotube–polystyrene composites printed on film substrate. Appl Phys Lett 103(15):153902–153904. doi:10.1063/1.4824648

Weber J, Potje-Kamloth K, Haase F et al (2006) Coin-size coiled-up polymer foil thermoelectric power generator for wearable electronics. Sens Actuators A 132:325–330. doi:10.1016/j.sna.2006.04.054

Yang Y, Xu GD, Liu J (2014) A prototype of an implantable thermoelectric generator for permanent power supply to body inside a medical device. ASME J Med Devices 8:014507-6. doi:10.1115/1.4025619

Three-Axis Contact Force Measurement of a Flexible Tactile Sensor Array for Hand Grasping Applications

Yancheng Wang, Kailun Xi, Deqing Mei, Zhihao Xin and Zichen Chen

Abstract This chapter explores the development of a flexible tactile sensor array based on pressure conductive rubber for three-axis force measurement in grasping application. The structural design, testing principle, fabrication process, and characterization of the tactile sensor array are presented. Three-axis force measurement performances for all nine sensing units have been carried out. The full-scale force measurement ranges of the tactile sensor array in x-, y- and z-axis are 5, 5, and 20 N, respectively. The corresponding sensitivities in x- and y-axis are 0.838 and 0.834 V/N, respectively. In z-axis, the sensor array has two sensitivities: 0.3675 V/N for 0–10 N and 0.0538 V/N for 10–20 N measurement ranges. Then, the tactile sensor array has been mounted onto a human hand finger and used to measure the real-time 3D contact forces during grasping application. Results showed that the developed tactile sensor array features high sensitivities and has the potential for real-time tactile images for gripping positioning and 3D grasping force feedback control in grasping applications.

Keywords Three-axis force · Flexible tactile sensor array · Pressure conductive rubber · Grasping

1 Introduction

Recently, tactile sensor arrays have been widely utilized in medical surgery and robotic manipulations. For prosthetic hand grasping application, tactile sensor arrays with high flexibility and real-time 3D contact force measurement abilities are

Y. Wang (✉) · K. Xi · D. Mei · Z. Xin · Z. Chen
Key Laboratory of Advanced Manufacturing Technology of Zhejiang Province, College of Mechanical Engineering, Zhejiang University, Hangzhou 310027, China
e-mail: yanchwang@zju.edu.cn

Y. Wang · D. Mei · Z. Chen
State Key Laboratory of Fluid Power and Mechatronic Systems, College of Mechanical Engineering, Zhejiang University, Hangzhou 310027, China

© Zhejiang University Press and Springer Science+Business Media Singapore 2017
C. Yang et al. (eds.), *Wearable Sensors and Robots*, Lecture Notes in Electrical Engineering 399, DOI 10.1007/978-981-10-2404-7_6

generally required. They can provide real-time distributed contact force measurement; generate tactile images for grasping positioning, object's shape and/or surface texture recognition (Tiwana et al. 2011).

For the development of tactile sensor array, there are several types of sensing principles, including piezoresistive (Zhang et al. 2015), piezoelectric (Dahiya et al. 2009), capacitive (Wang et al. 2014; Liang et al. 2015), optical (Hoshino and Mori 2008), organic filed effect transistors (OFETs). Among these sensing principles, piezoresistive tactile sensor array based on pressure conductive rubber, usually features good resolutions and linearity, and low cost fabrication, has been widely utilized. INASTOMER pressure conductive rubber, using it as the sensing material, has been utilized to develop the pressure sensor for normal force and slip detection (Xi et al. 2015). In this paper, we also utilized this conductive rubber as the sensing material to develop a flexible tactile sensor array for 3D contact force measurement.

For 3D contact force measurement during robot grasping application, several attempts have been made in the past two decades. A novel flexible capacitive tactile sensor array with the capability of measuring both normal and shear force distribution using polydimethylsiloxane (PDMS) as a base material was developed (Lee et al. 2008). This sensor array features high sensitivity for micro-force measurement, but cannot be utilized for large-scale force detection, especially in grasping application. Liu et al. (2009) designed a new 3D tactile sensor by using pressure-sensitive electric conductive rubber (PSECR) as the sensing material for plate normal and shear forces measurement. The developed sensor has a relatively large dimension and may not suit for the wearable applications. In this study, we will develop a flexible tactile sensor array with low spatial resolution and small dimensions, which can be easily mounted on the human hand for contact force measurement in grasping application and become one goal of this research.

In this paper, we developed a flexible tactile sensor array with the capability of measuring both normal and shear contact forces using INASTOMER pressure conductive rubber for hand grasping applications. The detailed structural design, working principle, and fabrication of the sensor array are presented, followed by the experimental characterization of the three-axis force measurement performance. Then, this sensor array is mounted on a finger for 3D contact force measurement during grasping application. Finally, results and discussions are conducted.

2 Sensor Array Design and Fabrication

2.1 Design and Fabrication

Previously, we have developed a flexible tactile sensor array based on pressure conductive rubber with the capabilities to measure 3D forces (Xi et al. 2015). This sensor array can be mounted onto the hand finger for distributed grasping force

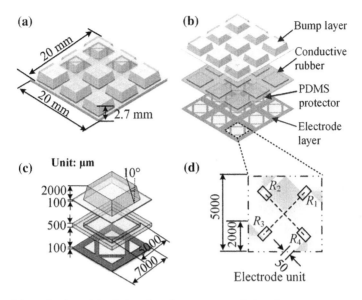

Fig. 1 **a** Schematic view of the designed tactile sensor array, **b** three layers structure, **c** exploded view of the sensing unit, **d** five electrodes generate four resistors

measurement. Figure 1a shows the schematic view of the designed flexible sensor array, it has $3 \times 3 (=9)$ sensing units. Each unit consists of three main layers: bottom electrode layer, top PDMS bump, and between them are the pressure conductive rubber chips, which is surrounded by a thin-layer of PDMS protector, as shown in Fig. 1b. The overall dimensions of the sensor array are 20 mm $\times$ 20 mm $\times$ 2.7 mm, and the sensing unit has the dimensions of 5 mm $\times$ 5 mm. Thus, spatial resolution defined as the distance between two adjacent units is 7 mm. The shape and dimension of the PDMS bump is critical for the transmission and detection of the 3D forces. In the sensor array design, the rectangular shape of the PDMS bump with dimensions of 5 mm $\times$ 5 mm $\times$ 2 mm was utilized.

For the sensing unit, five electrodes are designed on the flexible electrode layer and underneath the pressure conductive rubber chip. These five electrodes generate four resistors, as shown in Fig. 1d, can be used for 3D forces measurement. When normal force was applied on the bump, the PDMS bump will be compressed, which leads to the reduction of the thickness of the conductive rubber. Then, the resistance of these resistors will be increased. When shear force applied, it will generate a torque on the conductive rubber. As a result, the resistances of two resistors will be increased and the other two will be decreased.

By using the process developed in (Xi et al. 2015), the tactile sensor array was successfully fabricated as shown in Fig. 2. In Fig. 2a, five electrodes were fabricated on the flexible electrode layer. Then, rubber chips with the dimensions of 5 mm $\times$ 5 mm were bonded onto the electrode layer, as in Fig. 2b. In Fig. 2c, a thin-layer of PDMS protector was fabricated and surrounded the rubber chips.

Fig. 2 The photographs for the fabrication of the tactile sensor array: **a** the electrode layer, **b** rubber chips bonded onto the electrode layer, **c** PDMS protector layer surrounded the rubber chips, **d** final fabricated tactile sensor array

Finally, a PDMS bump layer was manufactured and glued onto the rubber chips. Figure 2d shows the final fabricated tactile sensor array with 9 sensing elements.

2.2 Basic Operating Principle

The designed tactile sensor array could be utilized for three-axis forces measurement. The working principle of 3D force measurement has been described in (Xi et al. 2015). After applying an external force on the sensing unit, we can obtain four voltage signals measured by the resistance change of four resistors, and then decompose the external force into three components: F_x, F_y, and F_z.

For a sensing unit, we have derived the equations to calculate the F_x, F_y and F_z, as in (Xi et al. 2015):

$$F_x = k_1 V_1 - k_2 V_2 - k_3 V_3 + k_4 V_4 \qquad (1)$$

$$F_y = k_1 V_1 + k_2 V_2 - k_3 V_3 - k_4 V_4 \qquad (2)$$

$$F_z = \lambda(k_1 V_1 + k_2 V_2 + k_3 V_3 + k_4 V_4) \tag{3}$$

where k_1, k_2, k_3, k_4, and λ are the calibration coefficients for four sensing resistors, they are determined by experimental measurements (Xi et al. 2015; Liu et al. 2009).

3 Experimental Setup and Procedure

3.1 Scanning Circuit

In order to measure exerted three-axis force, the scanning circuit matching the proposed tactile sensor array needs to be designed. Figure 3 shows the schematic diagram of the readout circuit for this 3×3 sensor array. It can measure and record the voltage outputs of the sensor array. For the row scanning, a 16-channel multiplexer (ADG1206) controlled by the MCU (STC89S52) provides the driving positive voltages (+5 V) to $12(= 3 \times 4)$ electrodes from sensing units. In the vertical direction, all 3 analog output voltage signals of each column are connected to the operational amplifiers with reference resistors R_{ref}, and these voltages will be converted to the digital signals by a high-precision 12-bit ADC chip (ADS 7950). Scanning requests are made by PC through a RS232 interface.

Data acquisition and communications are transferred to the PC through RS232 interface for visualization and data analysis. For this scanning circuit, the working serial baud rate was set at 19,200 bit/s. Thus, the scanning speed for the designed tactile sensor array is about 24 frames per second, this speed is a little slow but sufficient for real-time 3D grasping force measurement.

Fig. 3 Schematic view of the readout circuit for the tactile sensor array

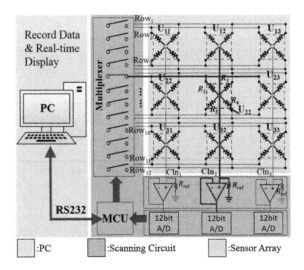

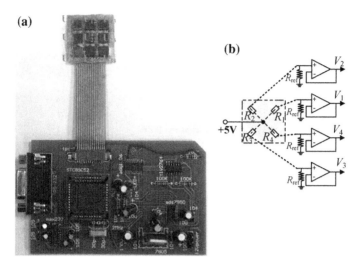

Fig. 4 **a** Photography of the sensor array connected with the designed scanning circuit. **b** Readout circuit for four resistors in each unit

Figure 4b shows the fabricated tactile sensor array, which is connected with the developed scanning circuit through a pin board connector. The distance between the sensor and scanning circuit can be extended by the flexible flat cable (FFC), and depends on the specific applications. For each unit, schematic view of the readout circuit for each resistor is shown in Fig. 4b.

3.2 3D Force Measurement Calibrations

To study the 3D force measurement performance, we have developed a visualization force measurement system, as shown in Fig. 5a. The system includes a 3-axis manual linear motion stage (Newport 460P, USA), a 3-axis force load cell (Interface 3A120, USA), related scanning circuit, and PC with the matching Matlab program to record data and real-time display. The 3-axis linear stage with a fixed round plastic bar (Fig. 5b) was utilized to apply the normal and tangential forces on the sensing unit. The plastic bar with a diameter of 5 mm at the bottom, contacted with the top surface of the bump unit via acrylic foam tape. The 3-axis linear motion stage, which can be positioned precisely within 0.001 mm was used to generate minute force change. The Interface 3D force load cell with resolution of 0.01 N in each axis was used to measure the applied force as the datum and for comparison.

During the experiment, the linear stage with the round plastic bar applied an external force on the sensing unit. Then, the purchased force sensor measured the generated force as the datum. At the same time, the original four voltage signals of

Fig. 5 Experimental setup to study the sensing performance of the sensor array: **a** overview, **b** a round plastic bar used to apply the normal and shear forces

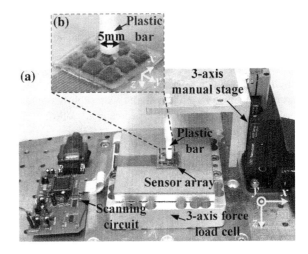

four resistors were recorded, and using Eqs. (1)–(3) to decompose the F_x, F_y, and F_z, and compared to that of the purchased force sensor.

3.3 Grasping Force Measurement for the Tactile Sensor Array

After 3D force measurement calibration, we have mounted the sensor array on the hand finger for grasping force measurement. As shown in Fig. 6a, the sensor array was worn on a human hand finger and connected with the sensing circuit via the FFC. Then, the hand grasped a plastic cup of water. The detailed close-up view of the sensor array contacted with the plastic cup was shown in Fig. 6b.

During the experiment, the scanning circuit measured the original voltages which were generated in each resistor, and then these voltages were used to

Fig. 6 Measurement of the grasping force when the tactile sensor array mounted onto a human hand finger for grasping of a cup of water: **a** overall view, **b** close-up view. (The authors declare that they have no conflict of interest. Research involved human participant compliance with ethical standards.)

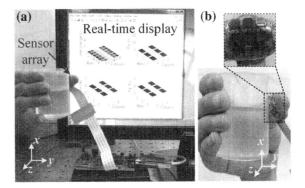

decompose the F_x, F_y, and F_z in each unit for the sensor array. The original voltage signals and 3D force components were transmitted to the computer and real-time displayed on the computer screen.

For the comparison, fast grasping and slow grasping were conducted. In each condition, four phases were conducted. Phase 1, the hand was not contacted with the plastic cup; Phase 2, the hand was slightly contacted and grasped the cup with the increasing of the grasping force, and followed with the result that the deformation of the plastic cup was increased; Phase 3, the hand applied an even greater force on the cup and lasted for about 15 s; Phase 4, the hand was releasing until the hand was not contacted with the cup. During these phases, the contact forces between the cup and finger were recorded by the scanning circuit and real-time displayed on PC screen.

4 Results and Discussions

4.1 3D Force Measurement of the Sensing Unit

For the 3D force calibration of sensing unit, the force measurement ranges for z-axis is 20 N and for x- and y-axis are 5 N, respectively. Figure 7 shows the measured original voltages of four resistors in the sensing unit. According to the reported test results in the tangential voltage differences at various F_z values at 0–20 N ranges do not change significantly. Thus the tangential forces' sensitivities are perceived to be

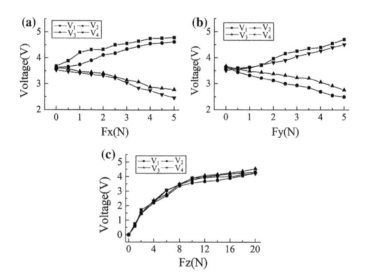

Fig. 7 Original measured voltages of four resistors when applying **a** shear force in x-axis, **b** shear force in y-axis, and **c** normal force in z-axis

stable. For x- and y-axis forces measurement, we firstly applied a constant normal force (about 10 N) on the top surface of the bump, then applied the transverse forces though moving 3-axis manual stage. In Fig. 7a, when shear force was applied on the x-direction, two voltages of the resistors on one side were increased as the increasing of the applied forces. The other two voltages of the resistors in the other side were decreased at the same time. Same results were observed when the force applied in the y-direction, as in Fig. 7b. For the normal force applied in z-direction, four original voltages are increased as the external force increased from 0 to 20 N, as shown in Fig. 7c. These four voltages generated by four resistors in one sensing unit have almost the same characteristics, and that demonstrated the developed tactile sensor array has the uniform resistors fabrication.

According to Eqs. (1)–(3), three-axis forces are calculated and shown in Fig. 8. In Fig. 8a, the calculated voltage was increased linearly as the increasing of the applied forces in x-direction. The fitting curve for the measured point is $y = 0.838x + 0.148$. The slope of this line can be used to characterize the sensitivity in x-direction of the sensing unit. Here, the sensitivity in x-direction is 0.838 V/N. In the same way, the sensitivity in y-direction is 0.834 V/N, as shown in Fig. 8b.

For the calculated voltages, when normal force applied, two distinguished sections can be observed. For 0–10 N, the fitting curve has the slope of 0.369. While for 10–20 N, the fitting curve's slope is about 0.054. That means two sensitivities of the sensing unit for z-axis force measurement could be obtained: 0.3675 V/N for 0–10 N force measurement range and 0.0538 V/N for larger force measurement

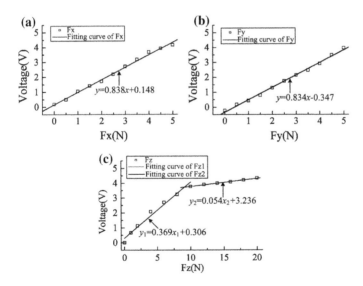

Fig. 8 Calculated voltages in three-axis of the sensing unit when applying **a** shear force in x-axis, **b** shear force in y-axis, and **c** normal force in z-axis

(10–20 N). That is to say, the developed tactile sensor array can be used for small contact normal force measurement with a relatively high sensitivity and larger normal force detection with a relatively low sensitivity (0.0538 V/N). Results obtained in this study also show that the calculated voltage components: V_x, V_y, and V_z have a relatively good linear response under applied forces. Thus, the developed tactile sensor array could be utilized for 3D force measurement by monitoring the voltages of four resistors in each sensing unit.

4.2 Real-Time 3D Forces Measurement in Hand Grasping Application

As described above, the fabricated tactile sensor array features high flexibility and can be easily worn on the human wrist or fingers. In this study, we have mounted this sensor array onto a human hand finger for three-axis grasping forces measurement during the hand grasping of a cup of water. Figure 9 shows the measured F_x, F_y, and F_z in five sensing units in grasping. During the process, we found four sensing units in the corners were not contacted with the plastic cup and almost had no forces readout. From the force curves in Fig. 9a, four phases were observed. In

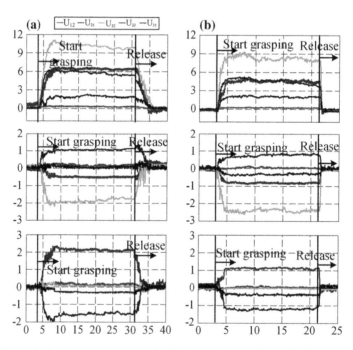

Fig. 9 Three-axis forces measurement results during **a** slow grasping and **b** fast grasping

Fig. 10 Three-axis force
measurement when **a** slightly
and **b** tightly grasped during
slow grasp condition

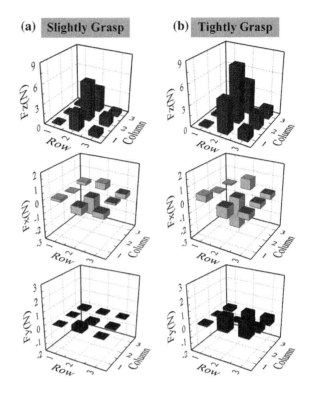

Phase 1, before the hand contact with the cup, there was no force measured. In Phase 2, after the hand was contacted with the cup and increasing the grasping force, we can see that the measured three-axis forces were increased quickly. Phase 3 is the stable grasping phase, in which phase the measured contact forces were almost the same and lasted for 22 s. In Phase 4, the grasping force was reduced gradually until the hand was not contacted with the cup. The measured three-axis forces were gradually decreased unto 0 N. Same results were observed in fast grasping experiment, as shown in Figs. 9b and 10.

5 Conclusions

In this study, we developed a flexible tactile sensor array based on INASTOMER pressure conductive rubber for 3D contact force measurement during hand grasping application. The structural design, fabrication, and characterization of the tactile sensor array were conducted. The sensing unit has the full scale of force measurement range of 5, 5, 20 N for the x-, y- and z-axis, respectively. The sensitivities for x- and y- axis forces measurement are characterized as 0.838 V/N and 0.834 V/N, respectively. For z-axis, the sensor array has two sensitivities: higher sensitivity with the value of 0.3675 V/N for small contact (0–10 N) and lower sensitivity of 0.0538 V/N for larger gripping force measurements (10–20 N). After mounting the sensor array on the finger, it can be used for real-time 3D contact forces measurement during grasping application. The contact force images between the object and sensor array were obtained and visualized in different grasping speeds. Results demonstrated that the developed tactile sensor array can be utilized to measure the real-time 3D contact forces for grasping applications and/or object shape even surface recognition.

To enhance the sensing performance of developed tactile sensor array, spatial resolution needs to be improved and structural optimization will be conducted in future work. More experimental studies will be performed to characterize the dynamic sensing performances of the tactile sensor array.

Acknowledgments The author would like to acknowledge the supports from the National Basic Research Program (973) of China (2011CB013300), National Natural Science Foundation of China (51105333), Science Fund for Creative Research Groups of National Natural Science Foundation of China (51221004), and the Fundamental Research Funds for the Central Universities (2015QNA4004).

References

Dahiya RS, Valle M, Lorenzelli L (2009) Bio-inspired tactile sensing arrays. In: Proceedings of bioengineered and bioinspired systems IV. Dresden, Germany, SPIE: 73650D-9. doi:10.1117/12.821696

Hoshino K, Mori D (2008) Three-dimensional tactile sensor with thin and soft elastic body. In: Proceedings of the IEEE workshop on advanced robotics and its social impacts, ARSO, pp 1–6. doi:10.1109/ARSO.2008.4653601

Lee H-K, Chung J, Chang S-I et al (2008) Normal and shear force measurement using a flexible polymer tactile sensor with embedded multiple capacitors. J Microelectromech Syst 17(4):934–942. doi:10.1109/JMEMS.2008.921727

Liang G, Wang Y, Mei D et al (2015) Flexible capacitive tactile sensor array with truncated pyramids as dielectric layer for three-axis force measurement. J Microelectromech Syst 24(5):1510–1519. doi:10.1109/JMEMS.2015.2418095

Liu T, Inoue Y, Shibata K (2009) A small and low-cost 3-D tactile sensor for a wearable force plate. IEEE Sens J 9(9):1103–1110. doi:10.1109/JSEN.2009.2026509

Tiwana MI, Shashank A, Redmond SJ et al (2011) Characterization of a capacitive tactile shear sensor for application in robotic and upper limb prostheses. Sens Actuators A 165(2):164–172. doi:10.1016/j.sna.2010.09.012

Wang Y, Xi K, Liang G et al (2014) A flexible capacitive tactile sensor array for prosthetic hand real-time contact force measurement. In: IEEE international conference on information and automation, pp 937–942. doi:10.1109/ICInfA.2014.6932786

Xi K, Wang Y, Mei D et al., 2015. A flexible tactile sensor array based on pressure conductive rubber for three-axis force and slip detection. In: IEEE international conference on advanced intelligent mechatronics, pp 476–481. doi:10.1109/AIM.2015.7222579

Zhang T, Jiang L, Wu X et al (2015) Fingertip three-axis tactile sensor for multifingered grasping. IEEE Trans Mechatron 20(4):1875–1885. doi:10.1109/TMECH.2014.2357793

A Novel Silicon Based Tactile Sensor with Fluid Encapsulated in the Cover Layer for Prosthetic Hand

Ping Yu, Chun-xin Gu, Wei-ting Liu and Xin Fu

Abstract Tactile sensing is indispensable for prosthetic hand to obtain information about contact forces and thus improve grasp stability. To reduce the burden of the hand's control and data acquisition (DAQ) system, we have developed a simple silicon-based tactile sensor containing a small number of sensing elements and owning large active sensory area in our previous work. In this paper, we incorporate a novel cover layer containing fluid in the sensor design to improve the uniformity of sensitivity from point to point in the whole receptive field. Experimental results show that with the cover layer, the normalized sensitivities of the tactile sensor are 1, 0.89151, 0.87736, and 0.82075 for four different indenting locations, which are 0, 1, 2 and 3 mm away from the center of the top surface, respectively. While without the cover layer, the normalized sensitivities of the tactile sensor are just 1, 0.8055, 0.42283, and 0.24101 for the four indentation locations, respectively.

Keywords Cover layer · Fluid · Tactile sensor · Piezoresistive · Silicon

1 Introduction

With the development of intelligent robotics and prosthetic hands, tactile sensors are indispensable for obtaining information about contact forces in dexterous manipulation and exploration (Yousef et al. 2011). Many kinds of tactile sensors based on different sensing mechanism, such as piezoresistive (Beccai et al. 2005), piezoelectric (Dahiya et al. 2009), capacitive (Liang et al. 2015), optical (Iwasaki et al. 2015) and magnetic (Yu et al. 2014), have therefore been developed. Among

P. Yu · C. Gu · W. Liu (✉) · X. Fu
The State Key Lab of Fluid Power Transmission and Control, School of Mechanical Engineering, Zhejiang University, Hangzhou 310027, China
e-mail: liuwt@zju.edu.cn

© Zhejiang University Press and Springer Science+Business Media Singapore 2017   81
C. Yang et al. (eds.), *Wearable Sensors and Robots*, Lecture Notes in Electrical Engineering 399, DOI 10.1007/978-981-10-2404-7_7

these sensors, silicon based piezoresistive tactile sensors are widely utilized for their high reliability, high sensitivity, low signal-to-noise ratio, and ease to realize fine spatial resolution by silicon microelectromechanical system (MEMS) technology (Beccai et al. 2005). Thus, silicon piezoresistors were selected as the sensing elements of our sensor.

For silicone based tactile sensors in prosthetic hand application, soft cover is indispensable. It plays an important role in protecting the fragile silicon structure from damage and enhancing compliance for better hand-to-object contact (Beccai et al. 2008). Besides these functions, the cover acts as a force transmitting medium, and thus its mechanical properties will affect the sensing characteristics of the sensor (Vásárhelyi et al. 2006). Some of the effects are disadvantageous and needed to be reduced, while others are advantageous and can be utilized. The effects of cover layer on tactile sensor have been discussed by several groups. Shimojo (1997) analyzed the mechanical spatial filtering effect of an elastic cover. When two concentrated loads P1 and P2 were applied on top surface of the cover, the pressure distributions for P1 and P2 under the cover may overlap. This overlap decreases the sensor's spatial resolution, just like the function of a low pass spatial filter. Simulation and experimental results show that the spatial filter gain decreased rapidly with the increasing cover thickness, while affected slightly by the Young's modulus and Poisson's Ratio. Cabibihan et al. (2014) investigated effects of the cover layer on the shape recognition and found that the thicker cover results in the worse shape discriminating ability, for instance, the flat, curved, and braille surfaces can be discriminated from one another with the 1 mm and 3 mm thick cover, but not with the 5 mm thick cover. Vásárhelyi et al. (2006) measured the receptive field and sensitivity of a miniature force sensor underlying an elastic layer and found that with increase in thickness of the cover layer the receptive field is broadened while the sensitivity is decreased. In this paper, we introduce fluid to the cover layer design in order to improve the uniformity of sensitivity from point to point in the whole receptive field of the tactile sensor. To our knowledge, no paper has discussed this effect brought by the cover layer.

Overall, our design challenge was to develop a highly reliable, easily fabricated, and compliant tactile sensor that contains a small number of sensing elements but owns large active sensory area for prosthetic hands. In previous work (Gu et al. 2014), we have developed a simple prototype with the novel structure of combining a steel sheet and a silicon piezoresistive gauge. The gauge was fixed on the center of the rear surface of the steel sheet in order to measure external applied forces by detecting defamation of the steel sheet. This structure has enlarged receptive field of the silicone gauge because the two-side clamped steel sheet can strengthen the diffusion ability of the valid mechanical signal, which was discussed in detail by our another previous work using a three-dimensional finite element analysis (FEA) model (Gu et al. 2015). However, the sensitivity is not constant in the whole area of receptive field, but varies with the indenting location, which will introduce measurement deviation when the contact point between the hand and object changes in actual application. To solve this problem, in this paper, the novel cover layer containing fluid was incorporated in the sensor.

2 Methods

2.1 Sensor Design

The whole sensor is composed of two substantial parts, as illustrated in Fig. 1a, sensory structure and its cover layer. The sensory structure consists of a silicon piezoresistive gauge, a metal framework, a flexible printed circuit board (FPCB), and three gold wires. The metal framework has two supporting feet and one square steel sheet, forming a two-side clamped plate structure. The silicon piezoresistive gauge, consisting of two symmetrical arranged resistors with the initial resistance $R_0 \approx 3.4$ KΩ for each one, is fixed on the rear surface of the square sheet and connected to the FPCB by golden wires. The detail description of the sensory structure was in (Gu et al. 2014).

The cover layer consists of elastomer framework and silicon oil. The elastomer framework is made of Polydimethylsiloxane (PDMS), whose mechanical characteristics are relatively close to that of human skin (Lee et al. 2011). The silicon oil is filled in the cavity of the elastomer framework.

The sensor dimensions are shown in Figs. 1 b, c. To fit the size of fingertip, the overall dimensions of the whole sensor are set as 15 mm × 15 mm × 10 mm. The height of the cover layer is 5 mm, while the silicon oil is 3 mm. The thickness of both the roof and four side walls of the elastomer framework is 2 mm.

When a normal force is applied on top of the cover layer, the cover layer deforms, and the silicon oil in the cavity is compressed accordingly, by Pascal's law, the pressure is transmitted equally to the entire interface of the metal

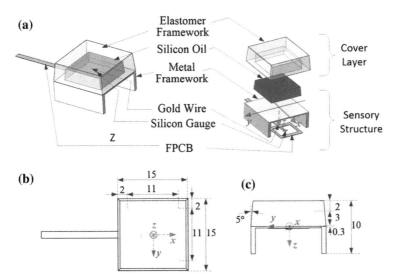

Fig. 1 Conceptual diagram of the proposed tactile sensor. **a** Schematic view. **b** Top view. **c** Lateral view

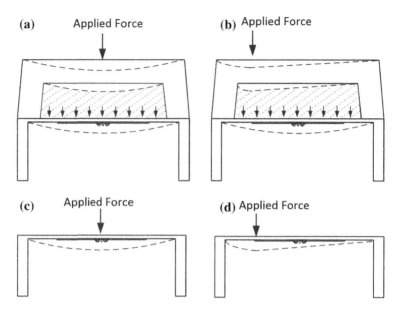

Fig. 2 Principle of operation to measure normal force. **a** Response of the tactile sensor with cover layer to normal force applied on the center region and **b** left region. **c** Response of the tactile sensor without cover layer to normal force applied on the center region and **d** left region

framework and the oil. As a result, forces on different location will cause the same deformation of the steel sheet, as shown in Fig. 2a, b. For comparing, the normal force was directly applied on the top surface of sensory structure [tactile sensor without cover layer developed by our previous work (Gu et al. 2014)], and the deformations are shown in Fig. 2c, d. It can be found that when the indenting location is shifted, the deformation of the steel sheet changes.

2.2 Fabrication

The whole tactile sensor is fabricated easily by combining the traditional machining process and the MEMS technology. The detail fabrication process of the sensory structure has been described in our previous work (Gu et al. 2014). In brief, the metal framework is grinded from a steel block (17-4PH) with Young's modulus as 197 GPa and Poisson ratio as 0.3, while the silicon gauge is fabricated with an SOI wafer based on the dry etching technology and fixed on the rear surface of the steel sheet by glass sintering process.

To fabricate the elastomer framework, a stainless concave mold that has the same shape as the external structure of the framework and a raised mold that has the same shape as the internal structure of the framework are manufactured by milling machine, and PDMS is utilized to duplicate the shapes of both these two molds.

Liquid-state PDMS (Sylgard 184, A: B = 10:1 in weight) is poured into the concave mold and then covered by the raised mold carefully. The redundant PDMS is squeezed out by tightening the screws used to connect the two molds, followed by vulcanization of the PDMS at 80 °C for 3 h. After that, the PDMS is peeled off from the concave mold, and the elastomer framework is fabricated.

To seal the silicon oil between the elastomer framework and steel sheet, the framework is firstly placed upside down, and then the silicon oil (H201) is poured into the cavity of the framework, at last the fabricated sensory structure is placed above the oil and glued to the elastomer framework by epoxy adhesive(Super XNo.8008).

Figure 3a shows the fabricated sensory structure, the same as we shown in our previous work (Gu et al. 2014). Figure 3b shows the fabricated tactile sensor, which is composed of sensory structure and cover layer. The silicon oil is well sealed in a

Fig. 3 **a** Photograph of the fabricated sensory structure (Gu et al. 2014).
b Photograph of the fabricated tactile sensor with cover layer.
c The developed sensor is mounted on a fingertip

closed space surrounded by the internal surfaces of the elastomer framework and the top surface of the metal framework. The fabricated sensor can be stably mounted on the fingertip by inserting the two supporting feet into two grooves in the fingertip, as shown in Fig. 3c.

2.3 Sensor Calibration

In order to demonstrate and investigate the ability of the sensor to measure normal force, especially the improvement of the uniformity of sensitivity from point to point in the whole receptive field, preliminary characterization experiments have been performed on both the tactile sensor with cover layer and the sensor without cover layer, respectively, as shown in Figs. 4a, c. Several points which are along y-axis and with spacing 1 mm between adjacent two points have been chosen as indenting locations for testing, as shown in Figs. 4b, d. The schematic of the calibrating system is shown in Fig. 5a and consists of a loading structure, signal

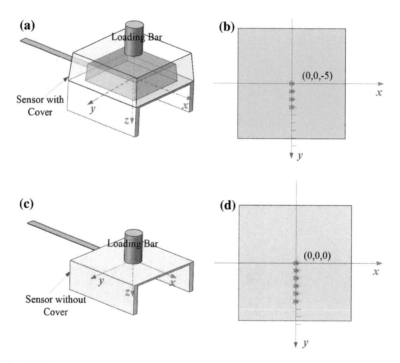

Fig. 4 a Calibrating method for the tactile sensor with cover layer. **b** View of the top surface of the tactile sensor with cover layer, where the *red* "*asterisk*" represents the projection of each indenting location. **c** Calibrating method for the tactile sensor without cover layer. **d** View of the top surface of the tactile sensor without cover layer, where the *red* "*asterisk*" represents the projection of each indenting location

Fig. 5 a Schematic of the calibrating system; **b** Photograph of the loading structure (① the reference sensor, ② the loading bar, ③ the fabricated tactile sensor, ④ the fingertip, ⑤ the link block, ⑥ the three-axial motion platform

conditioning electronic circuitry, data acquisition (DAQ), motion conditioner, and a PC workstation for data analysis. The photograph of the loading structure is shown in Fig. 5b. A copper loading bar is screwed to the final part of a reference six-axial force sensor (Nano43), whose measurement range is from −36 to 36 N with precision as 0.0078 N for normal force. The tactile sensor is mounted on a servo-controlled three-axial motion platform (M-VP-25XA-XYZL, Newport) with max movement range of 25 mm in each direction and can be automatically driven with the lowest speed of 1 um/s. By the movement of three-axial motion platform in x- and y-axes, different indenting locations can be oriented, and by the movement in z-axis, the tactile sensor can be pressed on the bottom surface of the loading bar, generating a uniform normal force on the top surface of the tactile sensor. The force value can be obtained by the reference sensor.

For the measurement of the resistance, a Wheatstone bridge circuit is set up for each piezoresistor (initial resistance $R_0 \approx 3.4$ KΩ) independently, which produces an output voltage proportional to the fractional resistance change $(R - R_0)/R_0$ in the piezoresistors, as described in Gu et al. (2014).

Using a graphical user interface developed by SignalExpress (Version 2011, National Instrument), the operator can impose a translation to the three-axial motion platform, while sampling the signals from the tactile sensor and the reference sensor through a DAQ card (NI 6343 from National Instruments).

Experiment procedure is as follows: (1) The three-axial motion platform is driven to enable the alignment between the axis of loading bar and the top surface center of the tactile sensor, while keeping the bottom surface of the bar slightly above the top surface of the tactile sensor; (2) The motion platform is driven again to move in the positive direction along z-axis with speed of 0.01 mm/s to load a normal force on the tactile sensor. The movement continues until the applied force is increased to 5 N; (3) The motion platform is driven to move along the negative z-axis to break the contact between the loading bar and the tactile sensor, and then move along the positive y-axis for a displacement of 1 mm to find next target location. The same test as described in procedure (2) is performed for each target location marked in Fig. 4b, d.

3   Experimental Results and Discussions

The response of tactile sensor without cover layer is shown in Fig. 6a. The fractional resistance change $(R - R_0)/R_0$ of the piezoresistor grows linearly when the applied force increases from 0 to 5 N, no matter which location the force is applied. However, the slopes of the curves, extracted using a linear regression technique and represent the sensitivities of the tactile sensor, are not equal for different indenting locations. With the increase of distance between the indenting location and the top surface center of the tactile sensor, the sensitivity of the tactile sensor $S_{xyz}[(x, y, z)$ is the coordinate position of the indenting points] decreases rapidly, as shown in Table 1.

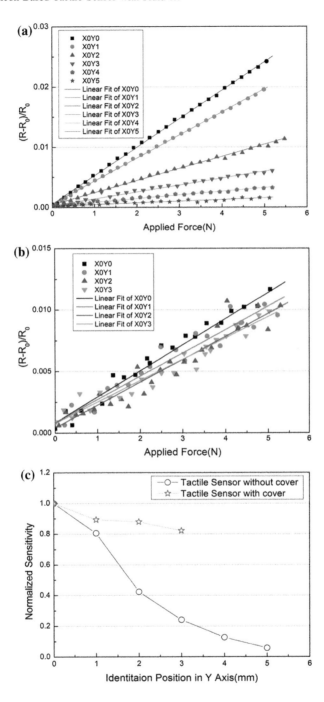

◀ **Fig. 6** Calibration results of tactile sensors. The fractional resistance change of the piezoresistor versus forces applied on different indention locations for **a** tactile sensor without cover layer and **b** with cover layer. **c** The normalized sensitivity of tactile sensors when external normal forces are applied on different indentation locations

Table 1 Resulting sensitivities when the normal force is applied on different indenting locations

Tactile sensor	Indenting location (x, y, z)	Sensitivity (S_{xyz})	Normalized sensitivity (S_{xyz}/S_{00z})
Without cover layer	(0, 0, 0)	0.00473	1
	(0, 1, 0)	0.00381	0.8055
	(0, 2, 0)	0.002	0.42283
	(0, 3, 0)	0.00114	0.24101
	(0, 4, 0)	0.0006	0.12685
	(0, 5, 0)	0.00027	0.05708
With cover layer	(0, 0, −5)	0.00212	1
	(0, 1, −5)	0.00189	0.89151
	(0, 2, −5)	0.00186	0.87736
	(0, 3, −5)	0.00174	0.82075

For tactile sensor with cover layer, the fractional resistance change $(R - R_0)/R_0$ also grows linearly with the applied force, as shown in Fig. 6b. Moreover, slopes of the curves (sensitivities) for four different indention locations are nearly equal. With the increase of distance between the indenting location and the top surface center of the tactile sensor, the sensitivity S_{xyz} decreases slowly, as shown in Table 1.

For comparing, the sensitivities of tactile sensor without cover layer and that of tactile sensor with cover layer are normalized (S_{xyz}/S_{00z}). The calculated normalized sensitivities are shown in Table 1 and the relationships between the normalized sensitivity and the indenting location are shown in Fig. 6c. In this figure, we can clearly find that the sensitivity of tactile sensor with cover layer decreases more slowly than that of tactile sensor without cover layer when the distance between the indenting location and the top surface center of the tactile sensor increases. For example, when the indenting location is 3 mm away from the top surface center of the tactile sensor, the normalize sensitivity of the tactile sensor with cover layer is 0.82075, while that of tactile sensor without cover layer is just 0.24101. This means that within the distance of 3 mm, the uniformity of sensitivity has been improved by 57.97 % by the novel cover layer.

4 Conclusions

We have developed a compliant tactile sensor with a novel cover layer. The cover layer, consisting of elastomer frame and silicon oil, improves the uniformity of sensitivity from point to point in the whole receptive field. Calibration of the fabricated sensor with cover layer shows that within 5 N, the normalized sensitivities are 1, 0.89151, 0.87736, and 0.82075 for four different indenting locations, which are 0, 1, 2 and 3 mm away from the center of the top surface, respectively. While that of tactile sensor without elastomer cover are 1, 0.8055, 0.42283, and 0.24101, respectively. Future work will focus on optimizing the structural dimensions of the cover layer to enhance the sensitivity of the sensor. Meanwhile, a better sealing method need to be found to avoid leakage of the fluid and enlarge the measurement range.

Acknowledgments Project supported by National Basic Research Program (973) of China (No.: 2011CB013303) and the Science Fund for Creative Research Groups of National Natural Science Foundation of China (No.: 51221004).

References

Beccai L, Roccella S, Arena A et al (2005) Design and fabrication of a hybrid silicon three-axial force sensor for biomechanical applications. Sens Actuate A-Phys 120(2):370–382. doi:10.1016/j.sna.2005.01.007

Beccai L, Roccella S, Ascari L et al (2008) Development and experimental analysis of a soft compliant tactile microsensor for anthropomorphic artificial hand. IEEE-ASME T Mech 13 (2):158–168. doi:10.1109/TMECH.2008.918483

Cabibihan JJ, Chauhan SS et al (2014) Effects of the artificial skin's thickness on the subsurface pressure profiles of flat, curved, and braille surfaces. IEEE Sens J 14(7):2118–2128. doi:10.1109/JSEN.2013.2295296

Dahiya RS, Metta G, Valle M et al (2009) Piezoelectric oxide semiconductor field effect transistor touch sensing devices. Appl Phys Lett 95(3):034105. doi:10.1063/1.3184579

Gu C, Liu W, Fu X (2014) A novel silicon based tactile sensor on elastic steel sheet for prosthetic hand. In: The 7th international conference on intelligent robotics and application, pp 475–483. doi:10.1007/978-3-319-13963-0_48

Gu C, Liu W, Fu X (2015) Three-dimensional finite element analysis of a novel silicon based tactile sensor with elastic cover. In: The 8th international conference on intelligent robotics and application, pp 402–409. 10.1007/978-3-319-22879-2_37

Iwasaki T, Takeshita T, Arinaga Y et al (2015) Shearing force measurement device with a built-in integrated micro displacement sensor. Sens Actuate A-Phys 221:1–8. doi:10.1016/j.sna.2014.09.029

Lee HK, Chung J, Chang SI et al (2011) Real-time measurement of the three-axis contact force distribution using a flexible capacitive polymer tactile sensor. J Micromech Microeng 21 (3):035010. doi:10.1088/0960-1317/21/3/035010

Liang G, Wang Y, Mei et al (2015) A modified analytical model to study the sensing performance of a flexible capacitive tactile sensor array. J Micromech Microeng 25(3):035017. doi:10.1088/0960-1317/25/3/035017

Shimojo M (1997) Mechanical filtering effect of elastic cover for tactile sensor. IEEE T Robot Autom 13(1):128–132. doi:10.1109/70.554353

Vásárhelyi G, Adám M, Vázsonyi E et al (2006) Effects of the elastic cover on tactile sensor arrays. Sens Actuate A-Phys 132(1):245–251. doi:10.1016/j.sna.2006.01.009

Yousef H, Boukallel M, Althoefer K (2011) Tactile sensing for dexterous in-hand manipulation in robotics—a review. Sens Actuate A-Phys 167(2):171–187. doi:10.1016/j.sna.2011.02.038

Yu P, Qi X, Liu W et al (2014) Development of a compliant magnetic 3-D tactile sensor with AMR elements. In: The 7th international conference on intelligent robotics and application, pp 484–491. doi:10.1007/978-3-319-13963-0_49

An Adaptive Feature Extraction and Classification Method of Motion Imagery EEG Based on Virtual Reality

Li Wang, Huiqun Fu, Xiu-feng Zhang, Rong Yang, Ning Zhang and Fengling Ma

Abstract With the aim to solve the problems such as low classification accuracy and weak anti-disturbances in brain–computer interfaces (BCI) of motion imagery, a new method for recognition of electroencephalography (EEG) was proposed in this work, which combined the wavelet packet transform and BP neural network. First, EEG is decomposed by wavelet packet analysis. Then, distance criterion is selected to measure the separable value of the feature frequency bands. Furthermore, the optimal basis of wavelet packet is attained by using a fast search strategy of "from the bottom to the top, from left to right." The classification feature is extracted by choosing the part wavelet package coefficient, which can attain higher classification evaluation value according to the optimal basis of wavelet packet. And then, the optimal bands are combined with BP neural network. The experimental results show that the proposed method can choose the feature bands of EEGs adaptively, and the highest classification accuracy is 94 %. The correctness and validity of the proposed method is proved. Lastly, establish the virtual robot in MATLAB and use the classification results to control the robot's arm motion.

Keywords Motion imagery EEG · Wavelet packet transform · Feature extraction · VR

1 Introduction

Brain–computer interface (BCI) is to establish a direct exchange channel of information and control between human and computer or other electronic devices, which is a new way to exchange information but not dependent on conventional channel

L. Wang (✉) · X. Zhang · R. Yang · N. Zhang · F. Ma
National Research Center for Rehabilitation Technical Aids, Beijing 100176, China
e-mail: cecilia_wl@126.com

H. Fu
101 Institute of the Ministry of Civil Affairs, Beijing 100070, China

© Zhejiang University Press and Springer Science+Business Media Singapore 2017
C. Yang et al. (eds.), *Wearable Sensors and Robots*, Lecture Notes in Electrical Engineering 399, DOI 10.1007/978-981-10-2404-7_8

(nerve and muscle). Its essence is through the brain electrical signal to identify people's intentions, so as to realize the man–machine communication (Virt 2006; Vaughan 2003; Xu 2007; Wolpaw et al. 2002). BCI has a broad application prospect in rehabilitation engineering, the disabled auxiliary control and other fields, and researchers pay more attention to it in the world (Yang et al. 2007). Human body's movement is controlled by the contralateral cerebral hemisphere. Imaging left–right hands movement or actually doing the action can change the activity of main cerebral sensorimotor area neuronal in the same way. The key to realize BCI system is that the patterns of motion imagery EEG identify accurately and quickly.

The characteristics of time varying, nonstationary, and individuation difference take great difficulties for the analysis of the EEG signal. The purpose of the feature extraction is to extract the eigenvector, which represent the thinking task from the EEG signal recorded by EEG amplifier. The commonly used methods of feature extraction are Power Spectral (Furstcheller et al. 2006), FFT (McFarland et al. 2006), SFT, AR (Burke et al. 2005), AAR (Furstcheller 2001), ICA, Bi-spectrum Estimation (Zhang et al. 2000), and Wavelet Analysis (Fatourechi et al. 2004), etc. FFT is limited in the spectrum analysis of the mutation signal. SFT is to average the feature of the signal during the time window. The shorter the signal is, the higher the time resolution is. However, the properties of time localization and frequency localization are contradicted. The AAR model parameters can reflect the condition of the brain better, but it is much more suitable for the analysis of stationary signal. Wavelet transform has lower time localization and higher frequency localization at low frequency, but higher time localization and lower frequency localization at high frequency which leads to the loss of feature information.

WPT is the extension of WP, it decomposes the low and high frequency information simultaneously at every grade decomposition for the extraction of abundant features. How to easily choose the optimal basis of wavelet packet to extract the frequency bands which embodies the individual characteristic is the key point of extracting personalized EEG features adaptively. A selection method of the optimal basis of wavelet packet based on entropy criterion is proposed by Coifman (1992). WPT is used to get the EEG signals of 8–16 Hz on C3 and C4 electrodes to calculate the frequency bands energy and Shannon entropy, which are adopted as the feature of imaging left–right hand movement EEG, the classification rate reach up to 87.14 % (Ren et al. 2008). As a kind of cost function, entropy can evaluate the concentrative degree of the energy distribution, but it cannot reflect the classification ability of the basis primely.

In recent years, virtual reality technology has been applied in some rehabilitation areas. In order to make patients immersed in this virtual world, VR can make the rehabilitation training full of entertainment, and stimulate patients in the task related to the training initiatively. It is vital for patients to encourage them and is significant for restore. In robot-assisted rehabilitation training, introduced VR can improve rehabilitation training effect after the stroke (Huang et al. 2005; Alma et al. 2002).

In this paper, the C3 and C4 channel of motion imagery EEG signals is respectively decomposed by using DB4 wavelet function. Then, use the distance criterion as the classification performance evaluation criteria of the optimal basis of wavelet packet. By using fast search strategy that from bottom to top, from left to right, we can obtain the optimal basis of wavelet packet and select the optimal basis, which features classification of partial bands wavelet packet coefficients of the corresponding to the high classification performance evaluation. There is classifier based on BP neural network to class motion imagine EEG signals. Finally, the classification results control arm movement of the robot of the virtual reality and verify the validity of the algorithm.

2 The Acquisition of Experimental Data

The experimental data of this paper is from the "BCI Competition 2003" data. The whole experimental data is composed of 280 experiments. The experimental data includes respective 90 times training samples and 50 times test samples of imagery left–right hand movement task. The data get from C3 and C4 channels of the international standard of 10–20 lead system. The sampling frequency is 128 Hz. It is implemented by 0.5–30 Hz bands pass filter. The experiment lasted 9 s each time. There are several minutes interval in every two experiments. During 0–2 s, the subject keeps relax. During 2–3 s, the displayer reminds the subject in preparation for performing the motion imagery tasks. During 3 s to 9 s, the subject starts to perform the relevant task which the displayer prompts.

3 Feature Extraction

3.1 Wavelet Packet Analysis

WPD is used to decompose EEG into wavelet subspace w_j by binary in order to achieve the purpose of improving the frequency resolution, and the decomposition process is shown in Fig. 1.

$$U_0^0 = U_3^0 + U_3^1 + U_3^2 + U_3^3 + U_3^4 + U_3^5 + U_3^6 + U_3^7, \tag{1}$$

Fig. 1 Decomposition of wavelet packet

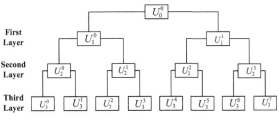

where U_j^n stands for the nth wavelet subspace in j scale ($n = 1, 2, 3, \ldots, 2^j - 1$, means the frequency factor). The orthogonal basis of the corresponding space is $u_{j,k}^n(t) = 2^{-j/2}u^n(2^{-j}t - k)$ (k is the translation factor) and it meets with the two-scale equation:

$$u_{j,0}^n = \sum_k g_0(k)u_{j-1,k}^i \quad (n \text{ is even}) \tag{2}$$

$$u_{j,0}^n = \sum_k g_1(k)u_{j-1,k}^i \quad (n \text{ is odd}), \tag{3}$$

where $j, k \in Z$, $n = 1, 2, 3, \ldots, 2^j - 1$, $g_0(k)$, and $g_1(k)$ are a pair of orthogonal filters meeting with $g_1(k) = (-1)^{1-k}g_0(1 - k)$. Let $f(t)$ be the function of $L^2(R)$ space, when the scale is sufficiently small and the sampling sequence of $f(t)$ approximate the factor $d_0^0(k)$ of U_0^0 space. Using the fast algorithm of orthogonal wavelet packet transform, wavelet packet coefficients of the jth stage, and the kth point can be obtained by Eqs. (4) and (5):

$$d_j^{2n}(k) = \sum_k g_0(m - 2k)d_{j-1}^n(m) \tag{4}$$

$$d_j^{2n+1}(k) = \sum_m g_1(m - 2k)d_{j-1}^n(m) \tag{5}$$

where $m \in Z$. The decomposition coefficients of jth level can be obtained, and so on, the wavelet packet decomposition coefficients of each level can be obtained from the digital signal $f(k)$. Each subspace's frequency bands of the jth level U_j^k are $\left\{\left[0, \frac{f_s}{2^{j+1}}\right]; \left[\frac{f_s}{2^{j+1}}, \frac{2f_s}{2^{j+1}}\right]; \left[\frac{2f_s}{2^{j+1}}, \frac{3f_s}{2^{j+1}}\right]; \ldots; \left[\frac{(2^j-1)f_s}{2^{j+1}}, \frac{f_s}{2}\right]\right\}$ after j level wavelet packet decomposition (f_s is the sampling rate of signal $f(t)$).

3.2   Distance Criterion

For the given orthogonal wavelet, the decomposition mode of signal which its length is $N = 2^L$ has at most 2^L kinds. That is the number of complete binary tree where its depth is L, and its value tends to be relatively large. In the complete binary tree, not all binary trees are valuable. Therefore, the problem of how to achieve the optimal decomposition of wavelet packet is generated.

In feature space, a model feature vector can be regarded as a point in this space, and the sample set of model is a point set of the space. Usually, the mean distance between classes of sample is larger, and the mean inner-class distance of sample is smaller, hence, the separability of class is better. Therefore, the distance between

samples is the most intuitive criterion, which study sample distribution and measures the separability of class.

Let c is the number of pattern class $(\omega_1, \omega_2, \ldots, \omega_c)$ and its combined feature vector
set is $\{x^{(i,k)}, i = 1, 2, \ldots, c, \quad k = 1, 2, \ldots, N_i\}$, which $x^{(i,k)} = (x_1^{(i,k)}, x_2^{(i,k)}, \ldots x_m^{(i,k)})^t$
is the number k in m-dimensional feature vector among ω_i, N_i is feature vector number
among ω_i. First, we calculate the average distance S_i between all feature vector among
ω_i. Its definition is:

$$S_i = \frac{1}{2} \frac{1}{N_i} \sum_{j=1}^{N_i} \frac{1}{N_i - 1} \sum_{k=1}^{N_i} \left\| x^{(i,j)} - x^{(i,k)} \right\|^2 \tag{6}$$

After seek average of S_i, $i = 1, 2, \ldots, c$, we can get the mean inner-class distance S_w.

$$S_w = \frac{1}{C} \sum_{i=1}^{c} S_i \tag{7}$$

Then, according to derivation, the mean inner-class distance is Eq. (8)

$$S_w = \frac{1}{C} \sum_{i=1}^{c} \sum_{l=1}^{m} \frac{1}{N_i - 1} \sum_{k=1}^{N_i} (x_l^{(i,k)} - \mu_l^{(i)})^2 \tag{8}$$

In Eq. (8) $\mu_l^{(i)} = \frac{1}{N_i} \sum_{k=1}^{N_i} x_l^{(i,k)}$, $\mu_l^{(i)}$ is the average of lth component in class ω_i
sample feature vector.

Besides, the average vector of class represents the class. The sample average
vector of class ω_i is $\mu^{(i)}$. The sample general average vector is μ. The mean distance
between classes of C class is defined as follows:

$$S_b = \frac{1}{c} \sum_{i=1}^{c} \left\| \mu^{(i)} - \mu \right\|^2 = \frac{1}{c} \sum_{i=1}^{c} \sum_{l=1}^{m} (\mu_l^{(i)} - \mu_l)^2 \tag{9}$$

In Eq. (9) $\mu_l = \frac{1}{c} \sum_{i=1}^{c} \frac{1}{N_i} \sum_{k=1}^{N_i} x_l^{(i,k)}$, μ_l is the lth component of general sample average
vector. The small mean inner-class distance and the big mean distance between
classes have good separability. Therefore, the definition of distance criterion is the
ratio between the mean inner-class distance and the mean distance between classes:

$$J_A = \frac{S_b}{S_w} = \frac{\frac{1}{c} \sum_{i=1}^{c} \sum_{l=1}^{m} (\mu_l^{(i)} - \mu_l)^2}{\frac{1}{c} \sum_{i=1}^{c} \sum_{l=1}^{m} \frac{1}{N_i - 1} \sum_{k=1}^{N_i} (x_l^{(i,k)} - \mu_l^{(i)})^2} \tag{10}$$

3.3 Adaptive Optimal Feature Selection

It is assumed that the wavelet packet coefficients of the training samples are uncorrelated with each other (wavelet transform has good correlation). We choose the wavelet packet basis for the value of Eq. (10). Specific algorithm is:

(1) Let c is the number of pattern class $(\omega_1, \omega_2, \ldots, \omega_c)$ and its training sample set is $\{\xi^{(i,k)}, \quad i = 1, 2, \ldots, c, k = 1, 2, \ldots, N_i\}$, which $\xi^{(i,k)} = (\xi_1^{(i,k)}, \xi_2^{(i,k)}, \ldots \xi_m^{(i,k)})^t$ is the kth feature vector which is m-dimensional among ω_i, N_i is feature vector number of ω_i. Do wavelet decomposition to $\xi^{(i,k)}$, the decomposed class is $j = 1, 2, \ldots, L$. The nth wavelet packet subspace of the grade j is $U_j^n (j = 1, 2, \ldots, L, n = 0, 1, 2, \ldots, 2^{j-1})$, and the wavelet packet coefficient in U_j^n is $\{x_p^{(i,k)}, p = 1, 2, \ldots P_j\}$

(2) Calculate the Pth wavelet packet coefficient $x_p^{(i,k)}$'s corresponding value J_{Ap} in U_j^n. The larger value J_{Ap} is, the better separability of characteristic component is.

(3) Find the wavelet packet basis which make $\sum\limits_{j,n} J_{Ap}$ maximum in library of wavelet packet, it is the optimal basis of wavelet packet. The number m wavelet packet coefficients corresponding value J_{Ap} is $J_{A1}, J_{A2} \ldots, J_{Am}$. It is sorted as $J_{A1}^* \geq J_{A2}^* \geq \cdots \geq J_{Am}^*$.

(4) The optimal basis is the wavelet packet basis which made the value of $\sum\limits_{j,n} J_{Ap}$ maximize in library of wavelet packet. The separable value of each node is written into the node. Using a fast search strategy to search the optimal basis, it is "from the bottom to the top, from left to right." Mark from the lowest layer nodes, all the upper node mark as father node, and lower node mark as child node. If the father node's separable value is higher than the children node, then mark the father node (delete child node at the same time). Otherwise, do not mark and so on. All the marked nodes form the optimal basis (Fig. 2).

(5) According to the selected optimal basis of wavelet packet, we do WPT to the signal mode which needs classification or recognition. The wavelet packet coefficients of $J_{A1}^* \sim J_{Ad}^* (d < m)$ are selected as feature vector.

Fig. 2 Adaptive optimal feature selection

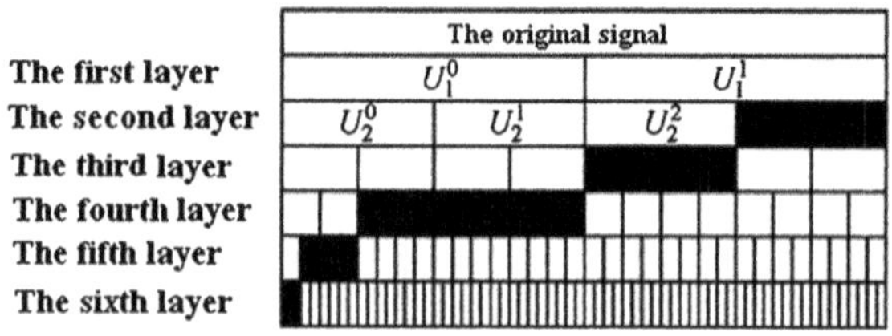

4 EEG Classification Method and VR Technology

The BP neural network has adaptive function, generalization ability, and strong fault tolerance. Hence, this paper uses BP neural network as the EEG nonlinear classifier and verifies the validity of the feature information. It constructs a three-layer BP neural network and the number of input layer neuron is equal to the eigenvector dimension. The number of hidden layer neuron is 12. The number of output layer neuron is 1.

The feature set which training BP network is extracted from the respective 90 times imagery left–right hand movement of EEG experiment data base on the optimal basis of wavelet packet by distance criterion. And then, the feature set which testing BP network is the respective 50 times imagery left–right hand movement of EEG experiment data. The selection of optimal basis wavelet packet is only in the training stage. The corresponding frequency bands of the selected optimal basis of wavelet coefficient constitute the training feature set. In the test stage it can be used directly. It can significantly reduce the computation of program and accelerate the speed of program.

BCI technology is a very valuable technology in rehabilitation project. It is combined with the virtual reality technology and can make the effect of rehabilitation training better. MATLAB Virtual Reality Toolbox provides an effective solution of visual operation and dynamic system in 3D virtual reality environment. Using the standard VRML technology, MATLAB and Simulink, we can generate 3D virtual scene. Through the Simulink interface, in 3D virtual reality environment, we can observe simulation of dynamic system. The classification results are connected to the rehabilitation robot model by the Simulink interface. This data can be used to control and manipulate the virtual rehabilitation robot model.

5 Experimental Results and Analysis

(1) Db4 wavelet belongs to Daubechies wavelet with excellent property, and it is a set of orthonormal compactly supported wavelet. And db4 wavelet function based on Eqs. (4) and (5) is chosen to decompose EEGs in 5 layers, and 32 wavelet subspaces the fifth layer are obtained.

(2) Fig. 3 shows the evaluation value of optimal basis of wavelet packet.

Due to the individual difference of EEG signals, the optimal basis of different subjects' EEG may be different. The method can automatically choose the optimal basis for each individual. The characteristic of this method is adaptively selecting feature frequency bands based on actual signal.

Capture the first four wavelet subspace coefficient as the classification features based on the results of distance criterion. The accuracy of classification is about 94 %. Figure 4 shows the classification results.

Fig. 3 The evaluation value
of optimal basis of wavelet
packet

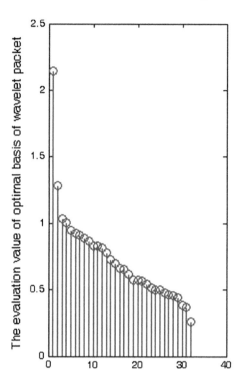

Fig. 4 BP neural network
classification results

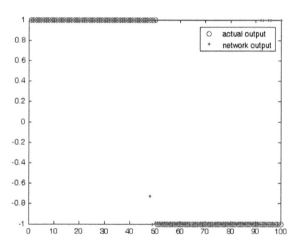

For the number of hidden neurons, the range is determined by the experience, and then several experiments are carried out. The number of hidden layer neuron effect the accuracy of the classification. At last the number of hidden neurons is 12. The process is shown in Table 1.

Table 1 The effect of the number of hidden layer neurons to the accuracy of the classification

Sequence	Number of hidden neurons	The accuracy of classification (%)
1	10	89
2	11	91
3	12	94
4	13	92

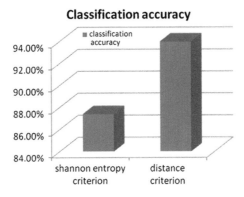

Fig. 5 The comparison of correct classification rates among two feature extraction methods

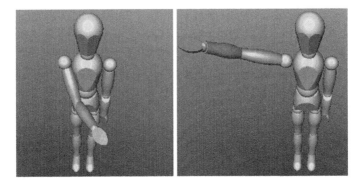

Fig. 6 The classification results control the motion of virtual robot

Figure 5 shows the comparison of correct classification rates among this paper's and Reference (Ren et al. 2008)'s feature extraction methods.

The classification results are connected to the rehabilitation robot model in VR through the Simulink interface. The classification results can be used to control and manipulate virtual rehabilitation robot model as Fig. 6. When the recognition is correct, the motion imagery EEG of left hand movement is used to control the robot arm up. The motion imagery EEG of right-hand movement is used to control the robot arm down.

6 Conclusion

According to the individual differences characteristics of imagery left–right hand movement EEG, this paper connect wavelet packet with the distance criterion. Use the criterion as the evaluating performance function of the optimal basis of wavelet packet. It is an adaptive extraction personalized characteristics method, which extract more abundant information. There is a classifier based on BP neural network to class motion imagery EEG signals. Finally, the classification results control arm movement of the robot in the virtual reality. It verifies the validity of the algorithm. It provides an effective way for the design and application of the BCI system for motion imagery EEG signal.

Acknowledgments Project supported by the National Natural Science Foundation of China (No. 51275101).

References

Alma SM, David J, Rares B et al (2002) Virtual reality–augmented rehabilitation for patients following stroke. J Phys Ther 82(9):898–915

Burke DP, Kelly SP, De Chazal P et al (2005) A parametric feature extraction and classification strategy for brain-computer interfacing. J IEEE Trans Neural Syst Rehabil Eng 13(1):12–17

Coifman RR, Hauser MVW (1992) Entropy based algorithms for best basis selection. J IEEE Trans IT 38(3):713–718

Fatourechi M, Mason SG, Birch GE (2004) A wavelet-based approach for the extraction of event related potentials from EEG. In: IEEE international conference on acoustics, speech, and signal processing. Montreal: ICASSP, pp 737–740

Furstcheller PG, Muller-Putz GR, Schlogl A et al (2006) 15 years of research at Graz university of technology: current projects. J IEEE Trans Neural Syst Rehabil Eng 14(2):205–210

Furstcheller P, Neuper G (2001) Motor Imagery and direct brain-computer communiction. Proc IEEE 89(7):1123–1134

Huang H, Ingalls T, Olson L et al (2005) Interactive multimodal biofeedback for task-oriented neural rehabilitation. In: Proceedings of the 2005 IEEE engineering in medicine and biology 27th annual conference, pp 2547–2550

McFarland DJ, Anderson CW, Muller KR et al (2006) BCI meeting 2005-workshop on BCI signal processing: feature extraction and translation. J IEEE Trans Neural Syst Rehabil Eng 14 (2):135–138

Ren LP, Zhang AH, Hao XH et al (2008) Study on classification of imaginary hand movements based on bands power and wavelet packet entropy. J Chin J Rehabil Theory Pract 14(2): 141–143

Vaughan TM (2003) Brain-computer interface technology: a review of the second international meeting. J IEEE Trans Neural Syst Rehabil Eng 11(2):94–109

Virt SJ (2006) The third international meeting on brain- computer interface technology: making a difference. J IEEE Trans Neural Syst Rehabil Eng 14(2):126–127

Wolpaw JR, Birbaumer N, McFarland DJ et al (2002) Brain- computer interface for communication and control. J Clin Neurophysiol 113(6):767–791

Xu AG, Song AG (2007) Feature extraction and classification of single trial motor imagery EEG. J J SE Univ 37(4):629–630

Yang BH, Yan BH, Yan GZ (2007) Extracting EEG feature in brain-computer interface based on discrete wavelet transform. J Chin J Biomed Eng 25(5):518–519

Zhang J, Zheng C, Xie A (2000) Bispectrum analysis of focal ischemic cerebral EEG signal using third-order recursion method. J IEEE Trans Biomed Eng 47(3):352–359

One-Handed Wearable sEMG Sensor for Myoelectric Control of Prosthetic Hands

Yin-lai Jiang, Shintaro Sakoda, Masami Togane, Soichiro Morishita and Hiroshi Yokoi

Abstract A novel sEMG (surface electromyography) sensor using polypyrrole-coated nonwoven fabric sheet as electrodes (PPy-electrode) is proposed for the disabled to control prosthetic limbs in daily life. The PPy-electrodes are sewed on an elastic band to guarantee closely contact to the skin thus to enable stable sEMG measurement with high signal-to-noise ratio. Furthermore, the sensor is highly customizable to fit for the size and the shape of the stump so that the disabled can wear the sensor by themselves. The performance of the proposed sensor is investigated by comparing with Ag/AgCl electrodes with electrolytic gel in an experiment to measure the sEMG from the same muscle fibers. The high correlation coefficient (0.87) between the sEMG measured by the two types of sensors suggests the effectiveness of the proposed sensor. The experiment to control myoelectric prosthetic hands showed that the disabled can use it with one hand to obtain sEMG signals for myoelectric control.

Keywords Surface electromyography · EMG electrodes · Polypyrrole-coated nonwoven fabric sheet · Prosthetic hand

1 Introduction

Benefitting from the advance of robotics and biomedical engineering, the research on EMG prosthetic hands has developed quickly in recent years to meet the demands of those who have lost their hands. Many dexterous prosthetic hands, such as the iLimb hand and the iLimb Pulse by Touch Bionics, the Bebionic hand by

Y. Jiang (✉) · S. Morishita · H. Yokoi
Brain Science Inspired Life Support Research Center,
The University of Electro-Communications, Tokyo 1828585, Japan
e-mail: jiang@hi.mce.uec.ac.jp

S. Sakoda · M. Togane · H. Yokoi
Department of Mechanical Engineering and Intelligent Systems,
The University of Electro-Communications, Tokyo 1828585, Japan

© Zhejiang University Press and Springer Science+Business Media Singapore 2017
C. Yang et al. (eds.), *Wearable Sensors and Robots*, Lecture Notes in Electrical
Engineering 399, DOI 10.1007/978-981-10-2404-7_9

RSL Steeper, and the Bebionic hand v2 by RSL Steeper (Belter et al. 2013), have been developed and commercially available. Surface electromyography (sEMG) has been widely studied for the control of prosthetic limbs since motion intentions can be identified based on the sEMG associated with the residual muscle contraction. Although many advanced algorithms are proposed to analyze sEMG signals, few of them are practical since they are based on the high-quality signals measured in labs which are not available for a disabled person in his/her daily life.

We have developed prosthetic hands with interactive wire-driven mechanism (Seki et al. 2013). An fMRI study with amputees has shown the adaptation process in the brain for control of the prosthetic hand (Kato et al. 2009). The prosthetic hand is controlled by sEMG signals with an online learning algorithm (Kato et al. 2006). However, for daily use of the prosthetic hand by a disabled person, the number of recognizable sEMG patterns is limited by the usability and stability of the sEMG electrodes. This study, therefore, introduces a sEMG sensor that can be worn one-handed for stable measurement of sEMG signals.

2   Methods

The sEMG sensor is an electrochemical transducer that detects biopotentials by using electrodes placed on the skin. Ag/AgCl electrodes, so-called wet electrodes utilizing an electrolytic gel to form a conductive path between skin and electrode, are currently most widely used in hospitals or laboratories. Despite the low electrode–skin impedance of Ag/Agcl electrodes contributes to a high S/N ratio, dry electrodes made from metals are always adopted in human robot interface since Ag/AgCl electrodes are disposable and not suited for long-term measurement. An important factor when using dry electrodes is their ability to maintain electrical contact with the skin since the electrical contact affects the electrode–skin impedance. Additionally, the end shape of the residual arm is different individually; it is hard for the existing metallic electrodes to maintain good electrical contact.

We therefore attempted to develop novel dry electrodes with soft conductive materials.

3   Results

The PPy-electrode sensor band and two disabled sEMG (surface electromyography) Engineering Surface electromyography sensor using polypyrrole-*coated nonwoven fabric sheet* (Japan Vilene Company, Ltd) as electrodes (PPy-electrode) is proposed for the disabled to control prosthetic limbs in daily life. Compared with other conductive polymer which has been studied as bioelectrical sensors (Tsukada et al. 2012), polypyrrole is more stable in the environment of daily life.

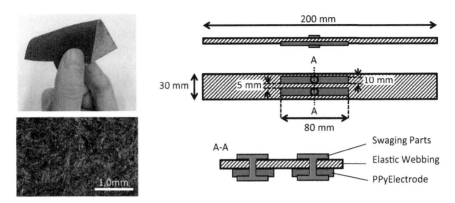

Fig. 1 Polypyrrole-coated nonwoven fabric sheet (*left*) and the PPy-electrodes sewed on the elastic webbing band (*right*)

As shown in Fig. 1, the PPy-electrodes are sewed on an elastic band to guarantee closely contact to the skin thus to enable stable sEMG measurement with high signal-to-noise ratio. Furthermore, the sensor is highly customizable to fit for the size and the shape of the stump so that the disabled can wear the sensor by themselves.

4 Results and Discussions

Figure 2 shows a 3-channel sEMG sensor band with six recording electrodes for differential amplification and a common ground electrode. The sensor band was manufactured for a 14-year-old girl with congenital right upper limb deficiency. The subject wearing the bands customized for her is shown in the right of Fig. 2. She was able to wear the sensor band one-handed by herself to get sEMG signals to control a two degree of freedom prosthetic hands developed by a previous study (Jiang, et al. 2014). With the myoelectric prosthetic hand, she tried to do some daily task, for

Fig. 2 PPy-electrode sEMG sensor band (*left*) and two subjects wearing sensor bands customized for them (*middle* and *right*)

example, moving a cup by holding the handle (Fig. 2). The experimental protocol was approved by the ethics board of The University of Electro-Communications.

To investigate the quality of the signals measured by the proposed sensor, simultaneous measurement of sEMG from the same muscle fibers with the PPy-electrodes that we developed and the conventional wet electrodes (Ag/AgCl electrodes with electrolytic gel), which are widely used in hospitals and laboratories, was carried out. The measurement settings and the results are shown in Fig. 3. The raw sEMG signals measured by the two types of electrodes and the root mean square (RMS) of the signals were nearly the same. The high correlation coefficient (mean = 0.87, SD = 0.065, $n = 12$) between the RMS values of them suggest the effectiveness of the proposed PPy-electrode sEMG sensor.

Surveys (Vasluian et al. 2013; Biddiss and Chau 2007) have suggested that appearance, function, comfort, control, and durability are the main factors affecting prosthesis acceptance and rejection. In order to provide the technology and opportunities necessary for the disabled to maximize their quality of life using prosthetic hands, the whole system, including the prosthetic robotic hand, the mount parts, and the sEMG measurement and analysis units, should be convenient to be used by the persons most of whom have only one hand.

In this study, a novel PPy-electrode is developed to enable highly usable and effective sEMG measurement in real-life settings. It is wearable with one hand, and the size and shape of the electrodes can be easily tailored to fit with the residual arms, guaranteeing the electrical contact with the skin. Multiple channel measurement of sEMG can be accomplished more conveniently by arranging the electrodes on a base cloth, which provides more options of sEMG measurement.

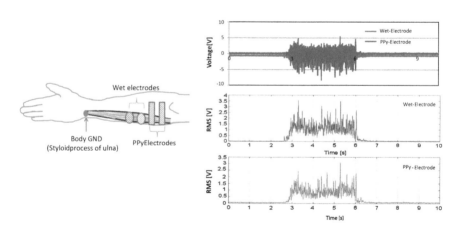

Fig. 3 Simultaneous measurement with wet electrodes and PPy-electrodes. Electrodes were placed along the same muscle fibers (*left*) and the sEMG signals measured were nearly the same (*right*)

Acknowledgments Project supported by JSPS KAKENHI Grant Number 25249025, "Brain Machine Interface Development" from Japan Agency for Medical Research and Development (AMED), and the Adaptable and Seamless Technology Transfer Program through Target-driven R&D JST(AS2524107P).

References

Belter JT, Segi JL, Dollar AM et al (2013) Mechanical design and performance specifications of anthropomorphic prosthetic hands: a review. J Rehabil Res Dev 50(5):599–618. doi:10.1109/PAHCE.2014.6849617

Biddiss EA, Chau TT (2007) Upper limb prosthesis use and abandonment: a survey of the last 25 years. Prosthet Orthot Int 31(3):236–257. doi:10.1080/03093640600994581

Jiang YL, Sakoda S, Hoshigawa S et al (2014) Development and evaluation of simplified EMG prosthetic hands. In: 2014 IEEE international conference on robitics and biomimetics, pp 1368–1383. doi:10.1109/ROBIO. 2014.7090524

Kato R, Yokoi H, Arai T et al (2006) Real-time learning method for adaptable motion-discrimination using surface EMG signal. In: IEEE/RSJ international conference on intelligent robots and systems, pp 2127–2132. doi:10.1109/IROS.2006.282492

Kato R, Yokoi H, Arieta AH et al (2009) Mutual adaptation among man and machine by using f-MRI analysis. Robot Auton Syst 57(2):161–166. doi:10.1016/j.robot.2008.07.005

Seki T, Nakamura T, Kato T, Morishita S et al (2013) Development of five-finger multi-DoF myoelectric hands with a power allocation mechanism. In: 2013 IEEE international conference on robotics and automation, pp 2054–2059. doi:10.1109/ICRA.2013.6630852

Tsukada S, Nakashima H, Torimitsu K (2012) Conductive polymer combined silk fiber bundle for bioelectrical signal recording. PLoS ONE 7(4):e33689. doi:10.1371/journal.pone.0033689

Vasluian E, de Jong IGM, Janssen WGM et al (2013) Opinions of youngsters with congenital below-elbow deficiency, and those of their parents and professionals concerning prosthetic use and rehabilitation treatment. PLoS ONE 8(6):e67101

Wearable Indoor Pedestrian Navigation Based on MIMU and Hypothesis Testing

Xiao-fei Ma, Zhong Su, Xu Zhao, Fu-chao Liu and Chao Li

Abstract Indoor pedestrian navigation (IPN) has attracted more and more attention for the reason that it can be widely used in indoor environments without GPS, such as fire and rescue in building, underground parking, etc. Pedestrian dead reckoning (PDR) based on inertial measurement unit can meet the requirement. This paper designs and implements a miniature wearable indoor pedestrian navigation system to estimate the position and attitude of a person while walking indoor. In order to reduce the accumulated error due to long-term drift of inertial devices, a zero-velocity detector based on hypothesis testing is introduced for instantaneous velocity and angular velocity correction. A Kalman filter combining INS information, magnetic information, and zero transient correction information is designed to estimate system errors and correct them. Finally, performance testing and evaluation are conducted to the IPN; results show that for leveled ground, position accuracy is about 2 % of the traveled distance.

Keywords Wearable indoor pedestrian navigation · MIMU · ZUPT · EKF · Hypothesis testing

1 Introduction

Indoor pedestrian navigation (IPN) system can track and locate indoor pedestrian, which can be widely used in location inside complex building, firefighters rescue at the fire scene, robot path planning, etc. The main difficulty is that some accurate navigation information cannot be used in indoor environment like the GPS. To

X. Ma · X. Zhao · F. Liu · C. Li
School of Automation, Beijing Institute of Technology, Beijing 100084, China

Z. Su (✉)
Beijing Key Laboratory of High Dynamic Navigation Technology, Beijing Information Science & Technology University, Beijing 100101, China
e-mail: jzys8000@163.com

© Zhejiang University Press and Springer Science+Business Media Singapore 2017
C. Yang et al. (eds.), *Wearable Sensors and Robots*, Lecture Notes in Electrical Engineering 399, DOI 10.1007/978-981-10-2404-7_10

solve this problem, preset-node pattern and autonomous pattern are two common solutions.

Preset-node pattern means some positioning nodes (like beacon node, base station) need to be arranged before location. Representative technologies include ultra-wideband wireless location (UWB) (Zampella et al. 2012), Bluetooth positioning (Bluetooth) (Kamisaka et al. 2012), WLAN positioning (WLAN) (Raitoharju et al. 2015), radio frequency identification positioning (RFID) (Zampella et al. 2013), ZigBee positioning, ultrasonic positioning, positioning infrared near-field electromagnetic positioning (NFER), etc. This method has a high precision, but requires high-density preset nodes in the navigation area. Not only high cost, it is unable to locate overall environment due to some blind zone in real time, and positioning may failure when some node cannot effectively serve.

Autonomous navigation is a method without pre-node. It is mainly used in MEMS inertial measurement unit (MIMU, mainly includes three-axis gyroscopes, three-axis accelerometer, even magnetic and pressure sensors) mounted on the pedestrian body (mostly feet) to obtain information on pedestrian movement, geomagnetism and others, then estimate attitude and location by dead reckoning or strapdown resolving (Foxlin 2005). This method is well adapted without preset beacon, but its accuracy is related to its running time and distance. At present, research on this problem is mainly on how to increase correction information in existing algorithm framework and improve information fusion filter (Chen et al. 2015).

This paper designs and implements a miniature wearable indoor pedestrian navigation system. The overall framework is inertial navigation systems based on MIMU, on this basis, a zero-velocity detector based on hypothesis testing is introduced for instantaneous velocity and angular velocity correction. Then an EKF integrating, INS information, and zero-velocity correction information are used to estimate the systematic error and corrected. Finally, performance testing and evaluation are conducted in an actual indoor environment.

2   Overall Framework

The indoor pedestrian navigation system outputs real-time position, velocity and attitude information, though a MEMS IMU to achieve basic functions of navigation and positioning. Meanwhile, using people's features of wearable and motional, the velocity, angular rate, and heading are corrected by the transient static detection trigger to suppress long-term drift of inertial devices, thereby improving positioning precision. The overall framework of the system is shown in Fig. 1, mainly including MIMU, SINS navigation module, zero-velocity detector, and EKF module.

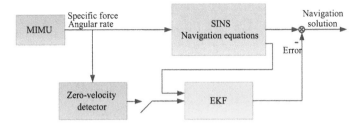

Fig. 1 The overall framework of IPN

2.1 MIMU and SINS Navigation Module

MIMU contains three accelerometers, three gyroscopes, three magnetometers, and even a pressure sensor. Fix it with human body, and the coordinate defines as shown in Fig. 2, coincide with front, right, and below side of the body. During navigation calculations, North-East Ground direction is selected as the navigation coordinate, expressed as n; carrier coordinate is expressed as b. The direction cosine matrix transformed from navigation frame to body frame defines as:

$$C_n^b = \begin{bmatrix} \cos\theta\cos\gamma & \cos\theta\sin\gamma & -\sin\theta \\ -\cos\varphi\sin\gamma + \sin\varphi\sin\theta\cos\gamma & \cos\varphi\cos\gamma + \sin\varphi\sin\theta\sin\gamma & \sin\varphi\cos\theta \\ \sin\varphi\sin\gamma + \cos\varphi\sin\theta\cos\gamma & -\sin\varphi\cos\gamma + \cos\varphi\sin\theta\sin\gamma & \cos\varphi\cos\theta \end{bmatrix},$$

$$(1)$$

where φ represents the yaw angle, θ represents the pitch angle, γ represents the roll angle.

In SINS navigation calculating, we must first determine the initial direction cosine matrix, which is the initial alignment of the system. Due to the low precision of MEMS gyroscopes we used cannot effectively sensitize the Earth's rotation rate, the initial alignment can be simplified as: estimate the pitch angle θ and roll angle γ with level accelerometer, and estimate the yaw angle φ from the output of the magnetometers.

First, estimate the pitch angle θ and roll angle γ with acceleration information in x, y, z directions:

$$\theta = \arcsin\frac{-a_x^b}{\sqrt{\left(a_x^b\right)^2 + \left(a_y^b\right)^2 + \left(a_z^b\right)^2}} \tag{2}$$

$$\gamma = \arctan\frac{a_y^b}{a_z^b}, \tag{3}$$

Fig. 2 Pedestrian coordinate schematic

where a_x^b, a_y^b, a_z^b are accelerometer output value after compensation in x, y, z directions, respectively.

Magnetic field strength on navigation coordinates is

$$\boldsymbol{B}^n = C_b^n \cdot \boldsymbol{B}^b \tag{4}$$

Simultaneous Eqs. (2), (3) and (4), yaw angle φ can be obtained as:

$$\varphi = \arctan \frac{B_z^n \cos \gamma - B_y^n \cos \gamma}{B_x^n \cos \theta + B_y^n \sin \theta \sin \gamma + B_z^n \sin \theta \sin \gamma} \tag{5}$$

The initial alignment is complete. Then we can calculate the navigation equations.

According to the three-axis angular rate measured, quaternion method is used to update the motion attitude of human body. Quaternion discrete differential equations can be written as

$$Q(\mathrm{k}+1) = \left\{\cos\frac{\Delta w_k}{2}I + \frac{[\Delta w]}{\Delta w_k}\sin\frac{\Delta w_k}{2}\right\}Q(\mathrm{k}), \tag{6}$$

where $Q(k) = [\,q_0 \quad q_1 \quad q_2 \quad q_3\,]^T$, I is the unit matrix, Δw_k is the vector sum of three axial angular increments.

Solve quaternion differential equation by Runge–Kutta method, the direction cosine matrix and corresponding attitude angle can be obtained as:

$$C_b^n = \begin{bmatrix} q_0^2 + q_1^2 - q_2^2 - q_3^2 & 2(q_1q_2 - q_0q_3) & 2(q_1q_3 + q_0q_2) \\ 2(q_0q_3 + q_1q_2) & q_0^2 - q_1^2 + q_2^2 - q_3^2 & 2(q_2q_3 - q_0q_1) \\ 2(q_1q_3 - q_0q_2) & 2(q_2q_3 + q_0q_1) & q_0^2 - q_1^2 - q_2^2 + q_3^2 \end{bmatrix} \tag{7}$$

$$\begin{cases} \varphi = \arctan\left(\frac{2q_2q_3 + 2q_0q_1}{2q_0^2 + 2q_3^2 - 1}\right) \\ \theta = -\arcsin(2q_1q_3 - 2q_0q_2) \\ \gamma = \arctan\left(\frac{2q_1q_2 + 2q_0q_3}{2q_0^2 + 2q_1^2 - 1}\right) \end{cases} \tag{8}$$

Compensate the gravity base on the update attitude and accelerometer output, then the acceleration in navigation coordinate system can be obtained, as well as the velocity and position information:

$$\breve{a}_k = C_{b_{k|k-1}}^n \cdot a_k^{\prime b} - [\,0 \quad 0 \quad g\,]^T \tag{9}$$

$$v_{k|k-1} = v_{k-1|k-1} + \breve{a}_k \cdot \Delta t \tag{10}$$

$$r_{k|k-1} = r_{k-1|k-1} + v_{k|k-1} \cdot \Delta t \tag{11}$$

2.2 ZUPT and Zero-Velocity Detector

Above results of inertial navigation will bring a cubic-in-time error drift, which is not conducive to long time navigation. Zero velocity update (ZUPT) is an effective method to restrain the error drift (Foxlin 2005). The basic principle is: When a person walks, their feet alternate between a stationary stance phase and a moving stride phase; we can detect the stationary phase to open zero speed update, that is, to feed an EKF with the measured errors in velocity, so that the EKF correct the speed errors in every step so as to restrain drift and ensure positioning accuracy.

Similarly, zero angular rate update (ZARU) can be done in the stationary state, which means to feed the EKF with measured errors in angular rate. This method can directly inhibit the gyro drift, which is useful to improve navigation accuracy, especially the yaw drift reduction.

It can be seen that zero velocity detector has a key role in the indoor pedestrian navigation. The common detection method extracts the acceleration and gyroscope data and process, then judge by comparing the evaluation value with the threshold size. These evaluation values include the acceleration moving variance, acceleration magnitude, angular rate energy, or their logical "AND" (Skog Nilsson and Händel 2010). In addition, SHOE (stance hypothesis optimal estimation) (Skog et al. 2010) is a generalized likelihood ratio test detector based on hypothesis testing. It was confirmed that above three common detectors can be derived from this detector based on different priori knowledge and its detection result is the best (Skog Nilsson and Händel 2010; Skog et al. 2010). Therefore, we choose SHOE detector.

Its mathematical formulas are characterized as follows:

let $y_k^a \in R^3$ and $y_k^\omega \in R^3$ denote the measured specific force vector and angular rate vector at time instant $k \in N$, respectively. The objective of the zero-velocity detection is to decide whether, during a time epoch consisting of N observations between the time instants n and $n + N-1$, the IMU is moving or stationary, given the measurement sequence $z_n \triangleq \{y_k\}_{k=n}^{n+N-1}$.

Then the stance hypothesis optimal estimation (SHOE) detector can be expressed as:

$$T(z_n) = \frac{1}{N} \sum_{k \in \Omega_n} \left(\frac{1}{\sigma_a^2} \left\| y_k^a - g \frac{\bar{y}_n^a}{\|\bar{y}_n^a\|} \right\|^2 + \frac{1}{\sigma_\omega^2} \|y_k^\omega\|^2 \right) < \gamma', \qquad (12)$$

where $T(z_n)$ is the test statistics of the detector and γ' is the detection threshold. $\sigma_a^2 \in R^1$ and $\sigma_\omega^2 \in R^1$ denote the accelerometer and gyroscope noise variance, respectively. $g \in R^1$ is the magnitude of the local gravity vector. $\bar{y}_n^a = \frac{1}{N} \sum_{k \in \Omega_n} y_k^a$ denotes the sample mean.

The tuning parameters for the detector are the window size N, the detection threshold γ', and the ratio $\sigma_a^2/\sigma_\omega^2$. The IMU is stationary if Eq. (12) is true.

2.3 The Extended Kalman Filter (EKF)

This section is mainly about designing a method of EKF, and about error estimation and correction. The state model of indoor pedestrian navigation is nonlinear while it can be applied with EKF after being linearization. Assume that the state space is linearized as:

$$\begin{cases} X_k = \Phi_k X_{k-1} + W_{k-1} \\ Z_k = H_k X_k + V_k \end{cases} \qquad (13)$$

In the equation set above, we select 15-dimensional error vector X_k as the state vector: $X_k = \begin{bmatrix} \delta r_k & \delta v_k & \delta \psi_k & \delta a_k^b & \delta \omega_k^b \end{bmatrix}^T$, where $\delta r_k, \delta v_k, \delta \psi_k, \delta a_k^b, \delta \omega_k^b$ are errors of position, velocity, attitude, accelerate, and angular rate, respectively. Φ_k is a 15×15 state transition matrix, which can be expressed as:

$$\Phi_k = \begin{bmatrix} I_{3\times3} & I_{3\times3} & 0_{3\times3} & 0_{3\times3} & 0_{3\times3} \\ 0_{3\times3} & I_{3\times3} & St & C^n_{b_{k|k-1}} & 0_{3\times3} \\ 0_{3\times3} & 0_{3\times3} & I_{3\times3} & 0_{3\times3} & -C^n_{b_{k|k-1}} \\ 0_{3\times3} & 0_{3\times3} & 0_{3\times3} & I_{3\times3} & 0_{3\times3} \\ 0_{3\times3} & 0_{3\times3} & 0_{3\times3} & 0_{3\times3} & I_{3\times3} \end{bmatrix} \tag{14}$$

where, $C^n_{b_{k|k-1}}$ is a state transition matrix. St is a skew-symmetric matrix of acceleration. It can be used to estimate pitch angle and roll angle. Its expression is as:

$$St = \begin{bmatrix} 0 & -a_{zk} & a_{yk} \\ a_{zk} & 0 & -a_{xk} \\ -a_{yk} & a_{xk} & 0 \end{bmatrix};$$ W_k is process noise, its variance matrix is $Q_k = E\left(W_k W_k^T\right)$.

In the measuring equation, we select $\mathbf{Z_k} = \begin{bmatrix} \Delta\varphi_k & \Delta\omega_k^b & \Delta v_k \end{bmatrix}$ as measurement, where yaw error $\Delta\varphi_k = \varphi_k - \varphi_{\text{compass}}$ (φ_{compass} is yaw angle measured by magnetometer), $\Delta v_k = v_{k|k-1} - 0_{3\times1}$, and $\Delta\omega_k^b = \Delta\omega_k^b - 0_{3\times1}$ are velocity error and angular rate error, respectively, when an zero velocity is detected. V_k is measuring noise and its variance matrix is $R_k = E\left(V_k V_k^T\right)$; H is a 7×15 matrix as follows:

$$H = \begin{bmatrix} [001] & 0_{1\times3} & 0_{1\times3} & 0_{1\times3} & 0_{1\times3} \\ 0_{3\times3} & I_{3\times3} & 0_{3\times3} & 0_{3\times3} & 0_{3\times3} \\ 0_{3\times3} & 0_{3\times3} & 0_{3\times3} & I_{3\times3} & 0_{3\times3} \end{bmatrix} \tag{15}$$

Next, we apply EKF to deal with error model mentioned above. The estimation of X_k can be acquired from the following equation after setting the appropriate initial parameters such as original state X_0 and covariance matrix P_0

$$\hat{X}_k = \hat{X}_{k/k-1} + K_k\left(Z_k - H_k\hat{X}_{k/k-1}\right) \tag{16}$$

where, $\hat{X}_{k/k-1}$ is one-step state prediction $\hat{X}_{k/k-1} = \Phi_k\hat{X}_{k-1} + W_{k-1}$; K_k is the Kalman gain which is calculated with $K_k = P_{k/k-1}H_k^T\left(H_kP_{k/k-1}H_k^T + R_k\right)^{-1}$; $P_{k/k-1}$ is the estimation error covariance matrix calculated with $P_{k/k-1} = \Phi_{k/k-1}P_{k-1}$ $\Phi_{k/k-1}^T + Q_{k-1}$; at last update the estimated covariance matrix with $P_k = (I - K_kH_k)\,P_{k/k-1}$.

Now, we finish the error estimation with EKF and can build a closed navigation loop when using it to revise error information in navigation.

Integrated framework is shown in Fig. 3. The overall structure is SINS + EKF + ZUPT + ZARU + COMPASS (Seco et al. 2010).

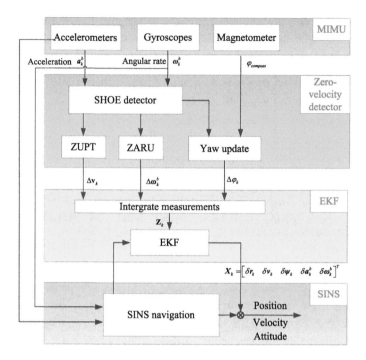

Fig. 3 Integrated framework

3 Experimental Test and Performance Evaluation

This section mainly talks about indoor pedestrian navigation system as a whole, and some experiments which verify its performance.

3.1 IMU Description

The system structure of the IMU is shown in Fig. 4, mainly composed of three MEMS gyroscopes, accelerometers, and magnetometer. The main performance parameters of sensors are shown in Table 1.

3.2 Parameter Tuning

(1) SHOE detector parameters

The parameters needed to be adjusted of SHOE detector are the window size N, the detection threshold γ', and the noise variance ratio $\sigma_a^2/\sigma_\omega^2$. The window size

Fig. 4 IMU and sensors description

Table 1 Performance parameters of sensors in IMU

	Accelerometer	Gyroscope	Magnetometer
Full scale (FS)	±10 g	±300°/s	±2 Gauss
Linearity	<0.2 % of FS	<0.1 % of FS	<0.2 % of FS
Bias stability	±4 mg	9.2°/h	±0.5 mGauss
Bandwidth	100 Hz	150 Hz	50 Hz

The dimension of the system is 45 × 35 × 30 mm. The weight is less than 100 g. Data output frequency is 100 Hz. The system can work continuously for longer than 1 h

N should be set as 5 or 10. Since the performance always degrades or stays unchanged as the window size increases, the adjustment of threshold value should be usually compared with external reference system (force-sensing resister installed on sole, video camera, etc.). The principle of adjustment is to make sure that stationary state can be detected at least once for most gait cycle rather than trying to detect all the samples in the time epoch when the IMU is stationary. The ratio $\sigma_a^2/\sigma_\omega^2$ reflects the disturbances in the information from the accelerometers versus them from the gyroscopes. It can be set to $\sigma_a = 0.02$ m/s^2 and $\sigma_\omega = 0.1°$/s, respectively, for example.

(2) EKF parameters

The parameters in EKF needed to be adjusted are covariance matrix Q of process noise, covariance matrix R of measuring noise, initial value P_0 of state estimation covariance matrix. These parameters exert a great influence on the whole algorithm. For this reason, it is necessary to adjust them repeatedly until filter achieves uniformly asymptotical stability.

In this paper, the EKF parameters can be set as $Q = \text{diag}\begin{bmatrix} 0_{1\times3} & 1 \times 10^{-4}_{1\times3} & 1 \times 10^{-4}_{1\times3} & 0_{1\times3} & 0_{1\times3} \end{bmatrix}$, where diag[*] means that * is diagonal line elements.

$R =\mathrm{diag}\begin{bmatrix} 1 \times 10^{-2} & 1 \times 10^{-2}_{1 \times 3} & 1 \times 10^{-4}_{1 \times 3} \end{bmatrix}$ is matched respectively with observation noise of each observed variable. The initial value of state estimation covariance matrix is $P = \begin{bmatrix} 0_{1 \times 3} & 0_{1 \times 3} & 0_{1 \times 3} & 1 \times 10^{-2}_{1 \times 3} & 1 \times 10^{-2}_{1 \times 3} \end{bmatrix}$.

3.3 Indoor Experiments

After the adjustment of system parameters, we performed indoor navigation experiments on the designed PDR.

In the indoor environment, a person with foot-mounted PDR walked along a rectangular route a circle at normal speed (5 km/h), as shown in Fig. 5, then back to initial position. The total distance is about 230 m, with 160 s total time consuming. In the figure, blue line represents trajectory estimated by PDR, and the red line is the true route.

As is shown in Fig. 5, estimated trajectory can basically form a closed-loop curve. Distance error between terminal point and initial point is 4.68 m, indicating that probable accuracy approximately reaches 2 % of the total walking distance.

We need to point out that on account of much the interference of magnetometer in the process of experiments which includes hard iron biases and soft iron distortions of the earth's magnetic field, yaw angle measured by magnetometer will become very unstable. As observed variable, it will bring the whole navigating estimation very great errors. Therefore, in practical application, we generally remove the yaw angle measured by magnetometer from the observed variables.

Measured data of gyroscope, accelerometer, and zero-velocity detection figure formed by SHOE detector are shown in Fig. 6. As you see, we can detect state of zero velocity at every step, which guarantees restraints of EKF on error drift of inertial navigation and also improve the accuracy of the indoor navigation.

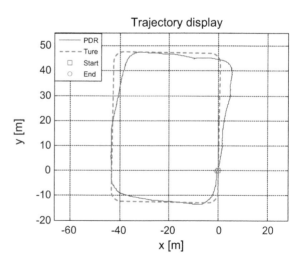

Fig. 5 Estimated trajectory in an indoor environment

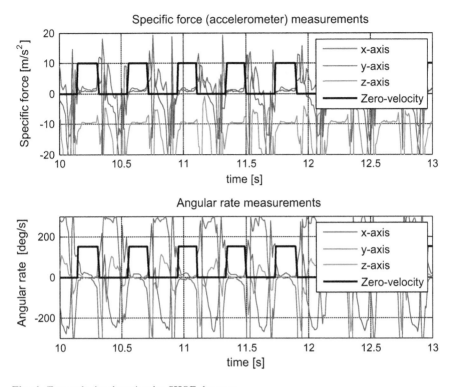

Fig. 6 Zero-velocity detection by SHOE detector

4 Conclusion

This paper designs and achieves a miniature of wearable indoor pedestrian navigation system in order to estimate the position and attitude of a person when walking indoor. With the purpose of decreasing the accumulated error caused by long-term drift of inertial device, a zero-velocity detector based on hypothesis testing is introduced to revise the instantaneous velocity and angular velocity of the PDR. This paper is also a design of Kalman filter with combinations of inertia navigation information, magnetic information, and the correction information of instantaneous zero state. The filter can achieve error estimation and correction. In the end, this chapter talks about the performance test and evaluation of the PDR. The results show that for level ground, position accuracy reaches 2 % of the whole walking distance.

Acknowledgments Project supported by the National Natural Science Foundation of China (No. 61471046).

References

Chen Z, Zou H, Jiang H et al (2015) Fusion of WiFi, smartphone sensors and landmarks using the Kalman filter for indoor localization. Sensors 15(1):715–732. doi:10.3390/s150100715

Foxlin E (2005) Pedestrian tracking with shoe-mounted inertial sensors. IEEE Comput Graphics Appl 25(6):38–46

Kamisaka D, Watanabe T, Muramatsu S et al (2012) Estimating position relation between two pedestrians using mobile phones. In: Pervasive Computing. Springer Berlin Heidelberg, pp 307–324

Raitoharju M, Nurminen H, Piché R (2015) Kalman filter with a linear state model for PDR + WLAN positioning and its application to assisting a particle filter. EURASIP J Adv Signal Process 2015(1):1–13. doi:10.1186/s13634-015-0216-z

Seco F, Prieto JC, Guevara J (2010) Indoor pedestrian navigation using an INS/EKF framework for yaw drift reduction and a foot-mounted IMU. In: 2010 7th Workshop on positioning navigation and communication (WPNC). IEEE, pp 135–143

Skog I, Nilsson JO, Händel P (2010) Evaluation of zero-velocity detectors for foot-mounted inertial navigation systems. In: 2010 International conference on indoor positioning and indoor navigation (IPIN). IEEE, pp 1–6

Skog I, Händel P, Nilsson JO et al (2010b) Zero-velocity detection—an algorithm evaluation. IEEE Trans Biomed Eng 57(11):2657–2666

Zampella F, De Angelis A, Skog I et al (2012) A constraint approach for UWB and PDR fusion. In: International conference on indoor positioning and indoor navigation (IPIN)

Zampella F, Jimenez R, Antonio R et al (2013) Robust indoor positioning fusing PDR and RF technologies: The RFID and UWB case. In: 2013 International conference on indoor positioning and indoor navigation (IPIN). IEEE, pp 1–10

Calibration Method of the 3-D Laser Sensor Measurement System

Qing-Xu Meng, Qi-Jie Zhao, Da-Wei Tu and Jin-Gang Yi

Abstract In order to improve the efficiency and adaptability of 3-D laser sensor measurement system, this paper proposed a calibration method based on structural parameters for the 3-D laser sensor measurement system. In this method, by scanning and measuring the structural calibration target with known structural parameters, the models are established in the laser scanning coordinate system and inertial measurement sensor coordinate system respectively, and the linear features are extracted then solve out the relative position and attitude by the constrains of structural parameters. The calibration results are carried out on the 3-D measurement system, which is used to conduct a simulation measurement experiment. In the experiment, the measured results' relative errors of lengths and angles of measured objects are less than 1.0 and 0.5 % on average, which indicate the accuracy of the calibration method. While the noise is increasing, the relative errors keep stability, which indicates the effectiveness of the calibration method.

Keywords Calibration · Laser sensor · Measurement · Structural parameters

Q.-X. Meng · Q.-J. Zhao (✉) · D.-W. Tu
School of Mechatronic Engineering and Automation,
Shanghai University, Shanghai 200072, China
e-mail: zqj@shu.edu.cn

Q.-J. Zhao · D.-W. Tu
Shanghai Key Laboratory of Intelligent Manufacturing and Robotics,
Shanghai 200072, China

J.-G. Yi
Department of Mechanical and Aerospace Engineering,
Rutgers University, Piscataway, NJ 08854, USA

© Zhejiang University Press and Springer Science+Business Media Singapore 2017
C. Yang et al. (eds.), *Wearable Sensors and Robots*, Lecture Notes in Electrical
Engineering 399, DOI 10.1007/978-981-10-2404-7_11

1 Introduction

3-D laser sensor measurement is the core technology of the large-scale measuring and terrain reconstruction. Therefore, it is very important to the stocks security, inventory costs reducing (Wang et al. 2013; Kim et al. 2009) and disaster relief (Huang and Wang 2012). Laser sensor-based measurement has characteristics of noncontact and wide range (Yu et al. 2012; Liu et al. 2013), which are suitable for measuring large target object. 3-D laser sensor measurement (Fojtik 2014; Diskin and Asari 2013) system consists of a variety of sensors and transmission platform. Faced with different actual measurement environment, the position and attitude of the different parts of the system have a variety of changes. Therefore, calibration of the measurement system is very necessary to ensure the reliability and accuracy of measurement results of the system.

In the field of laser sensors-based measurement, some domestic and foreign scholars have carried out some study on the calibration terms. Some of them seek the scanning position of the laser scanning line with the coaxial visible light source (Li et al. 2013) or silicon cells (Kang et al. 2008), then build the constrain equations of the homonymy points in object coordinate system and image coordinate system to calibrate the system. Some of them directly calibrate relative geometry relation of laser scanning device by building a standard template, such as a pyramid calibration target (Guo et al. 2011), flat paper target (Lao et al. 2011), or special aluminum plate (González and Riveiro 2011). These methods have a certain effect to improve the measurement accuracy. The features of these methods are that the features extraction is in the fitting planes, which results in errors in many directions (Li et al. 2012). Feature extractions based on linear fitting and structural parameters are easy to do and computation is small, but the precise fitting of lines is key points.

In this paper, for the condition that the inertial sensor real-time calibrates the position and attitude of the laser sensor in the 3-D laser measuring system, a calibration method based on structural calibration target is proposed. In this methods, through the way of features extraction in the obtained data combining the structural parameters, constrained equations are listed, then the relative position and attitude between inertial sensor and laser sensor are calibrated. Calibration experiment is conducted by the method proposed by this article, and then the calibrated results are used in the simulation measurement experiment, to prove the accuracy and reliability of the calibration method, which lay the foundation for the 3-D laser sensor measurement.

2 Theoretical Models

Through analyzing the three-dimensional laser measuring system and measurement method, the model of system coordinate transformation is presented in Fig. 1. $X_0Y_0Z_0$ is the base coordinate system of measurement system under any working

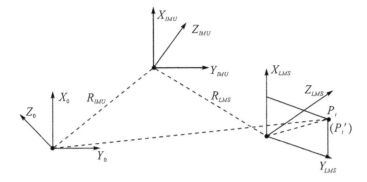

Fig. 1 Coordinate transformation model

situation. XYZ_{IMU} is the coordinate system of the inertial measurement sensor and XYZ_{LMS} is the coordinate system of laser scanning sensors. P_i is coordinate value in the laser scanning coordinate system and P'_i is the measured point's coordinate value in the basis coordinate system.

In order to unified operate data, measured target point $P_i(x_i, y_i, z_i)$ on the object relative to LMS coordinate system needs to transform to the basis $X_0Y_0Z_0$, also called measurement system coordinate system, and $P'_i(x'_i, y'_i, z'_i)$ is the measured target coordinate after conversion. The transformation from P_i to P'_i realize by the formula (1):

$$\begin{bmatrix} x'_i \\ y'_i \\ z'_i \end{bmatrix} = r_{\text{IMU}} + R_{\text{IMU}} \left[R \begin{bmatrix} x_i \\ y_i \\ z_i \end{bmatrix} + p \right] \tag{1}$$

r_{IMU} is the real-time coordinate values of inertial sensor under basis coordinate system, R_{IMU} is the rotation matrix corresponding to real-time angle of inertial sensor. R is the rotation matrix of laser sensor relative to inertial sensor and p is the displacement.

3 Calibration Methods

There is a fixed conversion between the inertial measurement sensors system and the laser scanner coordinate system. Under the inertial measurement sensors system the measured point can be described as:

$$\begin{bmatrix} r'_{\text{IMU}} \\ 1 \end{bmatrix} = [R|p] \begin{bmatrix} r_{\text{LMS}} \\ 1 \end{bmatrix} \tag{2}$$

Fig. 2 Diagram of calibration algorithm based on structured object

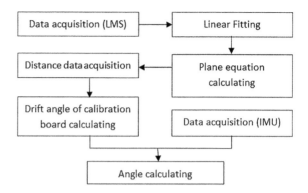

In the formula (2): r'_{IMU} is the coordinates in the IMU coordinate system, R and p are the rotation matrix and the translation vector from the inertial measurement sensors system to the laser scanner coordinate system. Because p can be measured and calculated after the sensors being assembled, so R is the key calibrate results, which can be calculated pitch angle α, roll angle β, and yaw angle γ (Fig. 2).

3.1 Calibration of Pitch Angle (α) and Roll Angle (β)

Three calibrated targets of different structural parameters were scanned, calculation model is built, which is shown in Fig. 3.

Three planes P_1, P_2, P_3 are scanned by the laser scanning sensor in the same position. The acquired point data is used to fit the linear equations by the way of RANSAC (Bazin and Pollefeys 2012; Li and Chen 2014), and set up the provided

Fig. 3 α and β calibration principle

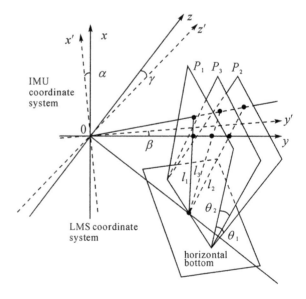

plane equations corresponding the lines. The normal vector of the plane which line l_i corresponds is (A_i, B_i, t_i), and A_i, B_i are fitted by the RANSAC method and t_i is the unknown number, $i = 1, 2, 3$. According to the calibration plate structure parameters, the following constraint Eq. (3) can be derived from the law of cosines:

$$\begin{cases} \cos\theta_1 = \dfrac{|A_1A_2 + B_1B_2 + t_1t_2|}{\sqrt{A_1^2 + B_1^2 + t_1^2}\sqrt{A_2^2 + B_2^2 + t_2^2}} \\ \cos\theta_2 = \dfrac{|A_1A_3 + B_1B_3 + t_1t_3|}{\sqrt{A_1^2 + B_1^2 + t_1^2}\sqrt{A_3^2 + B_3^2 + t_3^2}} \end{cases} \tag{3}$$

θ_1 and θ_2 are the structural parameters of the calibration targets. According to the constraints that the intersecting lines are unified of the three planes, the formula (4) can be list as follow:

$$\frac{B_1t_2 - B_2t_1}{B_1t_3 - B_3t_1} = \frac{A_2t_1 - A_1t_2}{A_3t_1 - A_1t_3} = \frac{A_1B_2 - B_1A_1}{B_1A_3 - B_3A_1} \tag{4}$$

The unknown number t_i is calculated. According to the constrain that the bottom of the calibrate target is horizontal, the pitch angle α_{LMS} and the roll angle β_{LMS} of LMS and the yaw angle γ_{LMS} of the laser sensor related the calibration target is calculated, shown as formula (5):

$$\begin{aligned} \alpha_{LMS} &= a\,\tan\frac{(B_1t_2 - B_2t_1)B_1 - (A_1t_2 - A_2t_1)A_1}{(A_2t_1 - A_1t_2)t_1 - (B_1t_2 - B_2t_1)B_1} \\ \beta_{LMS} &= a\,\tan\frac{(A_1B_2 - B_2t_1)A_1 - (B_1t_2 - B_2t_1)t_1}{(A_2t_1 - A_1t_2)t_1 - (B_1t_2 - B_2t_1)B_1} \\ \gamma_{LMS} &= a\,\tan\frac{t_1}{B_1} \end{aligned} \tag{5}$$

Combined the data $(\alpha_{IMU}, \beta_{IMU}, \gamma_{IMU})$ from inertial sensor, according the difference of the pitch angle and roll angle relative to horizontal plane respectively, the pitch angle α and the roll angle β between laser sensor and inertial sensor can be calculated as formula (6):

$$\begin{aligned} \alpha &= \alpha_{IMU} - \alpha_{LMS} \\ \beta &= \beta_{IMU} - \beta_{LMS} \end{aligned} \tag{6}$$

3.2 Calibration of Yaw Angle (γ)

The displacement of the transmit platform is d. The difference of the distance in the corresponding position of A and B is d_{AB}, which showed in the Fig. 4. AC is paralleled with the motion direction of the transmit platform. O and O' are the origin points of the IMU coordinate system before and after moving. In $\triangle ABC$, by

Fig. 4 γ calibration principle

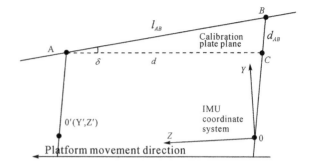

the law of cosines, the yaw angle γ_C between the calibration target and the motion direction of the transmit platform can be calculated by formula (7):

$$\cos \gamma_c = \frac{l_{AB}^2 + d^2 - d_{AB}^2}{2l_{AB}d} \tag{7}$$

Combined with the γ_{LMS} in the Chap. 3.1, $|\gamma_{\text{LMS}} - \gamma_C|$ is the yaw angle between laser sensor and motion direction of the transmit platform; the yaw angle between inertial sensor and the motion direction can be calculated by the coordinate values of O' in the inertial sensor coordinate system. With reference to the motion direction, the yaw angle can be settled as formula (8):

$$\gamma = \gamma_M - |\gamma_{\text{LMS}} - \gamma_c| = a \tan \frac{Y'}{Z'} - |\gamma_{\text{LMS}} - \gamma_c| \tag{8}$$

4   Experimental Analyses

As shown in Fig. 5, the 3-D laser measurement system is set up according to the proposed method. The system consists of 2-D laser scanning sensor, inertial measurement sensor, laser ranging sensor, structural calibration target, transmit platform, and drive devices. The measurement system obtain 3-D point cloud data by adopting stepper motor to drive the ball screw to realize the linear motion of the 2-D laser scanner of LMS511.

In this experimental platform, calibration experiment is conducted in order to get the relative position and attitude of the inertial measurement sensor and laser sensor. First, put the structural calibration target in the scene and keep the bottom horizontal. The surface points of different structural parameters calibration target are collected by laser sensor, and equations of the straight line are fit by RANSAC

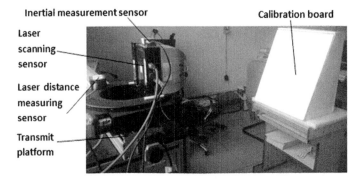

Fig. 5 3-D Laser measurement system

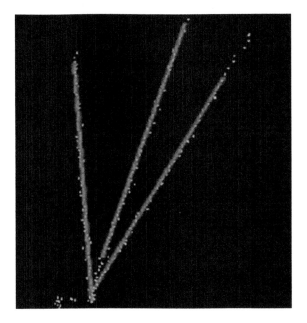

Fig. 6 Linear fitting results

using the collected points. The fitting result is shown in the Fig. 6. Calculate the angle between laser sensor and horizontal plane and the yaw angle between calibration plate and laser sensor. Second, the laser ranging sensors on the transmit platform is used to measure its distance to the calibration target in different position. The information of position and acceleration of IMU is adopted to calculate the angle between inertial sensor coordinate and motion direction. In the end, calculate the relative angle and coordinate transformation matrix of between laser sensor and inertial sensor according to the proposed calibration method.

Table 1 calibration result of initial values

Relative angle (RPY)/°	$[0.1243 \quad -3.7614 \quad -0.1339]$
Rotation matrix (R)	$\begin{bmatrix} 0.9978 & 0.0656 & 0.0023 \\ -0.066 & 0.9978 & 0.0022 \\ -0.0022 & -0.0023 & 0.9999 \end{bmatrix}$
Transmit vector (p)/mm	$[-84.50 \quad 57.05 \quad 112.75]$

Fig. 7 Generated 3-D point clouds

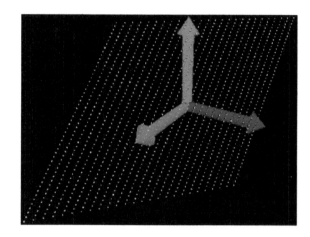

The rotation matrix and translation vector in Table 1 is the description of the relative position and attitude between laser sensor and inertial sensor, which are used to conduct a simulation measurement experiment.

In the experiment, the 51×26 three-dimensional spots are generated. Its shape is rhombus and has a certain geometric parameters as shown in Fig. 7. According to the scanning rules of the measuring system, random dimension noises are added to the laser spots. The mean of the noises is 0, and its amplitude variations from 1 to 7 mm. Its function is to simulate the error in the actual acquisition of the measure system after calibration, then to judge the performance of the calibration. In simulation experiments, the angle deformation of the rhombus is served as the angle relative error of the scanned object, side length deformation as the length relative error.

Shown as in the Fig. 8a, b are angle relative errors and length relative errors, respectively. Form the Fig. 8, it can be seen that the measurement errors are significantly reduced after calibrated and the angle relative error is less than 0.5 % stably and the length error is less than 1 % under the right noise, which verified the accuracy of the calibrated measurement system; the calibrated system is robust when noise level is creasing, which verified the effectiveness of the calibration method.

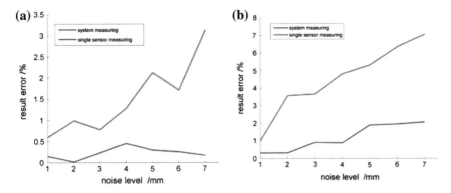

Fig. 8 Diagram of relative error, angle relative error (**a**), length relative error (**b**)

5 Conclusions

For the condition that the measurement system uses the inertial sensor to calibrate the position and attitude of laser sensor on real-time, this paper proposed a calibration method which is based on structural parameters to calibrate the relative position and attitude. The calibration method is on the basis of scanning the structural calibration targets and feature extraction, and then provided the computing models according to the different description of the same feature in the different coordinate systems, to figure out the rotation matrix and combining with the translation vector, the calibration is carried out. The simulation experiment verified the accuracy and the validity of the calibration method, which laid foundation for 3-D laser measurement.

Acknowledgments Project supported by the National Natural Science Foundation of China (No. 61101177).

References

Bazin J, Pollefeys M (2012) 3-line RANSAC for orthogonal vanishing point detection. J Intell Robot Syst, 4282–4287. doi:10.1109/IROS.2012.6385802

Diskin Y, Asari V (2013) Dense 3D point-cloud model using optical flow for a monocular reconstruction system. In: 2013 IEEE Applied imagery pattern recognition workshop (AIPR), 1–6. doi:10.1109/AIPR.2013.6749315

Fojtik D (2014) Measurement of the volume of material on the conveyor belt measuring of the volume of wood chips during transport on the conveyor belt using a laser scanning. In: International carpathian control conference, pp 121–124. doi:10.1109/CarpathianCC.2014. 6843581

González J, Riveiro B et al (2011) Procedure to evaluate the accuracy of laser-scanning systems using a linear precision electro-mechanical actuator. J Sci Measure Technol 6(1):6–12. doi:10. 1049/iet-smt.2011.0054

Guo Y, Du Z, Yao Z (2011) Calibration method of three dimensional (3D) laser measurement system based on projective transformation. J Measuring Technol Mechatron Autom 1:666–671. doi:10.1109/ICMTMA.2011.169

Huang G, Wang C (2012) Multiscale geostatistical estimation of gravel-bed roughness from terrestrial and airborne laser scanning. J IEEE Geosci Remote Sens Soc 9(6):1084–1088. doi:10.1109/LGRS.2012.2189351

Kang Y, Zhong R, Yu W (2008) Research of calibrating vehicle laser scanner's external parameters. J Infrared Laser Eng 37:249–253 (in Chinese)

Kim B, Koo J, Park B (2009) A raw material storage yard allocation problem for a large-scale steelworks. J Int J Adv Manuf Technol 41(9–10):880–884. doi:10.1007/s00170-008-1538-x

Lao D, Yang X et al (2011) Optimization of calibration method for scanning planar laser coordinate measurement system. J Guangxue Jingmi Gongcheng/Optics Precision Eng 19 (4):870–877. doi:10.3788/OPE.20111904.0870

Li H, Chen R (2014) Optimal line feature generation from low-level line segments under RANSAC framework. In: The 26th Chinese control and decision conference (2014 CCDC), pp 4589–4593

Li L, Zhang X, Tu D (2012) Joint calibration of 2D and 3D vision integrated sensor system. J Chin J Sci Instrum 33(11):2473–2479 (in Chinese)

Li L, Yan J, Ruan Y (2013) Calibration of vehicle-borne laser mapping system. J Chin Optics 6 (3):353–358 (in Chinese)

Liu Z, Zhu J et al (2013) A single-station multi-tasking 3D coordinate measurement method for large-scale metrology based on rotary-laser scanning. J Measur Sci Technol 24(10):1–9. doi:10.1088/0957-0233/24/10/105004

Wang H, Zhang D, He L (2013) Reserch on large open stockyard laser measurement methods. J Chin J Lasers 40(5):0508002–0508312. doi:10.3788/CJL201340.0508002

Yu M, Liou J, Kuo S (2012) Noncontact respiratory measurement of volume change using depth camera. J Eng Med Biol Soc, 2371–2374. doi:10.1109/EMBC.2012.6346440

Part II
Wearable Robots

Study on a Novel Wearable Exoskeleton Hand Function Training System Based on EMG Triggering

Wu-jing Cao, Jie Hu, Zhen-ping Wang, Lu-lu Wang and Hong-liu Yu

Abstract In order to improve the rehabilitation initiative of stroke patients with hand dysfunction, this paper proposes a novel wearable exoskeleton hand function training system based on myoelectric (EMG) triggering. EMG control system is designed and the exoskeleton hand function training device is modeled and tested to verify the motion capabilities. The experiment results show that subject can complete prehension and looseness independently. The maximum flexion degrees of fingers are in agreement with the theoretical values. The feasibility of the wearable exoskeleton hand function training system is validated.

Keywords Wearable · Exoskeleton · Rehabilitation training · EMG control

1 Introduction

Stroke is the result of a sudden disturbance of cerebral circulation (Kuang 2011). According to the investigation, there are about 10.36 million people who suffer from stroke over the age of 40 in China (Jia 2013). Among many sequelae of stroke, the hemiplegia accounts for the very great proportion. Because of the sophistication of hand joints, the recovery of hand function is the most difficult in all kinds of rehabilitation for hemiplegia. (Long 2007). Medical theory and practice proves that effective limb rehabilitation training is necessary to prevent muscle atrophy and remodel the nerve system of patients with hemiplegia. Therefore, it is significant to design a rehabilitation apparatus to help patients recover hand function (Zhang et al. 2012).

W. Cao · J. Hu · Z. Wang · L. Wang · H. Yu (✉)
University of Shanghai for Science and Technology, Shanghai 200093, China
e-mail: yhl98@hotmail.com

W. Cao
e-mail: caowujing414@126.com

© Zhejiang University Press and Springer Science+Business Media Singapore 2017
C. Yang et al. (eds.), *Wearable Sensors and Robots*, Lecture Notes in Electrical Engineering 399, DOI 10.1007/978-981-10-2404-7_12

With the development of sensor technology, material technology, control technology, and other related areas, research on finger rehabilitation exoskeleton trainer has made great progress (Liu 2008). In China, the only mature product is the mechanical arm robot triggered by muscle from the Hong Kong Polytechnic University. (Ho et al. 2011). In foreign countries, the intelligent hand function rehabilitation training robot named Golreha is designed by an Italian company called DROGENET and the exoskeleton hand robot based on electroencephalogram (EEG) control named ExoHand is developed by a Germany company called FESTO (Chen et al. 2011; Zhang 2011). However, the three products mentioned are large and medium-sized rehabilitation equipments, generally expensive and not appropriate for home use. To solve these problems and improve hand function rehabilitation training effect in daily life, this paper proposes and designs a wearable exoskeleton hand function training system based on EMG (electromyogram) (Wu et al. 2010).

Stroke patients with hand dysfunction can wear the exoskeleton hand function trainer proposed in this paper and do hand rehabilitation training anywhere according to the EMG signal of the forearm. Through optimizing structure and using proper materials, this hand function trainer can also be lighter.

2   Structure Design of Wearable Exoskeleton

This paper presents a light wearable exoskeleton hand function rehabilitation device, which mainly apply to stroke patients with hand dysfunction. This kind of patients can do hand rehabilitation training with the device anywhere and recover partly or fully from hand dysfunction.

The basic function of the mechanical structure of the device is to drive finger joints flex and extend within certain range. Moreover, the training apparatus must be also suitable for patients to wear. The finger rehabilitation training method of the design is the thumb movement bionic mechanism coordinate with other four fingers bionic mechanism. The thumb movement bionic mechanism fulfills the thumb rehabilitation training separately and the four fingers bionic mechanism drive the four fingers move simultaneously. The joint flexion and extension movement of metacarpal joint (MP) and proximal interphalangeal joint of four fingers (PIP) is realized by using chute bionic mechanism located in the back of the fingers, as shown in Fig. 1.

To simplify the model of four-finger movement bionic mechanism, the design takes an approximate line as the rotating shaft of the metacarpophalangeal joints of four fingers, which is similar to the straight line that goes through rotation centers of the metacarpophalangeal joint of the four fingers. The shaft of the mechanism and the fitting shaft of the four fingers' metacarpophalangeal joints are coaxial, so only one motor is needed to drive the four fingers flex and extend. This significantly reduces the weight and volume of the wearable exoskeleton hand function trainer.

Fig. 1 Bionic mechanism design of a joint **a** Flexion position; **b** Extension position

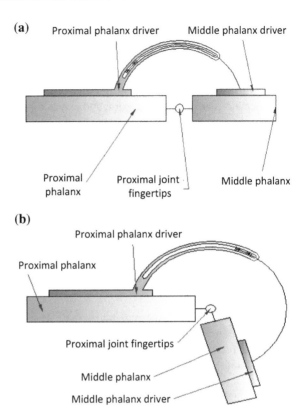

(**a**)   Proximal phalanx driver   Middle phalanx driver

Proximal phalanx   Proximal joint fingertips   Middle phalanx

(**b**)

Proximal phalanx driver

Proximal phalanx

Proximal joint fingertips

Middle phalanx

Middle phalanx driver

3 Kinematics Simulation Analysis

To verify whether the grabbing movement of the wearable exoskeleton hand function trainer is consistent with the normal people, motion model of SolidWorks is used to conduct motion simulation of the grabbing movement of the trainer. In the motion simulation analysis, index finger bionic movement mechanism is taken as the research subject (Li et al. 2011). The motion simulation curves of index finger metacarpophalangeal joint and proximal interphalangeal joint are extracted, comparing to the angular displacement curve obeying the cycloid motion law (Zhuang et al. 2006) and the 3-4-5 polynomial motion law (Yang et al. 2004; Gao. 2010), respectively. It indicates that the design is with high imitation by nature. Because of the fact that the finger lengths are different from people to people, the study of trajectory of the finger marker cannot stand for finger motion law of most people. Therefore, the angular displacement of index finger metacarpal joint and proximal interphalangeal joint are taken as the comparison targets in the simulation analysis. In this paper, the length of the index finger proximal phalanx is set to L1; the length of middle phalanx is set to L2; the length of the distal phalanx is set to L3; the angle of the metacarpophalangeal joint is set to θ_1; the angle of the proximal interphalangeal joint is set to θ_2 and the angle of the distal interphalangeal joint is set to θ_3. The forefinger motion analysis is modeled as shown in Fig. 2.

The common feature of angular displacement between cycloid motion law and 3-4-5 polynomial is that both the angular displacement represented by the tracks are continuous from 0 to 0 along with the time period, with no rigid impact. The two movement laws are similar to flexion and extension of human finger joints, and they are the most common laws used to study the movement of the fingers currently.

Angular displacement can be expressed according to the cycloidal motion law with time as Eq. 1:

$$\theta_a = \theta_0 \left(\frac{t}{t_0} - \frac{1}{2\pi} \sin \frac{2\pi}{t_0} t \right) \tag{1}$$

Angular displacement can be expressed according to the 3-4-5 polynomial motion law with time as Eq. 2:

$$\theta_b = 10\theta_0 \left(\frac{t}{t_0} \right)^3 - 15\theta_0 \left(\frac{t}{t_0} \right)^4 + 6\theta_0 \left(\frac{t}{t_0} \right)^5 \tag{2}$$

In Eq. 2, θ_0 represent the maximum angle of joint movement; t represent time variable; t_0 represent the time it takes for the angular displacement changes from 0 to maximum θ_0.

Simulation analysis is conducted to the simplified model of bionic finger movement mechanism in SolidWorks Motion, and set $t_0 = 2.5$s. Time-angular displacement curve of metacarpophalangeal joint angular displacement and angular displacement of the proximal interphalangeal joint can be get from the corresponding forefinger simulation model. According to the simulation, when $t = 0$ s, we can know $\theta_1 = 10.18$ $\theta_2 = 0.12$. So Eqs. 1 and 2 should be adjusted accordingly to Eqs. 3, 4, 5, and 6:

$$\theta_a = \theta_0 \left(\frac{t}{t_0} - \frac{1}{2\pi} \sin \frac{2\pi}{t_0} t \right) + 10.18 \tag{3}$$

Fig. 2 Finger motion analysis model

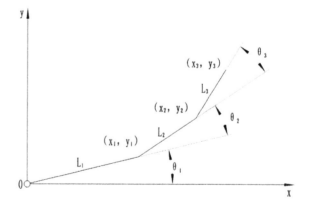

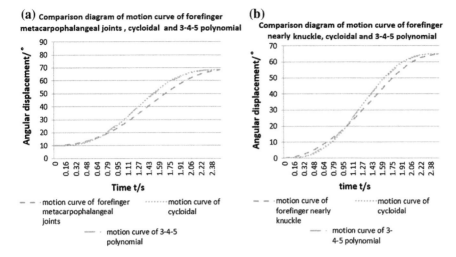

Fig. 3 Comparison diagram

$$\theta_b = 10\theta_0 \left(\frac{t}{t_0}\right)^3 - 15\theta_0 \left(\frac{t}{t_0}\right)^4 + 6\theta_0 \left(\frac{t}{t_0}\right)^5 + 10.18 \tag{4}$$

$$\theta'_a = \theta_0 \left(\frac{t}{t_0} - \frac{1}{2\pi} \sin \frac{2\pi}{t_0} t\right) + 0.12 \tag{5}$$

$$\theta'_b = 10\theta_0 \left(\frac{t}{t_0}\right)^3 - 15\theta_0 \left(\frac{t}{t_0}\right)^4 + 6\theta_0 \left(\frac{t}{t_0}\right)^5 + 0.12 \tag{6}$$

Compare the time (t)-angular displacement (θ_1) curve of the metacarpopha-langeal joint obtained from the simulation to Eqs. 3 and 4. The comparison chart is shown in Fig. 3a. Compare the time (t)-angular displacement (θ_2) curve of proximal interphalangeal joint to Eq. 5. The comparison chart is shown in Fig. 3b.

As can been seen in Fig. 3, the two curves are very consistent. As a result, the wearable exoskeleton hand function trainer can well simulate joints' movements of human grabbing, and has good wearable biomimetic.

4 Design of EMG Control System

Based on the EMG control system, active training can be achieved through EMG signal generated by patients' wishes (Zheng et al. 2003). In this kind of training, rehabilitation effect can be proved by observing actual hand movements.

Surface EMG (sEMG) is a complex biological signals generated by muscle contraction. Although the morphology of sEMG has great randomness and

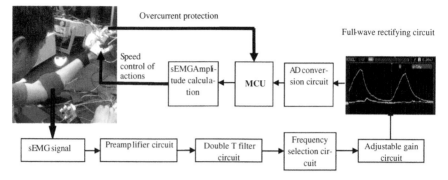

Fig. 4 EMG control system block diagram

instability, it contains large amounts of information about muscle contraction function. sEMG is an alternating voltage signal. Researches show that the greater the muscle strength is, the larger the amplitude of the muscle will be. There is a proportionality between them. In addition, there is a linear relation between the voltage amplitude of sEMG and muscular flaccidity and tension (Hou et al. 2007). In this paper, the grabbing action of the wearable exoskeleton hand function trainer is controlled by signals generated in the antagonistic muscle of patient's forearm using a pair of electrodes. In the EMG control system, weak EMG signal processing technology is used to collect and process the signal. The EMG control system designed is shown in Fig. 4.

In this paper, the forearm EMG signal is collected by surface electrode. The weak EMG signal is amplified by the preamplification circuit, and then 50 Hz power-line interference is eliminated through T filter circuit. Low-pass filtering and high-pass filtering are conducted to the signal by frequency selection circuit. Finally, gain adjustable circuit is used to regulate magnification to accommodate different EMG signal strength and obtain recognizable EMG signal. To apply EMG signal to control motors, EMG signal is transformed to positive amplitude by full-wave rectification circuit and then transformed to digital signals through digital analog convertor. Finally, the signal is analyzed and processed by MCU (Microprogrammed Control Unit). The EMG control system can control the speed of motion according to the EMG signal strength of the forearm. Meanwhile, the system has automatic overcurrent protection function.

5 Prototype Experiment

After finishing the design of the finger motion bionic mechanism and the EMG control system, we made the experimental prototype of the wearable exoskeleton hand function trainer according to the size of people aging from 18 to 26 conform to the GB standard (Chinese National Standards). The corresponding EMG triggering system and voice control system are designed. The prototype weighs 450 g. It is very portable. In order to prevent user's hand from friction of the mechanism, the biocompatible material liner is used in the contact surface. Meantime, the elastic cotton bandage is used to keep user's hand fix on the trainer. The following is the grabbing and holding experiment based on EMG triggering.

5.1 Experiment of Grabbing and Holding

The grabbing and holding experiment is designed to study grabbing and holding ability of the exoskeleton hand function trainer. When the thumb and four fingers extend to limit position, hand is in open state. When the thumb and four fingers flex to limit position, hand is in gripping state. In this experiment, the open and gripping motion of the hand is realized by the forearm EMG signal of the subject. In grabbing experiment, subject grabbed a cup from the table to the pad, as shown in Fig. 5. In holding experiment, subject holds the notebook and turns the wrist, as shown in Fig. 6.

We can see that subject can grab and hold stuff independently with the wearable exoskeleton hand function trainer From Figs. 5 and 6. It can be concluded that the trainer can well grab the cup and hold the notebook. The strength is appropriate that the things will neither slip off nor produce an oppressive feeling.

(a)   **(b)**   **(c)**

Fig. 5 Grabbing the cup

(a)   **(b)**   **(c)**

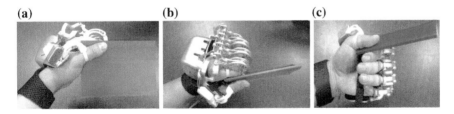

Fig. 6 Holding the book

Table 1 Maximum angle of joint flexion (°)

Angular	Times					
	1	2	3	4	5	Average
α	67.9	67.4	67.2	68.1	68.2	67.8
β	64.9	64.3	64.2	64.5	64.2	64.4

5.2 Experimental Measurements of Joint Activities

According to the grabbing and holding experiment, the maximum flexion angle of index finger metacarpophalangeal joint α and the maximum flexion angle of index finger proximal interphalangeal joint can be measured. The angles are shown in Table 1.

We know that the greatest flexion angles of index finger metacarpophalangeal joint and proximal interphalangeal joint are 90° from the human body physiology parameters. From the simulation analysis of grabbing motion, we achieve that the greatest flexion of index finger metacarpophalangeal joint is 68.5°, and the greatest flexion of index finger proximal interphalangeal joint is 65°. From Table 1, we know that the two angles are within the scope of the theory and are similar to the motion simulation.

6 Conclusion

Through the design of the wearable exoskeleton hand function trainer and the EMG control system, the wearable exoskeleton hand function training system based on EMG triggering is accomplished. In the grabbing motion simulation experiment, it is verified that the trainer can well-imitate human grabbing motion. In the experiment of grabbing and holding with prototype of wearable exoskeleton hand function trainer, the subjects can well grab and hold with the wearable exoskeleton hand function trainer. The maximum degrees of flexion of index finger metatarsophalangeal joint and proximal interphalangeal joint are within the scope of the theory and they are familiar with the result of simulation. The reasonability and feasibility of the wearable exoskeleton hand function trainer triggered by EMG is verified.

References

Chen X, Li Y (2011) Activities of daily living in patients with stroke sequelae guidance. Chin J Rehabil Theory Practice 17(9):825–829
Gao J (2010) Research and polynomial curve sine series series curve. Mech Transm 34(10):31–35
Hou W, Xu R, Zheng X (2007) Correlation between the size of grip strength and forearm muscle activity patterns of surface electromyography. Space Med Med Eng 20(4):265–268

Ho NSK,Tong K,Hu X (2011) An EMG-driven exoskeleton hand robotic training device on chronic stroke subjects. In: 2011 IEEE International conference on rehabilitation robotics. Switzerland, pp 1–5

Jia X (2013) The 40 over the age of stroke patients have been tens of millions. Beijing Evening News 27(2):0–10

Kuang G (2011) Comprehensive treatment of patients with post-stroke hemiplegia. China Modern Drug Appl 5(10):125–126

Long Y (2007) Wearing wrist hand orthosis conduct training on the effects of creeping paralysis of upper limb function in stroke patients. J Phys Med Rehabil 29(5):331–333

Liu Z (2008) Structure and function of the hand rehabilitation robot control system design. Huazhong University of Science and Technology, Wuhan, p 2008

Li L, Li H, Zhu X (2011) Based on solidworks motion's main work device simulation crane. Sci Technol Eng 11(5):1070–1072

Wu D, Sun X, Zhang Z (2010) Surface EMG analysis and feature extraction. China Tissue Eng Res Clin Rehabil 14(43):8073–8076

Yang Q, Zhang L, Ruan J (2004) The movement of a human finger joint research crawling process. China Mech Eng 15(13):1154–1157

Yangmei Zhang Y, Gao Y (2012) Advances in disuse muscle atrophy. J Phys Med Rehabil 34 (7):550–552

Zhang Z (2011) Underactuated three-link robot control strategy. Central South University, Changsha, p 2011

Zhuang P, Yao Z (2006) Suspension parallel robot trajectory planning based on the movement of cycloidal. Mech Design 23(9):21–24

Zheng X, Zhang J, Chen Z (2003) Research status myoelectric prosthetic hand. Chin J Rehabil Med 18(3):168–170

Dynamic Analysis and Design of Lower Extremity Power-Assisted Exoskeleton

Shengli Song, Xinglong Zhang, Qing Li, Husheng Fang, Qing Ye and Zhitao Tan

Abstract A new design of lower extremity power-assisted exoskeleton (LEPEX), is presented in this paper, which is used for transmitting the backpack weight of the wearer to the ground and enhancing human motion, with each joint driven by corresponding actuator. Primarily, a 7-bar human machine mathematical model is introduced and analyzed with different walking phases using Lagrange's Equations. Second, dynamic parameters, such as torque and power consumptions of each joint in the sagittal plane are obtained for human with 75 kg payload in his (her) back with different conditions, i.e., flat walking and climbing stairs. Afterwards, the actuator for each joint is chosen based on the torque and power consumptions, i.e., a passive actuator for each knee joint and an active actuator for each ankle and hip joint; as well as the structure of LEPEX. Last but not least, the designed LEPEX is simulated under ADAMS environment by using wearer's joint movement data, which is obtained from flat walking and climbing stairs experiments. Eventually, the simulation results are reported to witness the potentialities of the structure.

Keywords Kinematics · Lower extremity exoskeleton · Simulation · Structure

1 Introduction

Lower limb power-assisted exoskeleton (LEPEX) is famous for transmitting the backpack weight of the wearer to the ground and enhancing human motion, with each joint driven by corresponding actuator (Yang et al. 2000; Robert 2009; Homayoon 2008; Guizzo et al. 2005; Wang and Li 2011). The earliest mention of a device resembling an exoskeleton is a set of U.S. (Makinson et al. 1971; Rose et al. 1994; Hardiman 2012) patents granted in 1890 to Yagn (1890). In the late 1960s, a number of research groups in the United States began to do deep research in exoskeleton

S. Song · X. Zhang (✉) · Q. Li · H. Fang · Q. Ye · Z. Tan
PLA University of Science and Technology, Nanjing 210007, China
e-mail: zxl19900427aaa@zju.edu.cn

© Zhejiang University Press and Springer Science+Business Media Singapore 2017
C. Yang et al. (eds.), *Wearable Sensors and Robots*, Lecture Notes in Electrical Engineering 399, DOI 10.1007/978-981-10-2404-7_13

devices, and some researchers in the former Yugoslavia began to do such work at the same time. But the former focused on the military while the latter was for physically challenged persons (Aaron et al. 2008). In the recent years, some research findings have sprung up, among which the Berkeley Exoskeleton (BLEEX) (Zoss et al. 2005a, b; Kazerooni et al. 2005; Kazuo et al. 2009; Sunghoom et al. 2004), MIT Exoskeleton and Hybrid Assistive Leg (HAL) (Lee et al. 2002; Sakurai et al. 2009; Kawamoto et al. 2002a, b; Yoshiyuki 2010, 2011; Romero 2011) are the most outstanding representatives. To be specific, BLEEX has four actuated joints: hip flexion/extension, hip abduction/adduction, knee flexion/extension, and ankle flexion/extension. The exoskeleton is designed with hydraulic actuators installed in a triangular configuration with joint in the sagittal plane, so that moment arm corresponding to each joint varies with joint angle (Kazerooni et al. 2006; Zoss et al. 2006; Chu et al. 2005). MIT Exoskeleton (Grabowski and Herr 2009; Tomohiro et al. 2005) is quasi-passive one. The quasi-passive elements, i.e., spring and variable damper, were chosen based on the fact that energy consumption and storage happen alternatively during human walking cycle (Walsh et al. 2006a, b; Valiente et al. 2005; Walsh 2006). HAL-5 exoskeleton drives the joints at the hip and knee in the sagittal plane via DC motors and harmonic reducer installed in the corresponding joints, while ankle flexion/extension DOF is passive. The HAL system does not transfer loads to the ground surface, but simply augments joint torques at the hip, knee (Kawamoto et al. 2002a, b, 2003; Kawabata et al. 2009).

In this paper, the exact 7-bar mathematical dynamic model for human lower extremity is presented in Sect. 2. Dynamic performance including power and torque is analyzed based on the experimental data is revealed in Sect. 3. Mechanical design in each joint of LEPEX is included in Sect. 4. The dynamic simulation of the LEPEX structure is simulated under ADAMS environment by using wearer's joint movement data obtained from flat walking and climbing stairs experiments.

2 The Dynamic Model for Human Lower Extremity

The 7-bar human machine model (7-bar HMM) is used in the human gait kinematics and dynamics simulation in this paper. Compared with the 5-bar HMM (Mu et al. 2004; Fabio 2004), the 7-bar HMM (Furusho 1990) increases the ankle joint and foot component, and can also describe the interactive relationship between the feet and the ground in gait much more precisely. In the 7-bar HMM, the upper part of the body is simplified into a rigid element. The hip joint, the knee joint, and the ankle joint of the lower part of the body are connected through rigid components. So the dynamic model is used to imitate and simplify in human gait.

The 7-bar HMM is shown in Fig. 1. The symbols of the 7-bar HMM is shown in Table 1.

Fig. 1 The 7-bar HMM

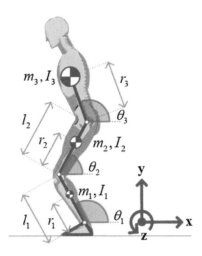

Table 1 The explanation for symbol in Fig. 1

Symbol	The meaning of the symbol
m_i	the mass of each rigid element
l_i	the length of element i
d_i	the distance from the center-of-mass to the joint
θ_i	absolute angle of the joint
I_i	the inertia of the element

It should be pointed out that absolute angle θ_i is convenient to describe the spatial coordinate of lower extremity exoskeleton joints, but the angle of the joint is usually expressed by the relative angle q_i. The relationship between θ_i and q_i can be expressed as:

$$q_i = M_{q\theta} \cdot \theta, \tag{1}$$

$$\text{where, } M_{q\theta} = \begin{bmatrix} 1 & 0 & 0 & 0 & 0 \\ -1 & 1 & 0 & 0 & 0 \\ 0 & -1 & 1 & 0 & 0 \\ 0 & 0 & -1 & 1 & 0 \\ 0 & 0 & 0 & -1 & 1 \end{bmatrix} \tag{2}$$

The whole cycle of the human gait can be simply divided into single leg support phase, double legs support phase, and ground collision model (Christopher et al. 1992; Wang et al. 2012).

2.1 Single Leg Support Model

The single leg support model is shown in Fig. 2. The system can be regarded as a multi-rigid-body and open-loop chain structure.

Set the end position coordinates of the support leg as (x_a, y_a), and the end position coordinates of swing leg as (x_b, y_b), then the center-of-mass coordinates of all rigid rods can be expressed as:

$$\begin{cases} x_{ci} = \sum_{j=0}^{i} (a_i l_j \sin \theta_j) + d_i \sin \theta_i + x_a \\ y_{ci} = \sum_{j=0}^{i} (a_i l_j con\theta_j) + d_i \cos \theta_i + y_a \end{cases}, \tag{3}$$

where, a_i is a rod parameter, if $i = 3$, $a_i = 0$; if $i = 1,2,4,5,6,7$, $a_i = 1$.

The potential energy and kinetic energy of the system are:

$$\begin{aligned} V &= \sum_{i=0}^{6} m_i g y_{ci} \\ &= \sum_{i=0}^{6} \left\{ m_i g \left[\sum_{j=0}^{i} ((a_j l_j \cos q\theta_j) + d_i \cos \theta_i + y_a) \right] \right\} \end{aligned} \tag{4}$$

$$KE = \sum_{i=0}^{6} \left[\frac{1}{2} m_i (\dot{x}_{ci}^2 + \dot{y}_{ci}^2) + \frac{1}{2} I_i \theta_i^2 \right] \tag{5}$$

Fig. 2 Single leg support model

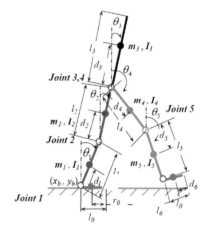

The formula (5) can be written as:

$$
\begin{aligned}
KE = & \sum_{i=0}^{6} \left[\frac{1}{2} (I_i + m_i d_i^2) \dot{\theta}_i^2 \right] \\
& + \sum_{i=0}^{6} \left\{ \frac{1}{2} m_i \left[\sum_{j=0}^{i} (a_j l_j \dot{q}_j \cos\theta_j) \right]^2 \right\} \\
& + \sum_{i=0}^{6} \left\{ \frac{1}{2} m_i \left[\sum_{j=0}^{i} (a_j l_j \dot{\theta}_j \sin\theta_j) \right]^2 \right\} \\
& + \sum_{i=0}^{6} \left\{ m_i d_i \dot{\theta}_i^2 \left\{ \sum_{j=0}^{i} \left[a_j l_j \dot{\theta}_j \cos(\theta_i - \theta_j) \right] \right\} \right\}
\end{aligned}
\tag{6}
$$

According to the Lagrange equation, the dynamic model equation of the system in the single leg support phase is derived as:

$$
T_\theta + T_{\text{ex}} = \frac{d}{dt} \left(\frac{\partial L}{\partial \dot{q}_i} \right) - \frac{\partial L}{\partial q_i}
\tag{7}
$$

Combining (7) with (3), (4), (5), the equation in matrix form is:

$$
T_\theta = M(\theta)\ddot{\theta} + C(\theta, \dot{\theta})\dot{\theta} + G(\theta) - T_{\text{ex}},
\tag{8}
$$

where, $M(\theta)$ is a 7×7 positive definite symmetric matrix, $C(\theta, \dot{\theta})$ is a 7×7 matrix with the centrifugal term, and Creole force, $G(\theta)$ is a 7×1 matrix with gravity, $\theta, \dot{\theta}, \ddot{\theta}$, and T_θ are all 7×1 matrix, They represent angle, angular velocity, angular acceleration, and torque of joint, respectively in the generalized coordinates. T_{ex} is the external force of the system.

2.2 Double Legs Support Model

The double legs support model is shown in Fig. 3. In this model, the double legs are in contact with the ground at the same time, so a closed chain constraint equation is introduced:

$$
\varphi(\theta) = \begin{bmatrix} f_1 \\ f_2 \end{bmatrix} = \begin{bmatrix} x_e - x_b - L \\ y_e - y_b \end{bmatrix} = 0,
\tag{9}
$$

where, (x_e, y_e) and (x_b, y_b) are the end of the supporting leg and swinging leg position coordinates, respectively. L is the distance between the two feet.

Fig. 3 Double legs support
model

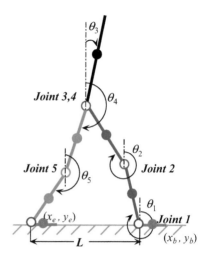

According to the Lagrange equation, the dynamic model equation of the system in the double legs support phase can be expressed as:

$$T_\theta + \boldsymbol{J}^T(\theta)\lambda = \mathrm{M}(\theta)\ddot{\theta} + \mathrm{C}(\theta, \dot{\theta})\dot{\theta} + \mathrm{G}(\theta) - T_{\mathrm{ex}}, \qquad (10)$$

where, λ is a 2×1 matrix with Lagrange operator, $\boldsymbol{J} = \partial\varphi/\partial\theta$ is a Jacobi matrix of 2×7.

The Jacobi matrix can be written as:

$$\boldsymbol{J}^T(\theta) = [\frac{\partial\dot{x}}{\partial\theta} \ \frac{\partial\dot{y}}{\partial\theta}], \qquad (11)$$

where, $\dot{x}$ is horizontal velocity of the joint, $\dot{y}$ is vertical velocity of the joint.

2.3 Ground Collision Model

When human is walking normally, the swing leg contacts with the ground, and the foot collides with the ground. In a very short period of time, the velocity has a great mutation, the acceleration becomes very large, and the transient collision force is generated. The collision force changes violently, therefore, the ground collision model is very complicated. In order to simplify the model, only the change of motion state is analyzed before and after the collision. The hypothesis for the ground collision model is:

The collision process is instantaneous, no rebound, or sliding.
After the collision, the supporting foot lifts immediately.
The collision force is a pulse function.

Because of the collision process is very short, the joint angle changes mutations, and the joint displacement does not change.

Let t^-, t^+ represent before and after the collision time, respectively, θ_c^-, θ_c^+ represent before and after the collision angle value, respectively. From the hypothesis, we can get that:

$$\begin{cases} t^- = t^+ \\ \theta_c^- = \theta_c^+ \end{cases} \tag{12}$$

Based on the Eq. (8), we can derive the Lagrange function of the ground collision model:

$$B_c u + (\frac{\partial E(\theta_c)}{\partial \theta_c})' \begin{bmatrix} F_x \\ F_y \end{bmatrix} = M(\theta)\ddot{\theta} + C(\theta,\dot{\theta})\dot{\theta} + G(\theta), \tag{13}$$

where, B_c is input matrix, u is the joint actuating torque, F_x, F_y are the vertical decomposition of the external collision force, E is the Jacobi matrix of Cartesian coordinate relative to generalized coordinate in the contact point caused by F_x and F_y.

Integrating the right side of Eq. (13), we have:

$$\int_{t^-}^{t^+} (M(\theta)\ddot{\theta} + C(\theta,\dot{\theta})\dot{\theta} + G(\theta))dt$$

$$= \int_{\dot{\theta}_c^-}^{\dot{\theta}_c^+} M(\theta)d\dot{\theta} + \int_{\theta_c^-}^{\theta_c^+} C(\theta,\dot{\theta})d\theta + \int_{t^-}^{t^+} G(\theta)dt \tag{14}$$

$$= \int_{\dot{\theta}_c^-}^{\dot{\theta}_c^+} M(\theta)d\dot{\theta} = M(\theta)(\dot{\theta}_c^+ - \dot{\theta}_c^-)$$

Integrating the left side of Eq. (13), we can get:

$$\int_{t^-}^{t^+} \left(B_c u + (\frac{\partial E(\theta_c)}{\partial \theta_c})' \begin{bmatrix} F_x \\ F_y \end{bmatrix} \right) dt$$

$$= B_c u(t^+ - t^-) + (\frac{\partial E(\theta_c)}{\partial \theta_c})' \begin{bmatrix} \int_{t^-}^{t^+} F_x\,dt \\ \int_{t^-}^{t^+} F_y\,dt \end{bmatrix} \tag{15}$$

$$= (\frac{\partial E(\theta_c)}{\partial \theta_c})' \begin{bmatrix} \int_{t^-}^{t^+} F_x\,dt \\ \int_{t^-}^{t^+} F_y\,dt \end{bmatrix}$$

The Lagrange function of ground collision model can be simplified as:

$$\mathbf{M}(\theta)(\dot{\theta}_c^+ - \dot{\theta}_c^-) = \left(\frac{\partial E(\theta_c)}{\partial \theta_c}\right)' \begin{bmatrix} \int_{t^-}^{t^+} F_x dt \\ \int_{t^-}^{t^+} F_y dt \end{bmatrix} \tag{16}$$

According to the hypothesis condition (1), we can get:

$$\left.\frac{\partial E(\theta_c)'}{\partial \theta_c}\right|_{\theta_c=\theta_c^+} = 0 \tag{17}$$

Combining Eqs. (15) and (16), we have:

$$\begin{bmatrix} \mathbf{M}_c(\theta_c) & -\left(\frac{\partial E(\theta_c)}{\partial \theta_c}\right)' \\ \left(\frac{\partial E(\theta_c)}{\partial \theta_c}\right)' & 0_{2\times 2} \end{bmatrix} \begin{bmatrix} \dot{\theta}_c^+ \\ \int_{t^-}^{t^+} F_x dt \\ \int_{t^-}^{t^+} F_y dt \end{bmatrix} = \begin{bmatrix} \mathbf{M}_c(\theta_c) & \dot{\theta}_c^- \\ 0_{2\times 1} \end{bmatrix} \tag{18}$$

To further simplify, then:

$$\begin{bmatrix} \dot{\theta}_c^+ \\ \int_{t^-}^{t^+} F_x dt \\ \int_{t^-}^{t^+} F_y dt \end{bmatrix} = \begin{bmatrix} \mathbf{M}_c(\theta_c) & -\left(\frac{\partial E(\theta_c)}{\partial \theta_c}\right)' \\ \left(\frac{\partial E(\theta_c)}{\partial \theta_c}\right)' & 0_{2\times 2} \end{bmatrix}^T \begin{bmatrix} \mathbf{M}_c(\theta_c) & \dot{\theta}_c^- \\ 0_{2\times 1} \end{bmatrix} \tag{19}$$

After the collision, the coordinate transformation occurs between the supporting leg and swing leg. The transform relation is:

$$\dot{\theta}_d^+ = D\dot{\theta}_t^- + \text{Const} = \Delta(\dot{\theta}_t^-), \tag{20}$$

where, θ_d^+ is the initial generalized angle of the heel.

Due to the constraint of the ground and the heel during the collision, we have:

$$\dot{\theta}_c^- = \begin{bmatrix} I_{5\times 5} \\ 0_{2\times 5} \end{bmatrix} \dot{\theta}_t^- \tag{21}$$

The initial generalized angle of the heel can be expressed as:

$$\dot{\theta}_d^+ = [D \quad 0_{i\times 2}]\dot{\theta}_c^+ = \Delta(\theta_t^-)\dot{\theta}_t^- \tag{22}$$

3 Dynamic Analysis

The purpose of the research is to design a reasonable LEPEX. Assume that human bears 75 kg load on his (her) back. We refer GB10000-88 human dimensions of Chinese adults (GB10000-88) and GB/T 17245-2004 Inertial parameters of adult human body (GB/T 17245-2004) to determine the parameters in the 7-bar HMM.

Walking on flat ground is the most common movement form. In addition, people often encounter various complex terrains, which is similar to climbing the stairs. So the torque of each joint during walking on flat ground and climbing the stairs are calculated. We choose the human experimental data by scholar Stansfield (Stansfield et al. 2001, 2006; Van 2003) for walking on flat ground and the experimental data by scholar Riener (2002) from Italy for climbing the stairs.

The experimental data are divided into single leg support phase, double legs support phase, and ground collision phase, then put the divided experimental data into the formula (9), (10), and (13), respectively. The torque of ankle joint, knee joint, and hip joint are obtained shown as Figs. 4 and 5.

In order to determine the forms of the joint drive, the change of joint power in a gait must be calculated. The joint power formula is:

$$P = T \cdot \frac{\mathrm{d}\theta}{\mathrm{d}t} \tag{23}$$

Fig. 4 The joint torque curve in a gait for walking on flat ground

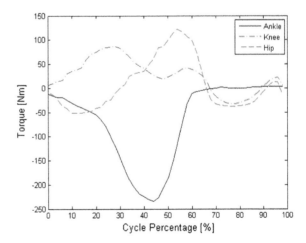

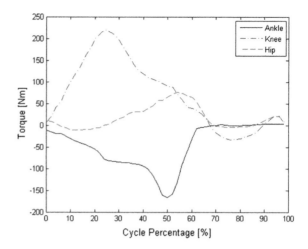

Fig. 5 The joint torque curve in a gait for climbing the stairs

4 The Design of LEPEX

4.1 *The Design of Ankle Joint*

In view of dynamic analysis above, we know that the ankle power has positive and negative value during walking on flat ground process and the average power is close to 0 (see Fig. 6), so some researches use spring assistance power for lower extremity exoskeleton (Aaron et al. 2008). Considering that the ankle joint torque is

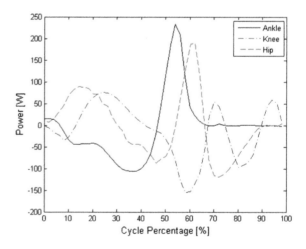

Fig. 6 The joint power curve in a gait for walking on flat ground

Fig. 7 The joint power curve in a gait for climbing the stairs

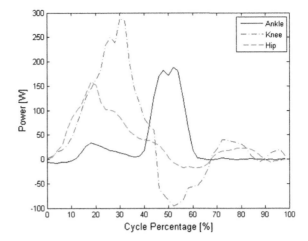

Fig. 8 The structure of ankle joint

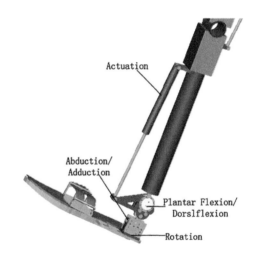

large in walking process (see Fig. 4) and the average power of ankle torque is positive in climbing the stairs process (see Fig. 7), the ankle joint takes the active actuator in the design of the exoskeleton. The ankle joint can be consisted by three motion axes, which are flexion/extension, adduction/abduction, and rotation motion (Channon et al. 1979). The designed ankle structure is shown in Fig. 8.

Fig. 9 The structure of knee joint

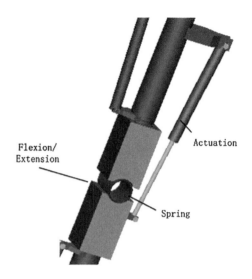

Flexion/
Extension

Actuation

Spring

4.2 The Design of Knee Joint

The knee joint torque is small (see Fig. 4) and the average power of knee joint is negative (see Fig. 6) in walking on flat ground process. Its main function is buffer and control, which we can use spring damper to realize them. The knee joint torque is large (see Fig. 5), the average power of knee is positive (see Fig. 7) and the power peak is also great (see Fig. 7) in climbing the stairs. So in the design of exoskeleton, active driver is introduced to realize climbing the stairs action. The human knee joint only has flexion/extension degree of freedom. The designed knee structure is shown in Fig. 9.

4.3 The Design of Hip Joint

The hip joint torque is small both in walking on flat ground process and climbing the stairs process (see Fig. 4). The hip power has positive or negative and the average power is positive (see Fig. 6) in walking on flat ground process. The average power is positive and very large (see Fig. 7) in climbing the stairs, thus, in the design of exoskeleton, active actuator is adopted. The human hip joint consist of three motion axes which are flexion/extension, adduction/abduction, and rotation motion (Xue et al. 2008). The designed hip structure is shown in Fig. 10.

Fig. 10 The structure of hip
joint

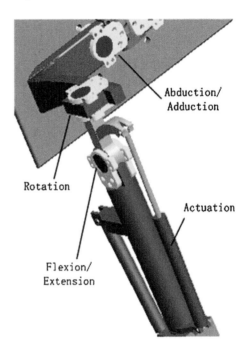

Abduction/
Adduction

Rotation

Actuation

Flexion/
Extension

5 ADAMS Simulation and Verification

To verify the correctness of the designed LEPEX, both simulations of walking on
flat ground and climbing stairs are performed under ADAMS, respectively. In these
simulations, the walker loads 75 kg-weight goods.

5.1 Build Prototype Models

Three steps are necessarily completed to build prototype models:

First, establish the human 7-bar HMM and the exoskeleton model by using
Pro/E according to GB10000-88 Chinese adult body size (GB10000-88). Then,
save the models as Parasolid (*. X_t) format, and import them into ADAMS. Third,
set the basic parameters of the models on the basis of the GB/T 17245-2004 inertial
parameters of adult human body (GB/T 17245-2004).

5.2 Obtain Experimental Human Gait Information

During walking, human gait is mainly generated by the movements of lower extremity in sagittal plane. The other movements of the lower extremity are basically used to keep balance during the movement. So, we only measure the motion information of human lower extremity joints in the sagittal plane. In the experiment, the human gait information of walking on flat and climbing stairs are measured, respectively. The experimental field is shown in Figs. 11 and 12. The angle data of each joint in a gait is shown in Figs. 13 and 14.

5.3 The Simulation and Verification

Simulations are performed under ADAMS (Liu et al. 2002; Liang et al. 2010) by using the joint angles obtained in the experiment above. The simulation results are shown in Figs. 15 and 16, from which we can see that human body models are able

Fig. 11 The experimental field of walking on flat

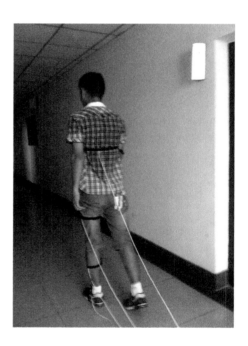

Fig. 12 The experimental
field of climbing stairs

Fig. 13 The experimental
angle curve of each joint in a
gait of walking on flat

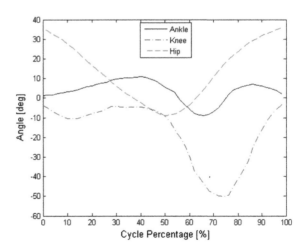

to walk naturally and smoothly in the power of the designed LEPEX. Figures 17
and 18 shows the torque curves of each joint, which is basically accord with the
dynamic modeling.

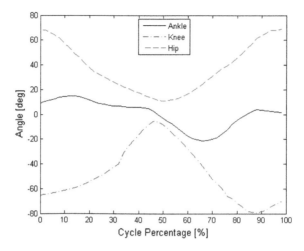

Fig. 14 The experimental angle curve of each joint in a gait of climbing stairs

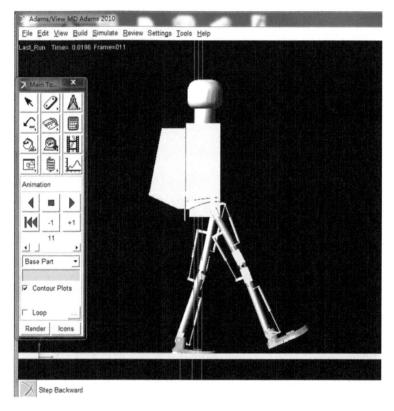

Fig. 15 Simulation for walking on flat under ADAMS

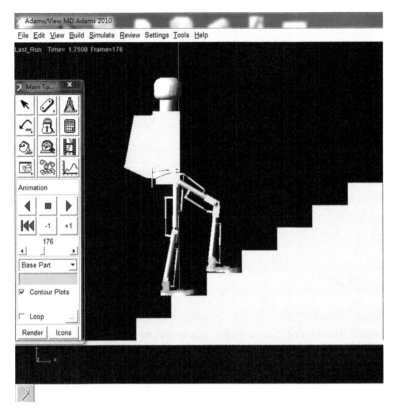

Fig. 16 Simulation for climbing stairs under ADAMS

Fig. 17 Torques curve of walking on flat under ADAMS

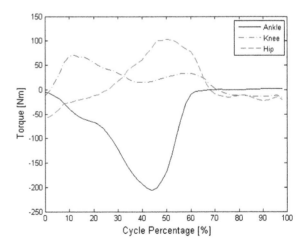

Fig. 18 Torques curve of climbing stairs under ADAMS

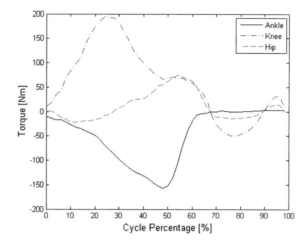

6 Conclusions

In this paper, dynamic analysis and design of the LEPEX are performed. A 7-bar human machine model of human walking is analyzed concerning different walking phases, i.e., single leg and double legs support phases, and ground collision phase using the Lagrange equations in the first place. Then, the flat walking and climbing stairs dynamic data, such as torque and power consumptions of lower limb joints is derived for human with 75 kg loads on his (her) back. The result shows that people in different walking modes, the joint's torque and power change apparently and instantaneously. The actuator for each joint is chosen based on the torque and power consumptions, i.e., a passive actuator for each knee joint and an active actuator for each ankle and hip joint. The structure of LEPEX is designed with combining characteristics of body joints and the functionality of exoskeleton. Eventually, the designed LEPEX is simulated in ADAMS environment by using a 75 kg person's joint movement data obtained in conditions of flat walking and climbing stairs. The simulation results show that the proposed LEPEX is reasonable and could achieve the initial objective of enhancing human motion.

References

Aaron MD, Hugh H (2008) Lower extremity exoskeleton and active orthoses: Challenges and state-of-the-art. IEEE T Robot 24(1):144–158

Bogue Robert (2009) Exoskeletons and robotic prosthetics: a review of recent developments. Ind Robot Int J (S0143-991x) 36(5):421–427

Yoshiyuki S. Centroid position detector device and wearing type action assistance device including centroid position detector device. USA, 20100271051A1. 28 Oct 2010

Channon M, Frankel VH (1979) The ball and socket ankle joint. J Bone Joint Surg 61B(1):85–89

Chu A, Kazerooni H, Zozz A (2005) On the biomimetic design of the Berkeley lower extremity exoskeleton (BLEEX). In: Proceeding of IEEE international conference robotics and automation. Barcelona, Spain, pp 4345–4352

DenBogert Van AJ (2003) Exotendons for assistance of human locomotion. Biomed Eng Online 2:17

Fabio Z (2004) Theoretical and experimental issues in biped walking control based on passive dynamics. Doctoral dissertation, Sapienza University of Rome, Rome

Furusho J (1990) Sensor-based control of a nine-link biped. Int J Robot Res 9(2):83–98

GB10000-88 (1988) human dimensions of Chinese adults

GB/T 17245-2004 (2004) Inertial parameters of adult human body

Grabowski AM, Herr H (2009) Leg exoskeleton reduces the metabolic cost of human hopping. J Appl Physiol 107(3):670–678

Guizzo E, Goldstein H (2005) The Rise of the Body Bots: Exoskeletons are strutting out of the lab-and they are carring their creators with them. IEEE Spectr (S0018-9235) 42(10):50–56

Hardiman http://davidszondy.com/future/robot/hardiman.htm [2012-10-6]

Hayashi T, Kawamoto H, Sankai Y (2005) Control method of robot suit HAL working as operator's muscle using biological and dynamical information. 2005 IEEE/RSJ international conference on intelligent robots and systems. IEEE, USA, pp 3455–3460

Homayoon K (2008) Exoskeletons for human performance augmentation. Springer handbook of robotics. Springer, Germany, pp 773–793

Katubedda K, Kiguchi K et al (2009) Mechanical designs of active upper-limb exoskeleton robots state-of-the-art and design difficulties. In: Proceedings of IEEE 11th international conference on rehabilitation robotics. Kyoto International Conference Center, Japan, pp 178–187

Kawabata T, Satoh H, Sankai Y (2009) Working posture control of robot suit HAL for reducing structural stress. In: Proceeding of the 2009 IEEE international conference on robotics and biomimetics, pp 2013–2018

Kawamoto H, Sankai Y (2002a) Power assist system HAL-3 for gait disorder person. In: Proceeding of international conference computing helping people special needs (ICCHP) (Lecture Notes on Computer Science), vol 2398. Springer-Verlag, Berlin, Germany

Kawamoto H, Sankai Y (2002b) Power assist system HAL-3 for gait disorder person. In: Proceedings of the 8th international conference on computers helping people with special needs, pp 196–203

Kawamoto H, Lee S, Kanbe S, Sankai Y (2003) Power assist method for HAL-3 using EMG-based feedback controller. In: Proceeding of IEEE international conference system man cybernetics pp 1648–1653

Kazerooni H, Steger R (2006) The Berkeley lower extremity exoskeleton. Trans ASME J Dyn Syst Measure Control 128:14–25

Kazerooni H, Racine J-L, Huang L, Steger R (2005) On the control of the Berkeley lower extremity exoskeleton (BLEEX). In: Proceedings of the 2005 IEEE international conference on robotics and automation. Barcelona, Spain, (4), pp 4353–4360

Lee S, Sankai Y (2002) Power assist control for walking aid with HAL-3 based on EMG and impedance adjustment around knee joint. In: IEEE/RSJ International conference intelligent robots and system. Lausanne , pp 1499–1504

Makinson BJ (1971) General Electric CO. Research and development prototype for machine augmentation of human strength and endurance, Hardiman I Project (General Electric Report S-71-1056). NY, Schenectady

Mu XP, Wu Q (2004) Development of a complete dynamic model of a planar five-link biped and sliding mode control of its locomotion during the double support phase. Int J Control 77 (8):789–799

Ning LIU, Jun-feng LI, Qing-yi FENG, Tian-shu WANG (2007) Underwater human model simulation based on ADAMS. J Syst Simul 19(2):240–243

Qing Liang, Xian-xi Song, Feng Zhou (2010) Modeling and simulationg of biped robot based on ADAMS. Comput Simul 27(5):162–165

Riener R, Rabuffetti M, Frigo C (2002) Stair ascent and descent at different inclinations. Gait Posture 15:32–34

Romero J (2011) Robot Suit [EB/OL]. HAL Demp at CES 2011. http://spectrum.ieee.org/automation/robotics/medical-robot-suit-hal-demo-at-ces-2011

Rose J, Gamble JG (1994) Human walking, 2nd edn. Williams and Wilkins, Baltimore, MD

Sakurai T, Sankai Y et al (2009) Development of motion instruction system with interactive robot suit HAL[C]. In: Proceedings of IEEE international conference on robotics and biomimetics. Robotics and Biomimetics (ROBIO), Japan, pp 1141–1147

Stansfield BW, Hillman SJ, Hazlewood ME, Lawson AM, Mann AM, Loudon IR, Robb JE (2001) Sagittal Joint Kinematics, moments, and powers are predominantly characterized by speed of progression. Not Age J Paed Orth 21:403–411

Stansfield BW, Hillman SJ, Hazlewood ME (2006) Regression analysis of gait parameters with speed in normal children walking at self-selected speeds. Gait Posture 23(3):288–294

Sunghoom Kim, George Arwar and Kazeooni H. High-speed communication network for controls with the application on the exoskeleton. In: Proceeding of American control conference Boston Massachusetts 2004: 355–360

Valiente A (2005) Design of a quasi-passive parallel leg exoskeleton to augment load carrying for walking. Master's thesis, Department of Mechanical Engineering Massachusetts Institute of Technology, Cambridge

Vaughan CL, Davis BL, O'Connor JC (1992) Dynamics of human Gait Kiboho, 2nd edn. Publishers Cape Town, Africa

Walsh CJ (2006) Biomimetic design of an underactuated leg exoskeleton for load-carrying augmentation. Master's thesis, Department of Mechanical Engineering, Massachusetts Institute of Technology, Cambridge, 2006

Walsh CJ, Paluska D, Pasch K, Grand W, Valiente A, Herr H (2006) Development of a lightweight, underactuated exoskeleton for loadcarrying augmentation. In: Proceeding of IEEE international conference on robotics and automation. Orlando, FL, pp 3485–3491

Walsh CJ, Pasch K, Herr H (2006) An autonomous, underactuated exoskeleton for load-carrying augmentation. In: Proceeding IEEE/RSJ international conference on intelligence robots system (IROS). Beijing, China, pp 1410–1415

Wang Yi-ji, Li Jian-Jun (2011) A device that can raise and improve the function of walking: power of lower limb exoskeleton system design and application. J Rehabil Theory Pract China 17 (7):628–631

Wang N, Wang J-H, Zhang M-W (2012) The gait of a human lower limb exoskeleton robot research status]. Chin J Orthop Clin Basic Res 4(1)

Xue Zhao-Jun, Jin Jing-na, Ming Dong et al (2008) The present state and progress of researches on gait recognition. J Biomed Eng 25(5):1217–1221

Yagn N (1890) Apparatus for facilitating walking, running, and jumping. US Patents 420 179 and 438 830

Yang Can-jun, Chen Ying, Lu Yu-xiang (2000) Intelligent man-machine integration system theory and its applied research. J Mech Eng 36(6):42–47

Yoshiyuki S (2011) Wearing type behavior help device calibration device and control program. USA, 2011000432A1. 06 Jan 2011

Zoss A, Kazerooni H, Chu A (2005a) Berkeley Lower Extremity Exoskeleton (BLEEX). In: 2005 IEEE/RSJ International conference on intelligent robots and system (3), pp 3132–3139

Zoss A, Kazerooni H, Chu A (2005b) On the mechanical design of the Berkeley lower extremity exoskeleton (BLEEX). IEEE, pp 3764–3370

Zoss AB, Kazerooni H, Chu A (2006) Biomechanical design of the Berkeley lower extremity exoskeleton (BLEEX). IEEE/ASME Trans Mechatron 11(2):128–138

Human Gait Trajectory Learning Using Online Gaussian Process for Assistive Lower Limb Exoskeleton

Yi Long, Zhi-jiang Du, Wei Dong and Wei-dong Wang

Abstract Human gait trajectory estimating and acquiring using human–robot interaction (HRI) is the most crucial issue for an assistive lower limb exoskeleton. The relationship between the HRI and the human gait trajectory is nonlinear, which is difficult to be obtained due to the complex dynamics parameters and physical properties of mechanism and human legs. A Gaussian process (GP) is an excellent algorithm for learning nonlinear approximation, as it is suitable for small-scale dataset. In this paper, an online sparse Gaussian process is proposed to learn the human gait trajectory, i.e., the increment of angular position of knee joints, where the input is the HRI signal and the output is the increment of angular position of knee joints. We collect the HRI signals and the actual angular position by using torque sensors and optical encoders, respectively. When collecting dataset, the subjects are required to wear the exoskeleton without actuation system and walks freely as far as possible. After purifying the dataset, a subspace of training set with appropriate dimensionality is chosen. The subspace will be regarded as the training dataset and is applied in the online sparse GP regression. A position control strategy, i.e., proportion- integration-differentiation (PID), is designed to drive the exoskeleton robot to track the learned human gait trajectory. Finally, an experiment is performed on a subject who walks on the floor wearing the exoskeleton actuated by a hydraulic system at a natural speed. The experiment results show that the proposed algorithm is able to acquire the human gait trajectory by using the physical HRI and the designed control strategy can ensure the exoskeleton system shadow the human gait trajectory.

Keywords Physical human–robot interaction · Human motion intent · Online sparse gaussian process · Exoskeleton

Y. Long · Z. Du · W. Dong (✉) · W. Wang
State Key Laboratory of Robotics and System, Harbin Institute of Technology,
Harbin 150001, China
e-mail: dongwei@hit.edu.cn

© Zhejiang University Press and Springer Science+Business Media Singapore 2017
C. Yang et al. (eds.), *Wearable Sensors and Robots*, Lecture Notes in Electrical
Engineering 399, DOI 10.1007/978-981-10-2404-7_14

1   Introduction

Lower limb exoskeletons are worn by human users as intelligent devices for performance assistance and enhancement. Generally, a lower limb exoskeleton is composed of two mechanical legs, which is similar to the real architecture of human legs. In recent years, the wearable robot has attracted interest of many researchers widely and much progress has been made, e.g., BLEEX (Kazerooni et al. 2005; Zoss et al. 2006; Kazerooni et al. 2006; Steger, et al. 2006), HAL (Sankai et al. 2011; Kawamoto et al. 2003; Satoh et al. 2009). The BLEEX uses the mechanical sensors placed on the mechanism to infer the human gait trajectory and utilize a control strategy, i.e., sensitivity amplification control (SAC), to drive the exoskeleton to achieve human exoskeleton coordination movement. The HAL uses biomedical signals collected from the human body to look for the actuation torque of joints. Although many kinds of control strategies are studied, the challenges of the human gait trajectory estimation limit the performance of the lower extremity exoskeleton continuously. Therefore, the acquisition of the human gait trajectory is a crucial issue for the human-exoskeleton coordination movement.

The human gait trajectory acquiring and estimating using physical human–robot interaction is a hot research topic in human–robot collaborative control especially for assistive exoskeleton. The accurate human gait trajectory prediction is able to help minimize hinder causing during the execution of these robots. If the human gait trajectory can be learned online, it is possible to improve the performance of exoskeleton to satisfy the users' need and help wearers walk easily (Aarno et al. 2008). There are two ways to acquire the human gait trajectory, i.e., biomedical information and physical HRI information. The former uses signals, e.g., EEG and EMG, or other biomechanical signals collected directly from human body to infer the human gait trajectory, which is used in many assistive robots and rehabilitation devices (Young et al. 2014; He et al. 2007; Kiguchi et al. 2009). The latter one uses mechanical sensors, e.g., force/torque sensors, kinematics and dynamics sensors, installed on the interaction cuffs not human body to collect the HRI, which is the resource to detect human gait trajectory to react to human beings behaviors in right ways (Lee et al. 2012).

The physical interaction signal is fundamentally affected by the physical properties of human limbs, exoskeleton's links, and the connection cuffs. Therefore, the appropriate relationship between the physical interaction force and the dynamics factors, e.g., angular position, velocity and acceleration, cannot be clearly determined (Tran et al. 2014). The human gait trajectory is typically time-varying, which cannot be represented by several states clearly as it is highly nonlinear (Li et al. 2014). In reality, the human gait trajectory can be supposed as a function of HRI force, actual position, and actual velocity and can be learned in real time with adopting RBFNN (Lewis et al. 1998; Ge et al. 2011). The Bayesian estimator and the graphical model of human-exoskeleton system are proposed to obtain the nonlinear relationship mapping between the EMG signal and the actuating torque, in which bio-signal model, inverse dynamics model and the exogenous disturbance model are used (Cheng et al. 2013). Chan et al. compared three kinds of algorithms, i.e., multilayer perceptron, differential evolution linear and differential evolution exponential to estimate human gait trajectory for an exoskeleton actuated by pneumatic artificial muscle and draw a conclusion that the approach, i.e., differential evolution linear, is considered as the most appreciate method to achieve the best overall performance (Chandrapal et al. 2013). The GP is applied to look for the relationship between the human-exoskeleton interaction and the actuating torque. The torque sensor is installed on the knee joint to measure torque signals, and the measured torque signals include actuating torque and actual human-exoskeleton interaction torque (Tran et al. 2014).

A GP is a general supervised learning method which is used widely in robotics control. An online sequential GP is applied to learn interface model to generate control algorithm for an inverted pendulum (Park et al. 2013). A local GP is designed to obtain the accurate model of robot with reduced computational cost. The performance of the GP is close to the standard GP and it is sufficiently fast for online learning (Nguyen et al. 2008). However, a general GP has a problem of computation complexity and time consuming. The GP regression algorithm is not useful for applications with large-scale datasets because it does not scale with the number of data points. The time complexity of training for kernel matrix is $O(n^3)$ and the prediction for mean and variance function is $O(n)$ and $O(n^2)$, respectively (Ranganathan et al. 2011). There are several approaches to solve the problem of computation complexity, i.e., sparse GP (SGP) and mixture of experts (ME) (Nguyen et al. 2008). In SGP, the whole input space is approximated by inducing inputs which is chosen from the original input space (Quiñonero et al. 2005). In ME, the whole input space is divided into smaller subspaces, however, the performance subjects to the number of experts for a particular dataset (Rasmussen et al. 2002).

In this paper, an online sparse GP is proposed to learn the human gait trajectory using the physical HRI signal measured by torque sensors directly representing human-exoskeleton interaction information. The training dataset is collected when the human user performs unconstraint motions without control. The training dataset is purified and a subspace is selected to reduce the sizes of input space. Based on the estimated gait trajectory learning of human, a position control strategy is designed to drive the exoskeleton system to follow the user's behaviors to achieve human-exoskeleton collaborative movement.

The remainder of this paper is organized as follows. The specific exoskeleton under study is given and the proposed online GP algorithm is explained in details in the second section. Experiments are performed and results analysis is presented in the third section. Conclusions are given in the final section.

2 Materials and Methods

2.1 Mechanical Architecture

Based on principles in the biological design, the designed exoskeleton is required to retain adaptability to multifunctionality of the human lower limbs. The enormous

Fig. 1 Prototype of lower limb powered exoskeleton. There are one active joint and four passive joints for each leg. θ_{knee} represents the active DoF in walking direction. All auxiliary facilities are packaged in the backpack. The hydraulic actuation system is used

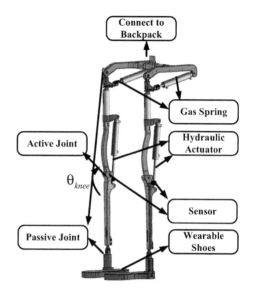

clinical gait analysis (CGA) data on human walking is an available powerful tool for an assistive lower limb exoskeleton. The CGA gives human limb joint angles, torques and powers for typical walking patterns, e.g., level walking (Andrew et al. 2005). Based on the CGA (Kirtley et al.), our designed lower exoskeleton is shown as Fig. 1. The exoskeleton is composed of two mechanical legs, each of which has five degrees of freedom (DoF), of which knee joints in sagittal plane are active while hip joints are passive by using gas spring.

The primary goal of an exoskeleton system control is to follow the gait trajectory of a user compliantly to help the user walk easily. To achieve human-exoskeleton collaboration, the exoskeleton system needs to imitate the control mode that human brain sends the command to actuate muscle to drive human limbs.

2.2 Online Sparse Gaussian Process for Gait Trajectory Learning

2.2.1 Gaussian Process Description

Machine learning can be classified into three categories, i.e., unsupervised learning, supervised learning, and reinforcement learning, which are widely used in robotics. Supervised learning techniques are used to obtain the nonlinear relationship between the input variables and the output variables and predict the new output for a new input. The supervised learning technique can be divided into two categories, i.e., nonparametric regression, e.g., neural network and parametric regression, e.g., GP.

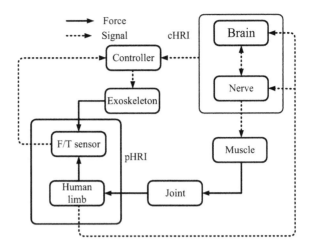

Fig. 2 Cooperative movement between human and exoskeleton, controller based on cHRI, or physical HRI needs to cooperate with the essential human movement protocol

Compared with other supervised learning techniques, e.g., local weight regression and neural networks, the GP is an effective and flexible tool for regression functions learning as the GP is capable of providing uncertainty estimations, which is adaptive to different situations (Seeger et al. 2004; Boyle et al. 2004; Aznar et al. 2009; Soh et al. 2012) (Fig. 2).

We define a training dataset $D = \{(x_i, y_i) \mid i = 1, 2, \cdots, n\}$, where x_i is the input variable and y_i is the output variable. The goal of a GP is to discover the latent function $f_i : x_i \rightarrow y_i$ which transforms the input vector x_i into a target value y_i. The obtained function can be expressed by $y_i = f_i(x_i) + \varepsilon_i$, where ε_i is Gaussian noise with zero mean and variance $\varepsilon_i \sim N(0, \sigma_n^2)$(Rasmussen et al. 2006). The GP regression is aimed at making inference about the relationship from the input variables to the output variables. The covariance function is frequently taken as the stationary anisotropic squared exponential covariance function

$$k(x_i, x_j) = \sigma_f^2 \exp(-\frac{(x_i - x_j)^2}{2w^2}), \tag{1}$$

where σ_f denotes the signal covariance and w is the width of the Gaussian kernel. The prediction $\mathbf{y}_*$ for a new input vector $\mathbf{x}_*$ can be obtained as below,

$$y_* = \mathbf{k}_*^T(\mathbf{K} + \sigma_n^2\mathbf{I})^{-1}\mathbf{y}$$
$$V(x_*) = k(x_*, x_*) - k_*^T(\mathbf{K} + \sigma_n^2\mathbf{I})^{-1}\mathbf{k}_* \tag{2}$$

where $\mathbf{k}_* = \mathbf{k}(\mathbf{X}, \mathbf{x}_*)$ and $\mathbf{K} = \mathbf{K}(\mathbf{X}, \mathbf{X})$ represent kernel matrix and Gram matrix for predictions, σ_n is the covariance of signal noise, $\mathbf{I}$ is an unity matrix and $k(x_*, x_*)$ can be given by (1).

The performance of a GP regression is dependent on the chosen kernel function and the hyper-parameters. The hyper-parameters of a GP are defined as

$\theta = [\sigma_n^2, \sigma_f^2, W]$ and their optimal value for a particular dataset can be derived by maximizing the log marginal likelihood using empirical Bayesian inference and its logarithmic form can be shown as follows (Li et al. 2015),

$$p(y|X, \theta) = -\frac{1}{2}\log|K + \sigma^2 I|$$
$$-\frac{1}{2}y^T(K + \sigma^2 I)^{-1}y - \frac{n}{2}\log(2\pi). \tag{3}$$

2.2.2 Architecture of Online Sparse Gaussian Process

As mentioned previously, the general GP has a problem of time complexity with respect to the training size. To address the problem, there are many new types of reported GPs. A newly $l_{1/2}$ regularization method is used to optimize the corresponding objective function to construct a sparse GP model (Kou et al. 2014). It is suitable to select an appreciate subspace of training dataset, to attain a sparse representation of the original GP (Zhu et al. 2014). There exist two kinds of criteria for selecting the subspace of training dataset, i.e., biggest entropy reduction based on information theory (Lawrence et al. 2003), and smallest Kullback–Leibler divergence between the variational distribution and the posterior distribution (Titsias et al. 2009). An efficient representation is used to update the Cholesky factor of the Gram matrix to reduce the kernel function computation complexity (Rasmussen et al. 2002). The simplest way is to select a subspace from the original dataset as the training dataset according to the real application and properties.

In this paper, the dataset is collected when the human user is attached to the exoskeleton robot and can move freely as far as possible, where the exoskeleton is not actuated. Since the lower limb is made of light-weight material, we do not take consideration of the effects caused by the weight of mechanism and friction. After purifying the training dataset, a subspace with appropriate dimensionality is chosen. The hyper-parameters are obtained offline using the chosen subspace in MATLAB. The designed online sparse GP for motion intention learning is shown in **Algorithm** 1 in **Appendix A**. As **Algorithm** 1 shows, the output is the increment of angular position $\Delta\theta_d(k)$ of active joint. The sent angular position to motion card from the high-level controller is represented as

$$\theta_d(k) = \theta_d(k-1) + \Delta\theta_d(k), \tag{4}$$

where $\theta_d(k)$ and $\theta_d(k-1)$ are the angular position of the kth and $(k-1)$th sampling interval respectively.

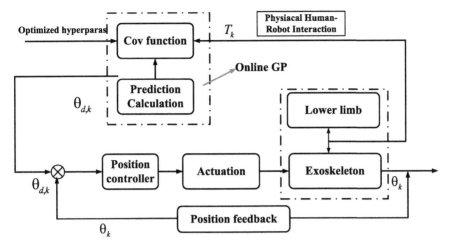

Fig. 3 Control diagram for the exoskeleton. The outside loop is mapping torque signal into joint trajectory and the inner loop is to control exoskeleton to follow the joint trajectory

2.3 Controller Design

The human gait trajectory can be inferred by the proposed online sparse GP. The learned human gait trajectory is regarded as the desired trajectory of the lower extremity exoskeleton. A simple position control strategy is designed to drive the exoskeleton to follow the human gait trajectory. The control system structure is shown as Fig. 3. As Fig. 3 shows, $\theta_{d,k}$ represents the human intention obtained by the online sparse GP, T_k is the human robot interaction information collected by torque sensors directly, θ_k is the actual angular position measured by the optical encoder. We divide the control software into two parts, i.e., high-level motion planner and low-level execution planner using motion control card.

3 Experiments and Results Analysis

3.1 Experiments Setup

In this section, the proposed method is examined through experiments. The experiments are performed through the designed exoskeleton system. The torque sensor is selected to measure physical HRI information and the actual angular position of mechanical system is measured using the optical encoder. The control software structure diagram is composed of two levels of controllers, as shown in Fig. 4. In high level controller, the gait trajectory of human lower limb is obtained by the proposed algorithm. The module of high-level controller includes the control module and the monitoring module. In low-level controller, a motion control card

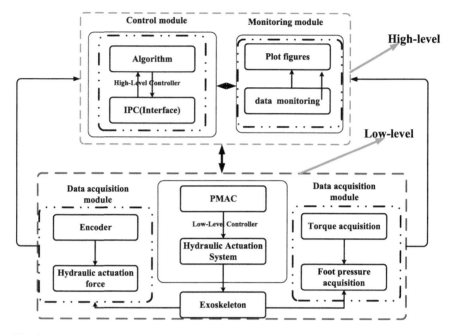

Fig. 4 Modularized structure of control software

called programmable multi axis controller (PMAC) is used to obtain the trajectory of high-level controller and send the control signals to hydraulic solenoid valves to actuate the exoskeleton system.

The data acquiring module includes four kinds of signals collection, i.e., angular position, physical HRI, hydraulic actuation force, and ground reaction force. They are transferred to data acquisition card through CAN, finally to the PC through USB. The signals from encoders are transferred directly to the PMAC. The central PC and the PMAC are connected through Ethernet.

For the sake of security, the whole experiment procedure can be demonstrated as Fig. 5 shows. If the user is ready he can start over the control system after system self-check to make sure all sensors are normal. During walking, the walking phases can be determined by ground reaction force to determine which leg should be actuated. With this exoskeleton robot, experiments for level walking are performed as shown in Fig. 6.

3.2   Results Analysis

During experiments, several kinds of data are collected by sensors in real time, i.e., ground reaction force signal used for gait phase identification, torque sensor signal for physical HRI, encoder signal for actual angular position, and force sensor signal for hydraulic cylinder information.

Fig. 5 Experiments
procedure of wearing
exoskeleton

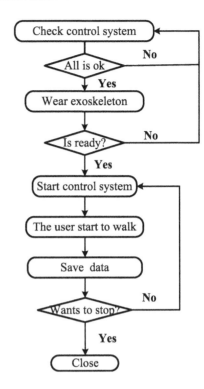

Fig. 6 Experiments for level
walking wearing the
exoskeleton system

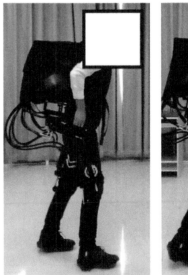

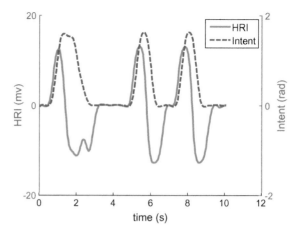

Fig. 7 Angular position of knee joint of left leg varies with respect to physical HRI

The mapping from the physical HRI to the angular position of joint in level walking is illustrated in Figs. 7 and 8 respectively. Figure 7 shows the corresponding relationship between the angular position of left knee joint and the physical HRI signal measured by the torque sensor while Fig. 8 shows that of right leg. When the physical HRI increases in the positive direction, the angular position of the knee joint become lager while the physical HRI varies in negative direction then the angular position will become smaller. The online sparse GP regression algorithm is applied to look for the relationship between physical HRI and the increment of angular position which is depicted in Figs. 9 and 10 for left leg and right leg, respectively. The output of online sparse GP includes two variables, i.e., the prediction mean and its variance. As Figs. 9 and 10 show, the mean of prediction will be regarded as $\Delta \theta_d(k)$.

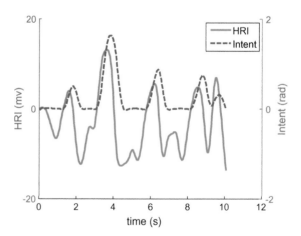

Fig. 8 Angular position of knee joint of right leg varies with respect to physical HRI

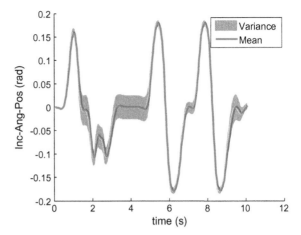

Fig. 9 The increment of angular position of knee joint with online sparse GP for left leg

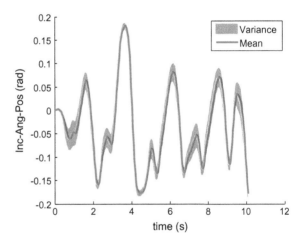

Fig. 10 The increment of angular position of knee joint with online sparse GP for right leg

As Fig. 3 shows, the position control strategy is developed to drive the exoskeleton follow the learned gait trajectory by the online sparse GP. The tracking performance is shown in Figs. 11 and 12 for left leg and right leg, respectively. As Figs. 11 and 12 show, the actual angular position curve tracks the commanded signal well. It is noted that the gap of tracking is caused by the software limit for the sake of safety. Therefore, the designed control algorithm can follow the human gait trajectory accurately.

Fig. 11 The tracking
performance of left leg with
position control strategy

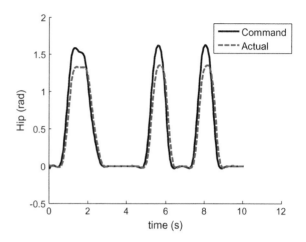

Fig. 12 The tracking
performance of right leg with
position control strategy

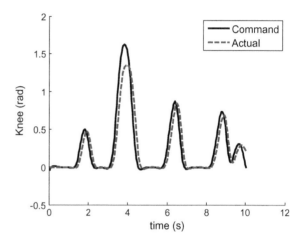

4 Conclusions

In this paper, an online sparse GP regression algorithm to learn the human gait
trajectory using the physical HRI signal. Experimental results show that the proposed
approach can improve overall performance, which achieves a good trade-off between
system performance and computation cost. The proposed algorithm has the potential
to be extended and employed in the field of intention recognition with physical HRI.

There are many interesting research points that we can explore in the near future.
The proposed online sparse GP should take consideration of effects on data subspace due to gravity of mechanism and friction. Since users' variety and external
disturbance are not considered in the implementation, some interaction control
strategy will be studied. In addition, some other intelligent control algorithms can
be looked into and integrated in the real-time control of the proposed system.

Appendix A: Algorithm 1

Algorithm: Online Sparse Gaussian Process for Learning Gait Trajectory

Input: Training data set $D = \{(T_i, \theta_i) | i = 1, 2, \cdots, n\}$

Remove points far away from the distribution and obtain the new purifying data set

$$S = \{(T_j, \theta_j) | j = 1, 2, \cdots, m\}, \; m < n$$

Select a subspace of data set $Q = \{(T_k, \theta_k) | k = 1, 2, \cdots, p\}$,

$$p \; \square \; m < n$$

Train the GP and **find** the optimal hyper-parameters $\tilde{\theta} = [\tilde{\sigma}_n^2, \tilde{\sigma}_f^2, \tilde{W}]$, save the optimal hyper-parameters.

Online Gaussian process, when a new input T_*

    for $i = 1, 2, \ldots, p$

        for $j = 1, 2, \ldots, p$

$$k_{ij} = \sigma_f^2 \exp(-\frac{(x_i - x_j)^2}{2w^2}) + \sigma_n^2 (x_i == x_j)$$

        end for

    end for

$$\mathbf{K} = \{k_{ij}\}$$

    for $i = 1, 2, \ldots, p$

$$ks_i = \sigma_f^2 \exp(-\frac{(T_* - x_i)^2}{2w^2}) + \sigma_n^2 (T_* == x_i)$$

    end for

$$\mathbf{Ks} = \{ks_i\}$$

$$k_T = \sigma_f^2 \exp(-\frac{(T_* - T_*)^2}{2w^2}) + \sigma_n^2 (T_* == T_*)$$

Output mean and variance of prediction

$$\bar{y}_* = \mathbf{Ks} * (\mathbf{K})^{-1} \mathbf{y}$$

$$V(x_*) = k_T - \mathbf{Ks} * (\mathbf{K})^{-1} * (\mathbf{Ks})^T$$

Angular position of joint is $\Delta\theta_d(k) = \bar{y}_*$.

References

Aarno D, Kragic D (2008) Motion intention recognition in robot assisted applications. Robot Auton Syst 56(8):692–705. doi:10.1016/j.robot.2007.11.005

Andrew C (2005) Design of the Berkley lower extremity exoskeleton(BLEEX). University of California, Berkeley

Aznar F, Pujol FA, Pujol M et al. (2009) Using Gaussian processes in Bayesian robot programming. In: Distributed computing, artificial intelligence, bioinformatics, soft computing, and ambient assisted living. Springer, Berlin, Heidelberg, pp 547–553. doi:10.1007/978-3-642-02481-8_79

Boyle P, Frean M (2004) Dependent Gaussian processes. In: Advances in neural information processing systems, pp 217–224

Chandrapal M, Chen XQ, Wang W et al (2013) Preliminary evaluation of intelligent intention estimation algorithms for an actuated lower-limb exoskeleton. Int J Adv Rob Syst 10:1–10

Cheng CA, Huang TH, Huang HP (2013) Bayesian human intention estimator for exoskeleton system. In: IEEE/ASME international conference on advanced intelligent mechatronics (AIM), pp 465–470. doi:10.1109/AIM.2013.6584135

Ge SS, Li Y, He H (2011) Neural-network-based human intention estimation for physical human-robot interaction. In: 8th International conference on ubiquitous robots and ambient intelligence (URAI), pp 390–395. doi:10.1109/URAI.2011.6145849

He H, Kiguchi K (2007) A study on emg-based control of exoskeleton robots for human lower-limb motion assist. In: 6th International special topic conference on information technology applications in biomedicine, pp 292–295. doi:10.1109/ITAB.2007.4407405

Kawamoto H, Lee S, Kanbe S et al (2003) Power assist method for HAL-3 using EMG-based feedback controller. In: Proceeding of IEEE international conference on systems, man and cybernetics, 2:1648–1653. doi:10.1109/ICSMC.2003.1244649

Kazerooni H, Racine JL, Huang L et al (2005) On the control of the Berkeley lower extremity exoskeleton (BLEEX). In: Proceeding of IEEE international conference on robotics and automation, pp 4353–4360. doi:10.1109/ROBOT.2005.1570790

Kazerooni H, Steger R, Huang L (2006) Hybrid control of the Berkeley lower extremity exoskeleton (bleex). Int J Robot Res 25(5–6):561–573. doi:10.1177/02783649060655

Kiguchi K, Imada Y (2009) EMG-based control for lower-limb power-assist exoskeletons. In: IEEE Workshop on robotic intelligence in informationally structured space, pp 19–24. doi:10.1109/RIISS.2009.4937901

Kou P, Gao F (2014) Sparse Gaussian process regression model based on ℓ 1/2 regularization. Appl Intell 40(4):669–681. doi:10.1007/s10489-013-0482-0

Lawrence N, Seeger M, Herbrich R (2003) Fast sparse Gaussian process methods: the informative vector machine. In: Proceedings of the 16th annual conference on neural information processing systems (EPFL-CONF-161319):609–616

Lee H, Kim W, Han J et al (2012) The technical trend of the exoskeleton robot system for human power assistance. Int J Precis Eng Manuf 13(8):1491–1497. doi:10.1007/s12541-012-0197-x

Lewis FW, Jagannathan S, Yesildirak A (1998) Neural network control of robot manipulators and non-linear systems. CRC Press

Li Y, Ge SS (2014) Human–robot collaboration based on motion intention estimation. IEEE/ASME Trans Mechatron 19(3):1007–1014. doi:10.1109/TMECH.2013.2264533

Li J, Qu Y, Li C et al (2015) Learning local Gaussian process regression for image super-resolution. Neurocomputing 154(284–295):2014. doi:10.1016/j.neucom.11.064

Nguyen-Tuong D, Peters J (2008) Local gaussian process regression for real-time model-based robot control. In: IEEE/RSJ International conference on intelligent robots and systems, pp 380–385. doi:10.1109/IROS.2008.4650850

Park S, Mustafa SK, Shimada K (2013) Learning based robot control with sequential Gaussian process. In: IEEE Workshop on robotic intelligence in informationally structured space (RIISS), pp 120–127. doi:10.1109/RiiSS.2013.6607939

Quiñonero-Candela J, Rasmussen CE (2005) A unifying view of sparse approximate Gaussian process regression. J Mach Learn Res 6:1939–1959

Ranganathan A, Yang MH, Ho J (2011) Online sparse Gaussian process regression and its applications. IEEE Trans Image Process 20(2):391–404. doi:10.1109/TIP.2010.2066984

Rasmussen CE, Ghahramani Z (2002) Infinite mixtures of Gaussian process experts. Adv Neural Inform Process Syst 2:881–888

Rasmussen CE, Williams CKI (2006) Gaussian processes for machine learning. MIT Press, Cambridge

Sankai Y (2011) HAL: Hybrid assistive limb based on cybernics. In: Robotics research. Springer Berlin Heidelberg, pp 25–34. doi:10.1007/978-3-642-14743-2_3

Satoh H, Kawabata T, Sankai Y (2009) Bathing care assistance with robot suit HAL. In: Proceeding of IEEE international conference on robotics and biomimetics (ROBIO), pp 498–503. doi:10.1109/ROBIO.2009.5420697

Seeger M (2004) Gaussian processes for machine learning. Int J Neural Syst 14:69–106. doi:10.1142/S0129065704001899

Steger R, Kim SH, Kazerooni H (2006) Control scheme and networked control architecture for the Berkeley lower extremity exoskeleton (BLEEX). In: Proceeding of IEEE international conference on robotics and automation, pp 3469–3476. doi:10.1109/ROBOT.2006.1642232

Titsias MK (2009) Variational learning of inducing variables in sparse Gaussian processes. In: International conference on artificial intelligence and statistics, pp 567–574

Tran H, Cheng H, Lin X et al (2014) The relationship between physical human-exoskeleton interaction and dynamic factors: using a learning approach for control applications. Sci China Inform Sci 57:1–13. doi:10.1007/s11432-014-5203-8

Kirtley C CGA Normative Gait Database. Available: http://www.clinicalgaitanalysis.com/data/

Young AJ, Kuiken TA, Hargrove LJ (2014) Analysis of using EMG and mechanical sensors to enhance intent recognition in powered lower limb prostheses. J Neural Eng 11(5):056021. doi:10.1088/1741-2560/11/5/056021

Zhu J, Sun S (2014) Sparse Gaussian processes with manifold-preserving graph reduction. Neurocomputing 138:99–105. doi:10.1016/j.neucom.2014.02.039

Zoss AB, Kazerooni H, Chu A (2006) Biomechanical design of the Berkeley lower extremity exoskeleton(BLEEX). IEEE/ASME Trans Mechatron 11(2):128–138. doi:10.1109/TMECH.2006.871087

Research on Bionic Mechanism of Shoulder Joint Rehabilitation Movement

Guo-xin Pan, Hui-qun Fu, Xiu-feng Zhang and Feng-ling Ma

Abstract In view of the clinical need of rehabilitation training to the human upper limb, the paper puts forward a novel exoskeleton device for shoulder rehabilitation. Based on the analysis of anatomy and biomechanics of shoulder joint, a novel bionic mechanism with 5° of freedom was proposed in the exoskeleton device. Then, the designs of mechanisms' scheme and mechanical structure to bionic mechanism were performed successively. The bionic mechanism of shoulder joint was optimized to match the physiological motion of anatomical center of rotation adaptively and improves the compatibility of human-machine kinematic chain. It is expected that the research will provide a reference method to the study of bionic mechanism in rehabilitation training related to other joints.

Keywords Rehabilitation robot · Rehabilitation training · Exoskeleton · Shoulder joint · Bionic mechanism

1 Introduction

It is seen that a major stroke is viewed by more than half of those at risk as being worse than death. Paralysis is caused due to complete loss of muscle function. It is proved that if they are under the process of rehabilitation for several months after stroke, their active range of motion as well as muscle strength can increase significantly and can promote the reorganization of brain function (Van et al. 2013; Pellegrino et al. 2012; Lew et al. 2012). Traditional treatments rely on the use of physiotherapy and on the therapist's experience (Masiero et al. 2009). The availability of such therapy programs, however, is limited by a number of factors such as the amount of a highly paid therapist's time they involve, and the ability of the

G. Pan (✉) · X. Zhang · F. Ma
National Research Center for Rehabilitation Technical Aids, Beijing 100176, China
e-mail: pan_guoxin@163.com

H. Fu
101 Institute of the Ministry of Civil Affairs, Beijing 100070, China

© Zhejiang University Press and Springer Science+Business Media Singapore 2017   181
C. Yang et al. (eds.), *Wearable Sensors and Robots*, Lecture Notes in Electrical
Engineering 399, DOI 10.1007/978-981-10-2404-7_15

therapist to provide controlled, quantifiable, and easily replicable assistance for complex movements. Consequently, rehabilitation robot that can quantitatively monitor and adapt to a patient's progress, and ensure consistency during rehabilitation may provide a solution to these problems, and has become an active area of research. The exoskeleton robots have been considered in the industry, military, and medical applications. In recent years, they have been applied in the areas of rehabilitation and power assistance for daily activities (Gopura et al. 2009). The robots can provide the parameters of the treatment and evaluate objectively and accurately, and have better operability and repeatability. Whether in stroke early or late, clinical studies confirmed that the treatment can significantly improve the movement function of the patient's upper limb (Burgar et al. 2011; Hsieh et al. 2011; Masiero et al. 2011; Liao et al. 2012; Mazzoleni et al. 2013).

In fact, many rehab-robots in the literature have been proposed for upper limb therapy and assistance. According to the mechanical structure of the rehab-robots, there are mainly three types that contact or interact with stroke patients. The first type is an endpoint-fixation system, such as MIT-Manus (Krebs et al. 1999), which can fix the distal part of UE of patients to guide the desired movements. That is, stroke patients can execute a task by the use of only forearm support. The second type is a cable suspension system, such as Freebal gravity compensation system (Stienen et al. 2007). It provides antigravity support for the UE during rehabilitation. The third type is an exoskeleton arm system, such as ARMin (Mihelj et al. 2007). But each one of them has several advantages and disadvantages. In this study, our rehab-robot is chosen to be of the exoskeleton type.

Among the human joints, the shoulder joint is an important one, as many human motions require its use. On the other hand, this joint is one of the most complexes, and therefore the design of exoskeleton devices for the shoulder joint is quite complex. Many of the exoskeletons described in the literature are mechanisms designed with seven degrees of freedom (DOF). In most of these mechanisms, the glenohumeral joint is modeled as a 3-DOF ball-and-socket joint and therefore, it does not include the translation of the glenohumeral joint and thus of the centers of rotation (McCormick 1970).

In this paper, we discuss the way in which the novel exoskeleton device for shoulder joint rehabilitation training is designed. In order to analyze the change of rotation axis of the human shoulder joint, the displacement of the center of rotation of human shoulder joint depending on the upper limb's posture is first analyzed from the anatomy point of view. After that, the novel shoulder joint mechanism is designed based on the DOF's analysis of the shoulder joint movement.

2 Anatomic Analysis of the Human Shoulder Joint

2.1 Human Shoulder Joint

A human upper limb mainly consists of the shoulder complex, elbow complex, and wrist joint. The shoulder complex shown in Fig. 1 consists of three bones: the

Fig. 1 Shoulder joint

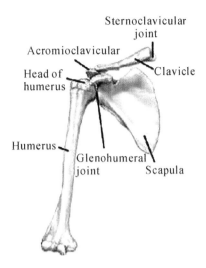

clavicle, scapula and humerus, and four articulations: the glenohumeral, acromio-clavicular, sternoclavicular, and scapulothoracic, with the thorax as a stable base (Martini 2003; Engin 1980). The glenohumeral joint is commonly referred to as the shoulder joint. The sternoclavicular joint is the only joint that connects the shoulder complex to the axial skeleton. The acromioclavicular joint is formed by the lateral end of the clavicle and the acromion of the scapula. The sternoclavicular joint is a compound joint, which has two compartments separated by articular disks. It is formed by the parts of clavicle, sternum, and cartilage of the first rib. In true sense, the scapulothoracic joint cannot be considered a joint because it is a bone-muscle-bone articulation, which is not synovial. It is formed by the female surface of the scapula and the male surface of the thorax. However, it is considered a joint when describing the motion of the scapular over the thorax.

2.2 Motion Analysis of Human Shoulder Joint

We made in-depth research analysis on the shoulder joint anatomy. In order to achieve the flexible movement of the human shoulder joint, the shoulder joint of the human body should have minimum 10 degrees of freedom, which includes: the glenohumeral joint should have 3 degrees of freedom; the scapulothoracic joint should have 3 degrees of freedom; the acromioclavicular joint should have 2 degrees of freedom; and the sternoclavicular joint should have 2 degrees of freedom (Kapandji 2011).

Basically, the shoulder complex can be modeled as a ball-and-socket joint. It is formed by the proximal part of the humerus (humeral head) and the female part of the scapula (glenoid cavity). However, the position of the center of rotation of

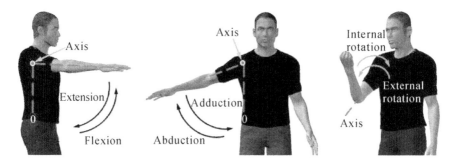

Fig. 2 Glenohumeral joint motions

shoulder joint is changing with the upper-arm motions. The main motions of the shoulder complex that are provided by the glenohumeral joint of shoulder complex are shoulder flexion/extension, shoulder abduction/adduction, and internal/external rotation (Fig. 2).

The motions of the scapulothoracic joint are scapula horizontal motion, scapula vertical motion, and scapula external rotation (Fig. 3). This paper will not describe in detail the motions of the acromioclavicular joint and the sternoclavicular joint. The DOFs and range of movement for each joint were studied, and the results of the research are shown in Table 1.

In a shoulder joint, the position of the center of rotation is changed according to the motions of a scapula and a clavicle depending on the posture of an upper limb (Kiguchi et al. 2011). Therefore, the difference between the position of the rotation axis of the shoulder joint mechanism and that of a human shoulder joint is generated when the user uses the power-assisted robot, which does not consider the characteristics of a human shoulder joint. Because of the position difference, the user feels uncomfortable and the user's body has to be adjusted in order to accommodate the ill effect. It is important to design an upper limb power-assisted exoskeleton robot considering the characteristics of a human shoulder joint so that the user is comfortable and the robot can perform proper power-assisted functions. In this study, the novel shoulder joint mechanism is designed based on the

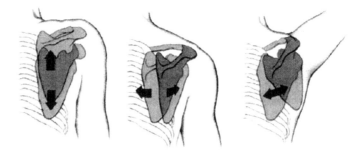

Fig. 3 Scapulothoracic joint motions

Table 1 Range of movement

Types of motion	Anatomical range	ExoRob's range
Glenohumeral joint		
Flexion	180°	150°
Extension	50°	30°
Abduction	180°	150°
Adduction	50°	30°
Internal rotation	110°	80°
External rotation	90°	60°
Scapulothoracic joint		
Horizontal motion	12 cm	8 cm
Vertical motion	12 cm	–
External rotation	60°	40°
Acromioclavicular joint		
Rotation(Front and back)	30°	–
Rotation(up and down)	90°	–
Sternoclavicular joint		
Horizontal rotation	90°	–
Coronal rotation	90°	–

anatomical analysis. In order to improve the flexibility and comfort levels of the bionic mechanism, in addition to the glenohumeral joint, we focus on adding feasibility of other DOFs in the mechanism design.

3 Design of the Bionic Mechanism of the Shoulder Joint Movement

3.1 DOFs of the Bionic Mechanism

In this study, the proposed bionic mechanism is modeled based on the concept of human shoulder joint articulations and movement to rehabilitate and to ease the shoulder joint motion of subjects. To meet the needs of rehabilitation training effectively, and meet reasonable engineering design, we simplify the motion of the bionic shoulder joint to reduce the complexity of the mechanical structure. Finally, the bionic mechanism of shoulder joint was designed with five DOFs:

- Glenohumeral joint (three DOFs: flexion/extension, abduction/adduction, internal rotation/external rotation);
- Scapulothoracic joint (two DOFs, horizontal motion, external rotation).

Because of the smaller influence of clavicle on the movement of the shoulder blade, the bionic mechanism of this study removes the movement freedom of the acromioclavicular joint and the sternoclavicular joint. Finally, the movement degree and the range of the bionic mechanism of the shoulder joint are obtained, as shown in Table 1.

Considering the safety of the robot users and to provide effective rehabilitation therapy as well as to assist in performing essential daily activities, such as eating, grasping, and bathing, preliminary studies on anatomical range (Hillman 2009; Rahman et al. 2009) of upper limb motion have been carried out to choose the suitable movable range for the proposed bionic mechanism. The movable range of proposed bionic mechanism is summarized in Table 1. Note that for technical simplicity, the range of motion of the bionic mechanism is kept to zero degree for shoulder joint extension. This should have minimal impact, as backward movement of the arm past the midline is seldom used in daily activities.

3.2 Principle Design of the Motion Mechanism

In the design of the bionic mechanism of the shoulder joint movement, the following factors are considered. First, to ensure that the safety of patients and to ensure that the performance of the mechanism are both reliable. Second, from the perspective of the experience of clinical rehabilitation medicine, bionic mechanism motion pattern should adapt to different conditions and to different recovery stages of patients, as far as possible to meet the needs of a majority of movements, so as to achieve the purpose of practicality. Finally, in the process of the design of bionic mechanism, the support and comfort of the patients were also considered. After the selection and configuration of the mechanism type, the five DOFs mechanism is designed. The DOFs model of each joint of the bionic mechanism is shown in Fig. 4. In the position of the elbow joint, seen from the picture, there is one DOF (elbow flexion-extension), which is used to support the upper limb.

The design of principle on the bionic mechanism is shown in Fig. 5. The rehabilitation movement of the shoulder complex is mainly composed of gear mechanisms. The bionic mechanism adopts serial mechanism. The movement of the scapulothoracic joint is realized by the gear rack mechanism and gear transmission mechanism. The movement of the glenohumeral joint is realized by two incomplete gear mechanisms and a motor direct driving mechanism.

4 Structure Design of the Bionic Mechanism

4.1 System Design of Exoskeleton

The 3D model of the shoulder rehabilitation system was designed, as shown in Fig. 6. The system mainly includes a lift support system, bionic mechanism of the shoulder joint, and forearm support system.

Fig. 4 DOFs model of bionic mechanism

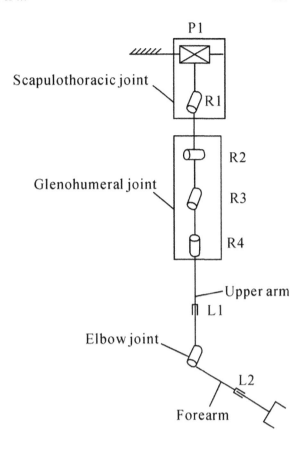

In the model, three DC servo motors assigned respectively to three degrees of freedom motion, drive the bionic mechanism of the glenohumeral joint.

4.2   Design of the Bionic Structure of the Scapulothoracic Joint Movement

The bionic mechanism of the scapulothoracic joint included in the horizontal linear motion mechanism and the external rotation is shown in Figs. 7 and 8a and b.

4.2.1   The Horizontal Motion Mechanism

As shown in Fig. 7, the bionic mechanism of the horizontal linear motion of the scapula includes the linear sliding mechanism and the gear rack drive mechanism. First, the linear sliding mechanism includes linear slide rail, linear slider, rack,

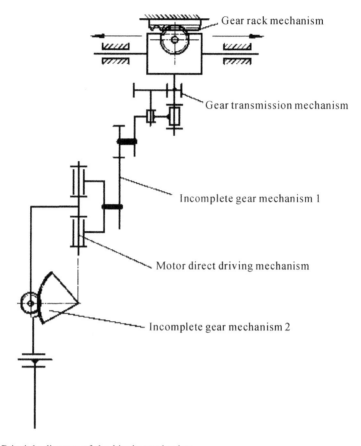

Fig. 5 Principle diagram of the bionic mechanism

gear 1, and so on. The gear 1 is always in a state of engagement with the rack. In addition, the trigger mechanism of gear rack transmission mainly comprises swing frame, rotating sleeve, sliding shaft, springs, sliding bracket, and a four-bar mechanism, which is connected with gear 1. The sliding bracket can automatically return to the middle position in the static state. The forearm support system is rotated to 90° from the vertical position to the horizontal position by the shoulder flexion motion. And when the abduction–adduction motion of the shoulder joint is performed in this position, the rotating sleeve will exert pressure on the swing frame's edge and drive the sliding shaft movement back and forth. The springs that are fixed on the sliding shaft can keep the shaft in the middle position. The axial movement of the sliding shaft can pull the sliding bracket movement back and forth along the support shaft, and drive the gear 1 simultaneously. The horizontal linear motion of the bionic mechanism of scapula is performed through the mesh movement of the gear 1 and the rack.

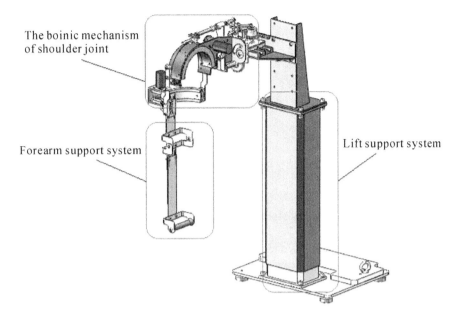

Fig. 6 The exoskeleton device system for shoulder rehabilitation

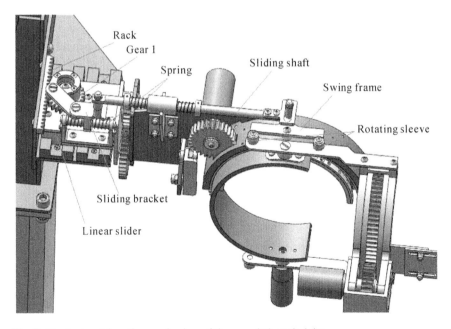

Fig. 7 The horizontal motion mechanism of the scapulothoracic joint

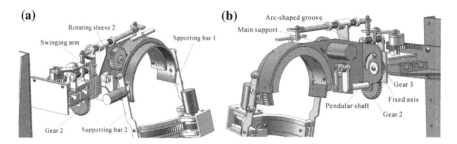

Fig. 8 The external rotation mechanism of the scapulothoracic joint: the *rear view* of the device (**a**), the *front view* of the device (**b**)

4.2.2 The External Motion Mechanism

As can be seen from Fig. 8a and b, the bionic mechanism of scapula external rotation motion includes the trigger mechanism of a swing and the gear transmission mechanism.

The trigger mechanism of a swing mainly includes supporting bar 2, rotating sleeve 2, a swinging arm, and a pendular shaft. The pendular shaft, which is installed on the main fixed seat, is driven to rotate the swing arm. First, the forearm support system is rotated to $0°$ from the horizontal position to the vertical position by the shoulder flexion motion. Then when the abduction-adduction motion of the shoulder joint is performed in this position, the rotating sleeve 2 will apply pressure on the swing arm's edge and trigger the swinging arm and the pendular shaft movement.

The gear transmission mechanism mainly comprises a main support, gear 2, gear 3, fixed axis, and so on. Gear 2 and the gear 3 are composed of the spur gear transmission mechanism. And gear 3 is a stationary gear, which does not rotate at all. Gear 2 is driven to rotate while the swinging arm is driven by the trigger mechanism. Gear 2 with its rotation will be swing around gear 3, and at the same time, it drives the main support rotation upward. The relative static between the fixed axis and the main support is realized through a guide pin, which is assembled in the arc-shaped groove that restricted the rotation range of external motion.

4.3 Design of the Bionic Structure of the Glenohumeral Joint Movement

Figure 9 shows the local amplification structure of the glenohumeral joint motion mechanism. It is composed of the flexion/extension movement mechanism, the abduction/adduction movement mechanism, and the internal/external rotation mechanism.

Fig. 9 The bionic
mechanism of the
glenohumeral joint

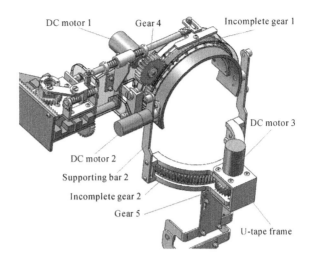

DC motor 1   Gear 4   Incomplete gear 1

DC motor 3

DC motor 2

Supporting bar 2

Incomplete gear 2

Gear 5

U-tape frame

(1) The flexion and extension movement mechanism mainly comprises main support, a pair of gears (gear 4, incomplete gear 1), DC motor 1, and so on. The rotation angle of the incomplete gear 1 meets the range of rehabilitation training. In case the shoulder joint is passively training, the DC motor 1 drives the gear 4 to rotate, and drives the imperfect gear 1 rotation, so as to realize the extension and flexion of the shoulder joint.

(2) The shoulder abduction–adduction mechanism is a rotation mechanism. When the supporting bar 2 of the mechanism is directly driven by the DC motor 2, the rehabilitation movement of the shoulder joint is realized.

(3) Internal and external rotation of the glenohumeral joint movement mechanism mainly comprises a DC motor 3, U-tape frame, incomplete gear 2, gear 5, and other parts. The incomplete gear 2 are arranged on the groove of the gear housing, and it meshes with the gear 5, which remain assembled in the U-tape frame. In addition, steel balls are installed on the upper and the lower planes of the gear housing by means of the ball brackets. The gear 5 is connected to the DC motor 3 directly. When the DC motor 3 directly drives the gear 5 and the incomplete gear 2 rotating, the internal and external rotating movement of the shoulder is performed.

According to the training data required through the clinical practice, the passive rehabilitation training of the shoulder joint can be carried out by driving the motor through the control program. The horizontal linear motion and the external rotation of the shoulder blade are two follow-up motion mechanisms. The two motions in the shoulder are flexion/extension and adduction/abduction. The upper limb under the condition of normal pituitary is driven the adduction/abduction motion to a certain angle. At the same time, the DC motor 2 triggers the rotation movement of the scapula bionic mechanism, which moves along with the human scapula external rotation. Under the condition of the upper limb being able to perform horizontal flexion, when the DC motor 2 drives the horizontal adduction and abduction motion of the exoskeleton device, the horizontal motion mechanism of the scapulothoracic joint is triggered, and it moves along with the scapula's linear motion of the human body. After this analysis it was found that the two mechanisms can effectively achieve the movement of shoulder joint rehabilitation training.

5   Conclusions

In recent years, the research of rehabilitation robot technology has been a research hotspot in the fields of rehabilitation engineering, mechanical design, electronic technology, automatic control, and other related fields. Exoskeleton rehabilitation robot is one of the most studied rehabilitation techniques, and many innovative research results have been obtained.

From the current points of view, in the global scenario, exoskeleton rehabilitation robot technology is still in its preliminary stage of development. The main reasons for this are as follows:

(1) Although many countries have supported the research and development work on the rehabilitation robot, there are only a very few practical and innovative achievements about it. Innovation in the field of multiple disciplines is an urgent need to improvise this technology.

(2) The transformation of the technological achievements of the exoskeleton rehabilitation robot has little to do with its widespread clinical application. Compared with traditional rehabilitation methods, the effectiveness of the rehabilitation robot in clinical rehabilitation is not yet confirmed in a credible manner.

(3) Interdisciplinary studies of rehabilitation medicine and engineering science are not deep and informative enough. They researchers involved in these studies have invested a lot of energy in this field but ignored the importance of effective collaboration with researchers in other disciplines.

The research concept of this paper is based on the discovery of clinical trials and the basic understanding of the rehabilitation robot technology. At present, there are many researches on the rehabilitation robot of upper limb and shoulder joint, but the match between the patients and the rehabilitation is not effective. The shoulder joint is the most complex joint in the human body; in the context of bone healing, this joint has not been studied effectively. In this paper, we have dealt in detail the design of the bionic mechanism of shoulder joint rehabilitation, and the comfort and safety involved in the rehabilitation training for patients.

In this paper, the bionic mechanism of shoulder rehabilitation training is studied in depth and the design of the mechanical structure of the exoskeleton device is discussed. The manufacturing of the mechanical prototype has been completed recently. The next work in this research is to carry out the function experiments using the exoskeleton device and optimize the mechanical structure through the analysis of the kinematics and dynamics on the bionic mechanism. The research idea in this paper can also provide reference for carrying out research on the designing of a rehabilitation robot to support other joints of the human body.

References

Burgar CG, Lum PS, Scremin AM et al (2011) Robot-assisted upper-limb therapy in acute rehabilitation setting following stroke: Department of Veterans Affairs multisite clinical trial. J Rehabil Res Dev 48(4):445–458. doi:10.1682/JRRD.2010.04.0062

Engin AE (1980) On the biomechanics of the shoulder complex. J Biomech 13(7):575–590. doi:10.1016/0021-9290(80)90058-5

Gopura RARC, Kazuo K (2009) Mechanical designs of active upper-limb exoskeleton robots: state-of-the-art and design difficulties. In: IEEE International conference on rehabilitation robotics, pp 178–187. doi:10.1109/ICORR.2009.209630

Hillman SK (2009) Interactive functional anatomy. J. Physiotherapy, New Zealand

Hsieh YW, Wu CY, Liao WW et al (2011) Effects of treatment intensity in upper limb robot-assisted therapy for chronic stroke: a pilot randomized controlled trial. J Neurorehabil Neural Repair 25(6):503–511. doi:10.1177/1545968310394871

Kapandji AI (2011) The physiology of the joints: the upper limb. People's Military Medical Press, Beijing, China, pp 22–74

Kiguchi K, Kado K, Hayashi Y (2011) Design of a 7DOF upper-limb power-assist exoskeleton robot with moving shoulder joint mechanism. In: Proceedings of IEEE international conference on robotics and biomimetics, pp 2937–2942. doi:10.1109/ROBIO.2011.6181752

Krebs HI, Hogan N, Volpe BT et al (1999) Overview of clinical trials with MIT-MANUS: a robot-aided neuro-rehabilitation facility. J Technol Health Care 7(6):419–423

Lew E, Chavarriaga R, Silvoni S et al (2012) Detection of self-paced reaching movement intention from EEG signals. J Front Neuroeng 5(13):2012. doi:10.3389/fneng.2012.00013

Liao WW, Wu CY, Hsieh YW et al (2012) Effects of robot-assisted upper limb rehabilitation on daily function and real-world arm activity in patients with chronic stroke: a randomized controlled trial. J Clin Rehabil 26(2):111–120. doi:10.1177/0269215511416383

Martini FH, Timmons MJ, Tallitsch RB (2003) Human anatomy. Prentice Hall, Pearson Education Inc (Chap. 8)

Masiero S, Carraro E, Ferraro C et al (2009) Upper limb rehabilitation robotics after stroke: a perspective from the University of Padua, Italy. J Rehabil Med 41(12):981–985. doi:10.2340/16501977-0404

Masiero S, Armani M, Rosati G (2011) Upper-limb robot-assisted therapy in rehabilitation of acute stroke patients: focused review and results of new randomized controlled trial. J Rehabil Res Dev 48(4):355–366. doi:10.1682/JRRD.2010.04.0063

Mazzoleni S, Crecchi R, Posteraro F et al (2013) Robot-assisted upper limb rehabilitation in chronic stroke patients. In: IEEE engineering in medicine and biology society, pp 886–889. doi:10.1109/EMBC.2013.6609643

McCormick EJ (1970) Human factors engineering, 3rd edn. McGraw-Hill, New York

Mihelj M, Nef T, Riener R (2007) ARMin II-7 DoF rehabilitation robot: mechanics and kinematics. In: IEEE International conference on robotics and automation, pp 4120–4125. doi:10.1109/ROBOT.2007.364112

Pellegrino G, Tomasevic L, Tombini M et al (2012) Inter-hemispheric coupling changes associate with motor improvements after robotic stroke rehabilitation. J Restor Neurol Neurosci 30 (6):497–510. doi:10.3233/RNN-2012-120227

Rahman MH, Saad M, Kenne JP et al (2009) Modeling and control of a 7DOF exoskeleton robot for arm movements. In: Proceedings of IEEE international conference on robotics and biomimetics, pp 245–250. doi:10.1109/ROBIO.2009.5420646

Stienen AHA, Hekman EEG, van der Helm FCT et al (2007) Freebal: dedicated gravity compensation for the upper extremities. In: IEEE international conference on rehabilitation robotics, pp 804–808. doi:10.1109/ICORR.2007.4428517

Van VP, Pelton TA, Hollands KL et al (2013) Neuroscience findings on coordination of reaching to grasp an object: implications for research. J Neurorehabil Neural Repair 27(7):622–635. doi:10.1177/1545968313483578

Reducing the Human-Exoskeleton Interaction Force Using Bionic Design of Joints

Wei Yang, Canjun Yang, Qianxiao Wei and Minhang Zhu

Abstract This paper presents a new method for design of lower body exoskeleton based on optimizing the human-exoskeleton physical interface to improve user comfort. The approach is based on mechanisms designed to follow the natural trajectories of the human hip and knee joints as flexion angles vary during motion. The motion of the hip joint center (HJC) with variation of flexion angle was experimentally measured and the resulting trajectory was modeled. Similarly, the knee joint rolling and sliding motion was calculated based on analytical knee joint model. An exoskeleton mechanism able to follow the hip and knee joints centers' movements has been designed to cover the full flexion angle motion range and adopted in the lower body exoskeleton. The resulting design is shown to reduce human-exoskeleton interaction forces by 25.5 and 85.5 % during hip flexion and abduction, respectively with bionic hip joint and to reduce human-exoskeleton interaction forces by 75.4 % during knee flexion with bionic knee joint. The results of interaction forces led to a more ergonomic and comfortable way to wear exoskeleton system.

Keywords Human-exoskeleton interaction · Bionic design · Hip joint center · Knee joint model

W. Yang · C. Yang (✉) · Q. Wei · M. Zhu
State Key Laboratory of Fluid Power Transmission and Control, Zhejiang University,
Hangzhou 310027, China
e-mail: zjuaway@163.com

W. Yang
e-mail: ycj@zju.edu.cn

Q. Wei
e-mail: wqx@zju.edu.cn

M. Zhu
e-mail: zoomingh@zju.edu.cn

© Zhejiang University Press and Springer Science+Business Media Singapore 2017
C. Yang et al. (eds.), *Wearable Sensors and Robots*, Lecture Notes in Electrical
Engineering 399, DOI 10.1007/978-981-10-2404-7_16

195

1   Introduction

With rapid progresses in mechatronics and robotics, anthropomorphic exoskeletons have been widely studied for rehabilitation applications and for general walking assistance. Key contributions in the area include lower extremity exoskeletons for post-stroke patient rehabilitation on treadmill [Lokomat (Hidler et al. 2009) and Lopes (Veneman et al. 2006)], wearable exoskeletons for paraplegic daily walking [HAL (Suzuki et al. 2007), and Indego (Farris et al. 2011)] and upper arm exoskeletons for upper body rehabilitation [IntelliArm (Ren et al. 2013)]. Although such exoskeletons can assist or guide motions of humans especially patients, there are potentials for discomfort and injuries if the designs are not compatible to human biomechanics (Wang et al. 2014). Without full sense of discomfort, paraplegic, or post-stroke patients may even suffer serious injuries during repeated rehabilitation where the comfort is far from ideal. To address such problems, we focus on the lower body and present a human biomechanics-based exoskeleton for providing support to the hip joint and knee joint in a way that is natural and based on human anatomical experimental data allowing the hip joint center and knee joint center to follow naturally occurring motions as flexion angle varies.

Traditional anthropomorphic exoskeleton designs are often based on assumptions that human anatomical joints are simplified to pin-and ball-and-socket jointed engineered designs to reduce kinematic complexity. It is usually the simplification that causes incompatibility of the exoskeleton's motion with human movements. Therefore, an understanding and quantification of anatomical joint center motion is necessary before designing exoskeletons joints. The hip joint center (HJC) and knee joint center are focused upon here before design of bionic exoskeleton joints which follow the motion of human anatomical joints.

In a pelvic anatomical coordinate system the motion of the HJC has been estimated previously using a functional method applied by calculating the center of the best sphere described by the trajectory of markers placed on the thigh during several trials of hip rotations (Leardini et al. 1999). However, the accuracy of the functional method is influenced by the hip motion range and researches show that the shape of the hip deviates from being spherical and becoming conchoidal (Menschik 1997) or aspherical (Rasquinha et al. 2011).

Similarly, the knee joint has a nonuniform geometry with varying articulating surfaces, and nonconstant rotation axis (Wang et al. 2014). O'Connor et al. (1989) present the crossed four-bar linkage knee joint model which consists of anterior cruciate ligament (ACL), posterior cruciate ligament (PCL), femur, and tibia. Wismans et al. (1980) take into account the knee joint surface geometry and build a three-dimensional analytical model. For standardization in a clinical joint coordination system, the knee joint is described with six-DOFs (Wu and Cavanagh 1995).

Some approaches have been utilized to realize alignment of the hip joint and knee joint motions of a human wearing an exoskeleton. Valiente (2005) designs the quasi-passive parallel leg with a cam and cam roller mechanism at the upper leg to realize hip abduction joint alignment. Because of the passive joint design, the friction caused by the mechanism leads to additional energy consumption by the human. To address this problem, Zoss et al. (2006) develop the Berkeley Lower Extremity Exoskeleton (BLEEX) with its flexion and abduction rotation axes intersecting at the human HJC which is seen to be fixed during flexion and abduction. Although these approaches contribute to realizing better hip joint alignment, dynamic motion of HJC due to human biomechanics is not accommodated for. Although some ergonomic knee joints are designed with passive or self-adjusting characteristics (Amigo et al. 2011), an analytical model considering a bio-joint (bionic joint) and the effect of exoskeleton on a human joint have not been well understood. This paper focuses on the joints alignment and develops an exoskeleton with compatible mechanisms which provide full coverage of HJC motions and knee joint motions. The remainder of this paper provides the following:

(a) This paper begins with analysis of the human HJC and knee joint center based on the experimental data and knee-exoskeleton analytical model, respectively.
(b) The results of the human HJC motion and knee joint motion are then employed to guide the design of a more biomechanically compatible exoskeleton joints.
(c) The validity of the compatible exoskeleton joints has been examined by studying human–machine interaction forces with comparison experiments between bio-joints exoskeleton and classical joints exoskeleton.
(d) Conclusions are made finally with description of bio-joints exoskeleton compatibility.

2 Method

To design exoskeleton bionic hip joint and knee joint, analysis of human corresponding anatomical joints is first studied based on motion capture experiment and analytical model. The mechanical joints are then designed applying the experimental data and analytical results.

2.1 Analysis of Human Anatomical Joints

2.1.1 Hip Joint Center Experimental Task

The experimental task of measuring the hip joint center has been designed consisting of static section and dynamic section to obtain its anatomical motion during normal walking. Figure 1 shows human hip coordinate systems and vectors with which both static section and dynamic section of experiments are conducted for anatomical HJC motion. The OptiTrack motion capture system (NaturalPoint, Inc.)

is used for measuring the motion of markers. Then the distances between the static HJC and markers can be calculated by a functional method. Different objective functions of the functional method have been compared and validated by Camomilla et al. (2006). In this research, the Spheric-4 (S4) algorithm is adopted for its high precision and repeatability. Details of the calculation are introduced by Gamage and Lasenby (2002).

Table 1 presents the results of six repeated experiments which were conducted on the same subject. Here r denotes distance between markers and HJC. With the static section results the dynamic section experiments are conducted and the optimal

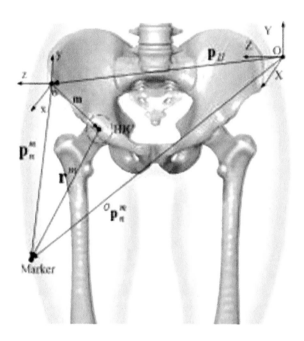

Fig. 1 Human hip coordinate systems and vectors

Table 1 Distances between markers and HJC (cm)

	r^1	r^2	r^3	r^4
1	27.38	30.30	39.95	39.86
2	27.74	30.41	40.21	40.11
3	26.78	29.50	39.18	39.14
4	27.90	30.50	40.26	40.13
5	27.28	29.80	39.57	39.42
6	28.36	30.85	40.79	40.56
AVE	27.57	30.23	39.99	39.87
STD	0.55	0.49	0.56	0.52

HJC motion of dynamic section can be calculated by minimizing $f(\mathbf{m})$ in Eq. (1) (Yan et al. 2014).

$$f(\mathbf{m}) = \sum_{m=1}^{M} \left[\|\mathbf{p}_n^m - \mathbf{m}\|^2 - (r^m)^2 \right], \tag{1}$$

where $\mathbf{p}_n^m$ denotes the distance between anterior superior iliac spine (ASIS) to marker and $\mathbf{m}$ is vector from ASIS to the HJC. Here $M = 4$ stands for markers number.

Figure 2 shows HJC position during thigh arc movements which implies that the HJC does not stand still during thigh arc movements. This phenomenon leads to bionic design of exoskeleton hip joints.

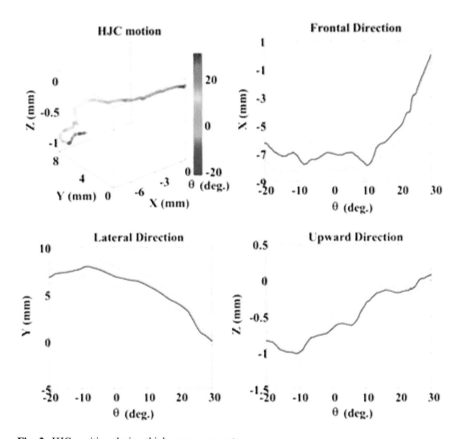

Fig. 2 HJC position during thigh arc movements

2.1.2 Knee Joint Analytical Model

MRI of an unloaded cadaver knee in sagittal plane is shown in Fig. 3a. The two white circles here represent for approximated geometries of femoral articular surfaces. To provide a continuous differentiable function, Lee and Guo (2010) present a more general bio-joint representation based on elliptical geometries as shown in Fig. 3b.

As shown in Fig. 3c, the tibia rolls and slides synchronously on the femur when the knee joint flexes. This motion leads to change of distance r from the tibia mass-center O to the contact point C. Thus, the simplified pin-joint design of exoskeleton knee joint is no more an appropriate choice, because of misalignment between human knee joint and exoskeleton knee joint. This misalignment will increase internal force to human knee joint and can even result in injury of knee joint. Figure 3d shows sketch of bio-joint design for the exoskeleton. The shank of exoskeleton is actuated through a cam mechanism located at the same position as initial contact point C_i. It is designed to adjust to the change of l from the tibia mass-center O to the contact point C.

$$Cp(\theta) = 1.078\theta^4 - 11.184\theta^3 + 26.542\theta^2 - 0.825\theta + 263.3, \qquad (2)$$

where θ is the flexion angle of human knee joint.

Fig. 3 Bionic knee joint illustration. **a** MRI of cadaver knee. **b** Bionic knee joint model. **c** Rolling/sliding contact. **d** Sketch of bio-joint design

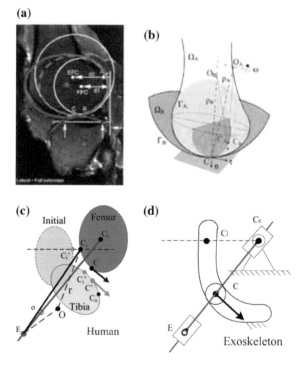

2.2 Design of Exoskeleton Joints

2.2.1 Hip Joint Design

To keep the mechanical HJC close to the anatomical one presented before, the sagittal, frontal, transverse, and rotation (SFTR) system is adopted which means the joint angles in the sagittal, frontal, transverse planes are measured. As shown in Fig. 4, based on the SFTR system, a new 3-DOF joint was constructed by translation of flexion/extension and abduction/adduction axes which were described in Y-X and Y-Z planes, respectively using polar coordinates. Both coordinates considered Y axis as polar axis. According to Fig. 4, the positions of the new intersection points, O_0 and O_1, could be expressed as Eq. (3).

$$\begin{cases} p_{O_0} = \rho_0 e^{i\alpha_0} \\ p_{O_1} = \rho_1 e^{i\alpha_1} \end{cases} \tag{3}$$

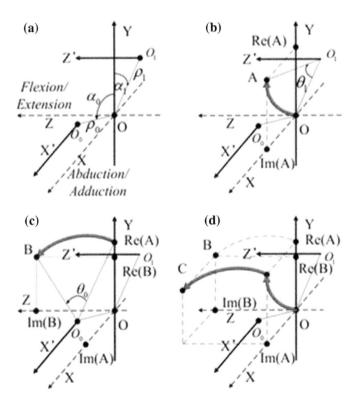

Fig. 4 Axes translation and corresponding HJC position

To determine ρ_0, ρ_1 α_0 and α_1, the root mean square of the distance between anatomical and mechanical HJC position was employed [shown in Eq. (4)] as a criterion with which the four design parameters could be obtained.

$$\overline{E} = \sqrt{\frac{1}{N}\sum_{i=1}^{N}[(O_{x_i} - C_{x_i})^2 + (O_{y_i} - C_{y_i})^2 + (O_{z_i} - C_{z_i})^2]}, \qquad (4)$$

where $(O_{x_i}, O_{y_i}, O_{z_i})$ and $(C_{x_i}, C_{y_i}, C_{z_i})$ are the position of the anatomical and mechanical HJC positions during the same hip joint arc movement. N is the sample number of motion capture system during arc movements, here $N = 700$. Applying the steepest descent method (Fletcher and Powell 1963), the minimum value of $\overline{E}$ was found to be 1.08 mm when $\rho_0 = 11.5$ mm, $\rho_1 = 4.1$ mm, $\alpha_0 = 278.2°$, and $\alpha_1 = 243.1°$. Figure 5 shows optimal mechanical HJC sphere and anatomical HJC.

2.2.2 Knee Joint Design

Figure 3d shows the sketch of bionic knee joint design and Fig. 6 shows CAD model of bio-joint mechanism. When the bio-joint flexes/extends, the roller in cam

Fig. 5 Optimal mechanical HJC sphere and anatomical HJC

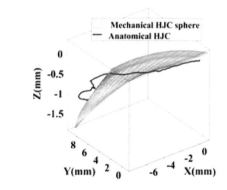

Fig. 6 Bio-joint mechanism CAD model

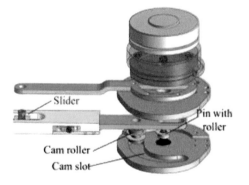

slot moves along the cam profile, which makes the roller in straight slot move along the straight slot at the same time. The motions of these two rollers lead to change of shank length, which is used to adjust to the human knee joint motion. In order to match the human knee joint flexion/extension ideally, the cam profile trajectory of bio-joint should be designed to follow the change distance of l presented in (1). Considering the actual mechanism size limitation, the constant term in (1) should be reduced and the r should be scaled down. Thus, cam profile is modified by: (1) Replacing the constant in (1) with a proper constant s on the basis of actual machining of cam profile as shown in (5a). (2) Scaling down the Variable Terms of Cp in (5a) by a coefficient λ as shown in (5b).

$$Cp(\theta) = 1.078\theta^4 - 11.184\theta^3 + 26.542\theta^2 - 0.825\theta + s, \qquad (5a)$$

$$Cp(\theta) = \lambda(1.078\theta^4 - 11.184\theta^3 + 26.542\theta^2 - 0.825\theta) + s, \qquad (5b)$$

where Cp is the distance from cam profile to knee joint center of rotation, $s = 35$ mm and $\lambda = 0.6$ considering sizes limitation of the knee joint mechanism. Figure 7 shows trajectories of cam profile described in (5a), (5b). The original and reduced cam profiles cannot support the human/exoskeleton because of the negative slopes near $\theta = 0°$. This will lead to upper link slipping from a high potential energy state while the human is standing. Therefore, the knee joint mechanism has been designed using the reduced cam profile as shown in Fig. 7.

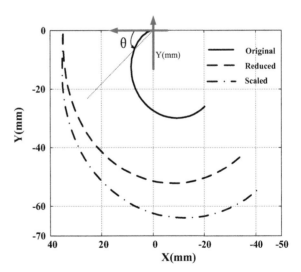

Fig. 7 Design of cam profiles

3   Experiments and Results

An anthropomorphic lower extremity exoskeleton with bionic hip joint and knee was designed and manufactured by implementing the optimal parameters. The exoskeleton hip and knee flexion/extension joint were driven by flat motors (Maxon Inc.) with harmonic gearboxes (CTKM Inc.). To keep balance and provide further manual support during walking training for rehabilitation, two crutches were also used. Figure 8a shows the exoskeleton structure being worn by the patient. The comfort of the human-exoskeleton physical interface is mostly evaluated by the interaction forces between the human and the exoskeleton. This mismatch between the human and exoskeleton joints gives rise to an interaction force, which is pressed onto the human soft tissues and reduces the wearing comfort. Therefore to assess the comfort quality of the resulting exoskeleton, these physical interaction forces can be compared with traditional design. This was done and the interaction forces between human legs and exoskeleton legs were measured by packaged force sensors consisting of two one-dimensional force sensors (Tecsis Inc.) as shown in Fig. 9. This study was approved by the Institutional Review Board of Zhejiang University. Informed written consent was obtained from all subjects.

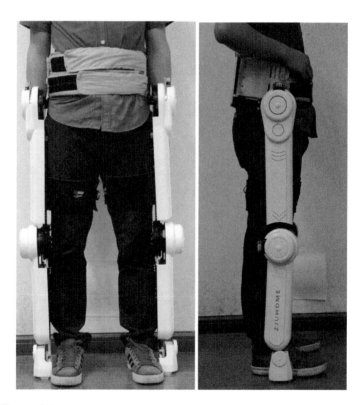

Fig. 8 The exoskeleton prototype

Fig. 9 Interaction force
measurement experiment

3.1 Human-Exoskeleton Interaction Force with Bionic Hip Joint

The hip flexion and abduction movements were repeated by the subject with the exoskeleton for 5 times. The exoskeleton hip joint flexion speed was set at 15°/s and abduction speed was set at 10°/s. Figure 10 shows the mean interaction force between the subject and exoskeleton during flexion and abduction movements driven by the exoskeleton. $F_{e\theta}$ compatible and F_{er} compatible mean normal and tangential interaction force with respect to connecting surface wearing exoskeleton with compatible hip joint. While $F_{e\theta}$ traditional and F_{er} traditional refer to normal and tangential interaction forces with respect to the contact surface as wearing the traditional hip joint exoskeleton. Obviously, both normal and tangential forces with compatible joint decreased during flexion movement. While only tangential forces with compatible joint decreased during abduction movement. The results showed the advantage of the bionic hip joint exoskeleton over the traditional one. However, the normal forces with the biocompatible (bionic and compatible) joint exoskeleton was close to the forces with the traditional joint during abduction movement. A reasonable explanation was that the abduction speed was slow and the anatomical HJC movement in the Z direction was not significant as shown in Fig. 2.

Fig. 10 Hip interaction force during flexion/abduction. **a** Hip interaction force during flexion. **b** Hip interaction force during extension

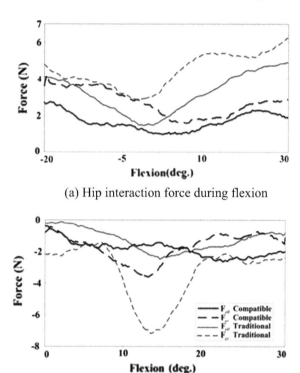

(a) Hip interaction force during flexion

(b) Hip interaction force during extension

3.2 Human-Exoskeleton Interaction Force with Bionic Knee Joint

Similarly, the knee flexion movements were repeated by the subject with the exoskeleton for 5 times. The exoskeleton knee joint flexion speed was set at 15°/s. Figure 11 is the knee interaction force experiment results. There were reductions of both normal and tangential interaction forces compared with interaction forces wearing traditional exoskeleton. Simplification of human knee joint as traditional pin-joint leads to additional relative movements during human-exoskeleton knee joint flexion. The relative movements cause additional friction forces through the bandage connecting human legs and exoskeleton legs. However, the exoskeleton bio-joint can well follow the human knee joint rolling and sliding during flexion leading to less relative movements. Hence, as experiment results show, the bio-joint can reduce the human-exoskeleton interaction force which is of great significance to reduce human internal joint load and make the exoskeleton compatible.

Fig. 11 Knee interaction
force during flexion

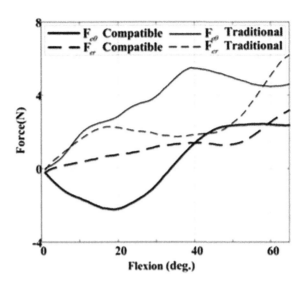

4 Conclusions

To realize biocompatible human-exoskeleton physical interfaces, new design methods are adopted and presented where a lower extremity exoskeleton with compatible hip and knee joints is designed. The compatibility of the new joints was validated by human–machine interaction force experiments compared with using an exoskeleton that with traditional joints. The design method can also be adopted as a reference for hip and knee replacement mechanical designs.

The key results of this research are summarized below:

(1) The mechanical hip joint is designed with its HJC best covering the anatomical one by translation of flexion/extension and abduction/adduction axes under the SFTR system. The RMS error of matching $\overline{E} = 1.08$ mm is low compared with the range of anatomical HJC motion which is about 10 mm. The human-exoskeleton interaction force experiments showed the force decreases by 25.5 and 85.5 % during hip flexion and abduction, respectively, when applying the new design method.

(2) The mechanical knee joint is also designed with its joint center best following the rolling and sliding of human knee joint center during knee flexion/extension. The human-exoskeleton interaction force experiments showed the force decreases by 75.4 % during knee flexion compared with interaction forces wearing traditional exoskeleton.

(3) The human-exoskeleton interaction forces of both hip joint and knee joint validate the compatibility of the exoskeleton with bionic design which will lead to a more ergonomic design method for human–machine systems.

Acknowledge This work was supported by the Science Fund for Creative Research Groups of National Natural Science Foundation of China under Grant 51221004. Manufacture was supported by Li-tong Lv and Kai-ge Shi. Thanks give to them.

References

Amigo LE, Casals A, Amat J (2011) Design of a 3-DoF joint system with dynamic servo-adaptation in orthotic applications. In: 2011 IEEE international conference on robotics and automation (ICRA). IEEE, pp 3700–3705. doi:10.1109/ICRA.2011.5980173

Camomilla V, Cereatti A, Vannozzi G, Cappozzo A (2006) An optimized protocol for hip joint centre determination using the functional method. J Biomech 39(6):1096–1106. doi:10.1016/j.jbiomech.2005.02.008

Farris RJ, Quintero HA, Goldfarb M (2011) Preliminary evaluation of a powered lower limb orthosis to aid walking in paraplegic individuals. IEEE Trans Neural Syst Rehabil Eng 19(6):652–659. doi:10.1109/tnsre.2011.2163083

Fletcher R, Powell MJ (1963). A rapidly convergent descent method for minimization. Comput J 6(2):163–168. http://dx.doi.org/doi:10.1093/comjnl/6.2.163

Gamage SSHU, Lasenby J (2002) New least squares solutions for estimating the average centre of rotation and the axis of rotation. J Biomech 35(1):87–93. doi:10.1016/S0021-9290(01)00160-9

Hidler J, Nichols D, Pelliccio M, Brady K et al (2009) Multicenter randomized clinical trial evaluating the effectiveness of the Lokomat in subacute stroke. Neurorehabilitation Neural Repair 23(1):5–13. doi:10.1177/1545968308326632

Leardini A, Cappozzo A, Catani F et al (1999) Validation of a functional method for the estimation of hip joint centre location. J Biomech 32(1):99–103. doi:10.1016/S0021-9290(98)00148-1

Lee KM, Guo J (2010) Kinematic and dynamic analysis of an anatomically based knee joint. J Biomech 43(7):1231–1236. doi:10.1016/j.jbiomech.2010.02.001

Menschik F (1997) The hip joint as a conchoid shape. J Biomech 30(9):971–973. doi:10.1016/S0021-9290(97)00051-1

O'connor JJ, Shercliff TL, Biden E, Goodfellow JW (1989) The geometry of the knee in the sagittal plane. Proc Inst Mech Eng [H] 203(4):223–233. doi:10.1243/PIME_PROC_1989_203_04

Rasquinha B, Wood G Rudan et al (2011) Ellipsoid fitting of dysplastic and early arthritic articular hip surfaces. Proc Orthop Res Soc, Paper, 844. (doi:10.1109/iembs.2007.4353484)

Ren YP, Kang SH, Park HS et al (2013) Developing a multi-joint upper limb exoskeleton robot for diagnosis, therapy, and outcome evaluation in neurorehabilitation. IEEE Trans Neural Syst Rehabil Eng 21(3):490–499. doi:10.1109/tnsre.2012.2225073

Suzuki K, Mito G, Kawamoto H, Hasegawa Y, Sankai Y (2007) Intention-based walking support for paraplegia patients with Robot Suit HAL. Adv Robot 21(12):1441–1469. doi:10.1163/156855307781746061

Valiente A (2005) Design of a quasi-passive parallel leg exoskeleton to augment load carrying for walking. Massachusetts Inst of Tech Cambridge Media Lab

Veneman JF, Ekkelenkamp R, Kruidhof R et al (2006) A series elastic-and bowden-cable-based actuation system for use as torque actuator in exoskeleton-type robots. Int J Robot Res 25(3):261–281. doi:10.1177/0278364914545673

Wang D, Lee KM, Guo J, Yang CJ (2014) Adaptive knee joint exoskeleton based on biological geometries. IEEE/ASME Trans Mechatron 19(4):1268–1278. doi:10.1109/TMECH.2013.2278207

Wismans JAC, Veldpaus F, Janssen J, Huson A, Struben P (1980) A three-dimensional mathematical model of the knee-joint. J Biomech 13(8):677–685. doi:10.1016/0021-9290(80)90354-1

Wu G, Cavanagh PR (1995) ISB recommendations for standardization in the reporting of kinematic data. J Biomech 28(10):1257–1261. doi:10.1016/0021-9290(95)00017-C

Yan H, Yang C, Zhang Y, Wang Y (2014) Design and validation of a compatible 3-degrees of freedom shoulder exoskeleton with an adaptive center of rotation. J Mech Des 136(7):071006. doi:10.1115/1.4027284

Zoss AB, Kazerooni H, Chu A (2006) Biomechanical design of the Berkeley lower extremity exoskeleton (BLEEX). IEEE/ASME Trans Mechatron 11(2):128–138. doi:10.1109/TMECH. 2006.871087

Development of a Lower Limb Rehabilitation Wheelchair System Based on Tele-Doctor–Patient Interaction

Shuang Chen, Fang-fa Fu, Qiao-ling Meng and Hong-liu Yu

Abstract Portable rehabilitation training devices are coming into normal families and becoming an important part of home rehabilitation. At the same time, the rehabilitation system combining virtual reality technology and tele-doctor–patient interaction and portable rehabilitation devices is a new research trend. In this paper, an intelligent rehabilitation training system which is one of the first product that make it possible for patients to do lower limb training at home is proposed. It includes an electric wheelchair with lower limb training function, a multivariate control module, a virtual reality training module and a tele-doctor–patient interaction module. This system can solve the shortcomings of large volumes of existing products. The lower limb training games module which is based on virtual reality technology make the rehabilitation procedure more interesting. The tele-doctor–patient interaction module enables patients to do lower limb training at home, meanwhile doctors can give assignments to patients based on the score of the last game to save more medical resources and time effectively.

Keywords Lower limb rehabilitation wheelchair · Virtual reality · Tele-doctor–patient interaction

1 Introduction

As is known to all that China has gradually entered the aging society. According to statistics, the number of stroke patients over the age of 40 in China has climbed to 10 million. The number of new stroke patients is coming out up to nearly 200 million people each year (Zhang 2011). Limb dysfunction seriously affects life quality of patients. In the rehabilitation of stroke patients, limb function recovery is

S. Chen · F. Fu · Q. Meng · H. Yu (✉)
Institute of Biomechanics and Rehabilitation Engineering, University of Shanghai for Science and Technology, Shanghai 200090, China
e-mail: yhl98@hotmail.com

© Zhejiang University Press and Springer Science+Business Media Singapore 2017
C. Yang et al. (eds.), *Wearable Sensors and Robots*, Lecture Notes in Electrical Engineering 399, DOI 10.1007/978-981-10-2404-7_17

the most difficult one. Modern rehabilitation theory and practice show that effective rehabilitation after stroke can speed up the recovery process, mitigate functional disability, reduce the high cost of potential long-term care needed and save social resources (Visintin et al. 1889).

A lot of lower limb rehabilitation equipment have been developed in recent years. Lokohelp is a lower limb training and assessment system based on BWSTT (Body Weight Support Treadmill Training). It improves patient outcomes by increasing therapy volume and intensity, providing task-specific training and increasing patient engagement (Freivogel et al. 2008). Flexbot is an intelligent lower limb training robot. It can do ambulation training in different body positions. Virtual reality scene is used to simulate walking state. HAL-5, designed by a Japan technology company called Cyberdyne, is an exoskeleton robot and can be driven by neural signals of brain. It can help users stand, walk, and climb upstairs (Kawamoto et al. 2003).

As we can conclude, existing lower limb rehabilitation devices can be divided into three kinds: BWSTT system, intelligent lower limb training robot, and assistive exoskeleton. BWSTT system mainly aimed for patients whose muscle strength are weak and do rehabilitation in hospitals. Also, the training procedure is quite boring (Hornby et al. 2005). In order to imitate human gait, intelligent lower limb training robot usually have large volume, and very expensive, not suitable for family use (Lv 2011). Lower limb exoskeleton is more portable than the other two kinds of devices and is designed for patients with stronger muscle strength. Because of its wearable feature, uncertain danger may occur during outdoor usage (Adam and Kazerooni 2005). Although the history of lower limb rehabilitation equipment is very short but it has become quite a trend. But there are only a few independent research and development of such products in China.

According to the requirement of the lower limb rehabilitation, a lower limb rehabilitation wheelchair with virtual reality games based on tele-doctor–patient interaction is designed. Mobile phone software is used for controlling of the wheelchair and virtual reality interactive system is added to the rehabilitation wheelchair to realize a variety of lower limb training. Also, a tele-doctor–patient interaction is designed for doctors to assign tasks for patients to do lower limb training at home. The structure of the system is as shown in Fig. 1.

2 Mechanical Design of Lower Limb Training Module

2.1 Preparing Manuscript

The lower limb training module is designed based on a multi-posture electric wheelchair so that it can be used as a common wheelchair while not doing rehabilitation training.

Training function is realized by four linear motors. As shown in Fig. 1, in the process of lying, motor 11 and motor 14 work together to realize the movement of the backrest and the legs. In the process of standing, motor 12, motor 11, and motor

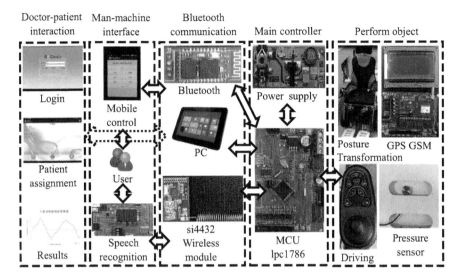

Fig. 1 Structure of a multi-posture intelligent wheelchair for rehabilitation

15 work together to realize the stand of the seat and the expend of the foothold. Motor 14 and motor 15 work together when doing lower limb rehabilitation trainings, realizing the flexion and extension of the knee joint of patient's as well as balancing trainings with virtual reality games, etc.

As shown in Figs. 2 and 3, the main control module 10 is responsible for the deployment of each module to work together. Users can select the lower limb training modes by touching tablet PC 5. Command is transmitted to the master control module 10 and converted to the corresponding command to realize posture transformations or lower limb training mode. In the process of training, pressure sensors 16 which are located in the footrests collect the state of motion in real time. The signal is transmitted via Bluetooth module to master control module, and then displayed on the tablet PC 5 to help nurses to know the training situation of the patients.

Fig. 2 Principle diagram of lower limb training structure

Fig. 3 Pressure sensors for lower limb training

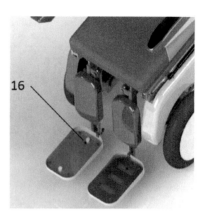

16

In order to know whether the structure meets the requirements of movement, the law of motion of the structure needs to be studied. In mechanical transmission, movement analysis of the mechanical structure is the foundation of the analysis of entire mechanical system. According to the movement of the driving link, displacement, velocity, and acceleration of a specific point in the entire mechanical drive system can be obtained. To solve the disadvantages that graphic method is of low precision and time-consuming when analyzing motion mechanism, analytic methods and computer are used to help the analysis. Analytic methods can not only conduct high precision analysis, but also draw a graph of the movement, helping improving design of the mechanical design (Lin and Wang 2003).

Because of the complexity of the wheelchair structure, vector equation analytic method is used in the analysis to obtain precisely the kinematical characteristic of the components in the process of posture change. Coordinate system is established as shown in Fig. 4. O is the origin of coordinate the system. OA and OB are the x axis and y axis of the coordinate system.

According to the relation between the vectors, we can get Eq. (1):

$$\overrightarrow{AE} + \overrightarrow{EB} = \overrightarrow{AB} \tag{1}$$

Project the vectors to the x axis and y axis and we can get Eq. (2):

$$\begin{cases} x: \left|\overrightarrow{AE}\right| \cos\theta_1 + \left|\overrightarrow{EB}\right| \cos\theta_3 = \left|\overrightarrow{AB}\right| \cos\theta_2 \\ y: \left|\overrightarrow{AE}\right| \sin\theta_1 + \left|\overrightarrow{EB}\right| \sin\theta_3 = \left|\overrightarrow{AB}\right| \sin\theta_2 \end{cases} \tag{2}$$

Eliminate θ_1 and solve Eq. (3):

$$\theta_2 = \theta_3 + \arccos\left(\frac{AB^2 + EB^2 - AE^2}{2\left|\overrightarrow{AB}\right|\left|\overrightarrow{EB}\right|}\right) \tag{3}$$

Fig. 4 Mathematical model coordinate system of the wheelchair

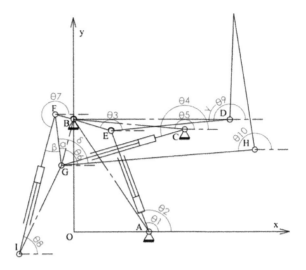

For A and B are fixed point on the wheelchair frame, we can get Eq. (4):

$$\theta_2 = \pi - \arctan\frac{\left|\overrightarrow{OB}\right|}{\left|\overrightarrow{OA}\right|} \tag{4}$$

$\left|\overrightarrow{AB}\right|$ mentioned above is a known variant, so θ_1 can be calculated according to the equations.

θ_4, θ_5 and θ_6 can be calculated in a similar way.

According to the geometric relationship between the vectors, expressions can be determined for each position parameter of the structure:

$$\theta_7 = \pi + \theta_6 + \alpha - \beta \tag{5}$$

$$\theta_8 = \theta_6 + \alpha - \beta \tag{6}$$

$$\theta_9 = \pi + \gamma \tag{7}$$

$$\theta_{10} = \pi + \theta_6 + \alpha - \delta \tag{8}$$

α, β, γ and can all be calculated according to the law of cosines:

$$\alpha = \arccos\left(\frac{BG^2 + GF^2 - BF^2}{2\left|\overrightarrow{BG}\right|\left|\overrightarrow{GF}\right|}\right) \tag{9}$$

When the position parameters of the structural components are known, the motion trails of the hinges of the components can be calculated. For example, the position parameter of E is as Eq. (10):

$$\begin{cases} E_x: -\left|\overrightarrow{EB}\right| \cos \theta_3 \\ E_y: \left|\overrightarrow{AE}\right| \sin \theta_1 \end{cases} \quad (10)$$

By computing the first derivative of (10) with respect to time, the velocity parameter equation of E can be obtained. By computing the first derivative of velocity parameter equation with respect to time, the acceleration parameter equation of E can be obtained. Kinematics parameters of other hinges can be calculated in the same way (Yuan and Zuomo 2001).

3   Design of Mobile Terminal Control Module Based on Bluetooth

3.1   Bluetooth Control Design

In order to control the wheelchair more conveniently, a special Bluetooth phone control module, which is suitable for families and rehabilitation centers, was developed. This module includes a Bluetooth communication module and a system of mobile phone software.

As shown in Fig. 5, there are several group of bottoms in the cell phone interface. Each group consisted of two bottoms, one is "open," the other is "close," correspondingly control the status of each posture. Users can touch the screen to select the posture that they want.

Fig. 5 Cell phone control interface

Fig. 6 Mathematical model coordinate system of the wheelchair

Intelligent control terminal includes four parts: Lpc1768 microcontroller control part, the Bluetooth module part, Power supply part, Photoelectric coupling switch part. A HC-05 Embedded Bluetooth serial interface communication module is used in the Bluetooth module part and is controlled by AT commands. Serial port parameters are set to 9600 bit/s and the passcode is 1234. Connection mode is set to any Bluetooth address link mode so that multiple phone manipulation can be realized (Lee et al. 2004).

To enrich the usability of the tele-doctor–patient interaction module, the control interface is transplanted to the tablet PC. The tablet PC interface is as shown in Fig. 6.

3.2 Control Module Experiment

In order to know how the Bluetooth control module works during lower limb training, mobile terminal experiment is conducted. The experiment platform includes lower limb training wheelchair module and Bluetooth control module. The Bluetooth receive module and the controller are installed in the wheelchair. The experimenter sits on the wheelchair, stands two meters, five meters, and eight meters away from the wheelchair while controlling it with the cell phone. All movements of the wheelchair are tested from sitting position to standing, lying, lower limb training, going forward, going backward, turning left and right. Every movement is tested 50 times, and all the results are recorded as shown in Table 1.

We can know from Table 1 that when experimenter sits and stands two meters away from the wheelchair, the success rates of Bluetooth control are almost 100 %. When experimenter stands five meters away from the wheelchair, the success rates are above 98 %, and when experimenter stands eight meters away from the wheelchair, the success rates reduces to 94 %. The success rate is related to operating distance. Therefore, Bluetooth control function is substantially reliable

Table 1 Experiment data of Bluetooth control system

Distance		Sit	Stand	Lie	Train	Forward	Backward	Left	Right
0 m	Success/times	50	50	50	50	50	50	49	50
	Rate/%	100	100	100	100	100	100	98	100
2 m	Success/times	50	50	50	49	50	50	50	49
	rate/%	100	100	100	98	100	100	100	98
5 m	Success/times	50	49	50	50	49	50	50	49
	Rate/%	100	98	100	100	98	100	100	98
8 m	Success/times	49	49	48	49	48	49	47	48
	Rate/%	98	98	96	98	96	98	94	96

and users are suggested to be at least four meters away from the wheelchair when operating for normal use.

4 Tele-Doctor–Patient Interaction Module Design

In order to make it possible for the patients to do lower limb rehabilitation at home to relieve the strain on medical resources, the tele-doctor–patient interaction module is designed. This module provides an interaction platform for doctors and patients and is consisted of doctor client, web server, and the patient client. The following sections focus on design of the virtual reality game and software of the interaction module.

4.1 Design and Realization of Virtual Reality Games

Patient client is a tablet PC which has following functions: controlling, doing rehabilitation training with virtual reality games, and interaction with doctors. Virtual reality games used in rehabilitation training is based on Unity 3D engine (Lange et al. 2010). Status of patients' lower limb is captured by the pressure sensors to be the input of the virtual reality games realizing the human–computer interaction function (Meng 2009).

There are two JHBM-100 kg pressure sensors in the both sides of the footrests. In standing mode, patients do active lower limb exercise by changing their center of gravity to control the bird to fly up and down. The pressure sensors change the anagogic pressure signal into an electrical signal. The more the patients' center of gravity shifts, the larger the pressure of the offset side of the foot will be and the smaller the pressure of the other side of the foot will be. If the pressure becomes small, the output voltage will become small and vice versa. Output voltage is amplified by the differential amplifier circuit and after A/D convert, it will be input to the microcontroller for processing. In the game, the airplane will fly to the side whose output voltage is larger. The larger the voltage is, the more the airplane will move. The interface of the virtual reality game is as shown in Fig. 7.

Fig. 7 Interface of starting
rehabilitation game

The information obtained by the pressure sensors is acted as the parameter in the game. Bombs and diamonds will be randomly generated in the game. Patients gain score by controlling the bird to capture the diamonds, and avoiding bombs. Patients can easily focus on the game and start rehabilitation training.

After each game, patient's movement in training will be saved as a line chart. As is shown in Fig. 8, relative position of patient's center of gravity is compared with the normal relative position which is the average value from 10 normal experimenter. A score is given calculated by Eq. (11).

$$S = \sum_{n=1}^{t} |F_0 - F_n|, \quad t = 1, 2, 3. \dots \tag{11}$$

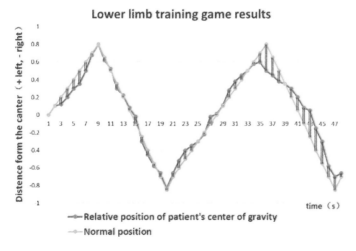

Fig. 8 Line chart of relative position of patient's center of gravity during training game

F_0 refers to the value of normal relative position at a certain time point while F_n refers to the one of patient. T refers to a time point, and S refers to the score of game.

This score quantify the result of game and reflects patient's rehabilitation progress. Also, it can be used as a recovery judgement.

4.2 Software Design

The software is a link between doctors and patients, so two interfaces are necessary for both patients and doctors, also, doctors should have permission to query patients' information. To satisfy the needs, JAVA is used to write the program. Login interface is as shown in Fig. 9, user can select a certain identity to login. After typing in user name and password, patient can login with their account. As can been seen in Fig. 10, patients can control the wheelchair, do lower limb training, see current position (Fig. 11) and check today's tasks assigned by doctor. After patients finish each training game, the score and the line chart of relative position of patient's center of gravity during training game will be sent automatically to the doctor to help detecting patient's recovery condition. Doctors' interface is as shown in Fig. 12. Doctors can manage their patients, check their training results, and assign everyday tasks with the system.

With the help of the tele-doctor–patient interaction module, patients can concentrate more on rehabilitation than worrying about going to rehabilitation center while they have mobility impairments. On the other hand, doctors can deal with more patients because the system can help save the score and the line chart of patient's movement which are the key to their recovery progress. This is also what makes the system different from existing products.

Fig. 9 Login interface

Fig. 10 Patient interface

Fig. 11 Current position

Fig. 12 Doctor interface

5 Conclusion

This study completed the structure design of a multi-posture intelligent rehabilitation wheelchair and successfully developed a module of mobile control and a module of Bluetooth communication which realize posture conversion and lower limb training control by a cell phone. In order to improve training effectiveness and stimulate patients' interests in training, Unity 3D engine and two pressure sensors were combined to realize rehabilitation game. Pressure sensors were set in the footplate of wheelchair to capture the information of patients' center of gravity as input signal to the game. A tele-doctor–patient interaction software is designed to enable doctors check patients' recovery condition and assign tasks remotely.

Experiments have been done to ensure the safety of the posture transformations and the reliability of the Bluetooth phone control module. And the results show that the wheelchair can fully fulfill the functions we expected. But mass of data is needed to show how the rehabilitation wheelchair can help the patients with lower limb dysfunctions and the efficiency of the system.

In order to improve the rehabilitation evaluation system to help the patients do lower limb rehabilitation according to their extent of dysfunction and their needs, a database of the patients' information should be established to help the design of the content and intensity of rehabilitation games. Also, we will continue doing clinical experiments on patients with lower limb dysfunction in the rehabilitation hospitals and more data will be taken to further verify the effectiveness of lower limb rehabilitation with the rehabilitation wheelchair.

Acknowledgments Project supported by the Shanghai Production-Study-Research Cooperation projects (No. 12DZ1941003, 12DZ1941004).

References

Adam ZH, Kazerooni AC (2005) On the mechanical design of the Berkeley lower extremity exoskeleton (BLEEX). In: IEEE Intelligent Robots and Systems Conference, August, Edmonton, pp 3465–3472

Freivogel S, Mehrholz J, Husak-Sotomayor TA, Schmalohr D (2008) Gait training with the newly developed "LokoHelp" system is feasible for non-ambulatory patients after stroke, spinal cord and brain injury. A feasibility study. Brain Inj 22:625–632

Hornby TG, Zemon DH, Campbell D (2005) Robotic-assisted, bodyweight-supported treadmill training in individuals following motor in-complete spinal cord injury. J Phys Ther 85(1):52–66

Kawamoto H, Lee S, Kanbe S, Sankai Y (2003) Power assist method for HAL—3 using EMG. Based feedback controller. Proceedings of the 2003 IEEE international conference on systems, man and cybernetics

Lange BS, Requejo P, Flynn SM et al (2010) The potential of virtual reality and gaming to assist successful aging with disabilityJ. Phys Med Rehabil Clinics N Am 21(2):339–356

Lee H-S, Chang S-L, Lin K-H (2004) A study of the design, manufacture and remote control of a pneumatic excavator. Int J Mech Eng Educ 32(4):345–361

Lin J, Wang X (2003) A comparative study of the institutional displacement analysis method. Mechanical Design and Manufacturing, pp 31–32

Lv C (2011) Research of a hemiplegia rehabilitation robot for upper-limb D. Shanghai Jiaotong University

Meng G (2009) Xiaohong of Scientific, such as bio-based virtual game's visuals anti—Applied Technology in Rehabilitation Exercise Training J Chinese. J Rehabil Med 24(2)

Visintin M, Barbeau H, Korner-Bitensky N, Mayo NE (1889) Anew approach to retrain gait in stroke patients through body weight support and treadmill stimulation. Stroke 29:1122–1128

Wang S, Mao Z, Zeng C et al (2010) A new method of virtual reality based on Unity 3D. In: 2010 18th international conference on geoinformatics, IEEE, pp 1–5

Yuan S, Zuomo C (2001) Mechanical principles. Higher Education Press, Beijing, pp 61–66

Zhang T (2011) Chinese Stroke Rehabilitation Treatment Guidelines. J Chinese J Rehabil Theory Practice 18(4):301–318

A Grasp Strategy with Flexible Contacting for Multi-fingered Hand Rehabilitation Exoskeleton

Qian-xiao Wei, Can-jun Yang, Qian Bi and Wei Yang

Abstract To recovering the functions of hand after stroke, many hand exoskeletons and their control methods are developed. However, less research involves in the multi-fingered grasping. There are two primary problems: the fingers are correlative in the movement and the contacting part, the human finger, is flexible. This paper presents a method, which takes not only all the fingers but also their mechanical impedance into a dynamic system. The method is divided into three levels. First level, grasping planning, the desired interface force of each finger is derived by the geometric and external force information of object. Second level, multi-fingered coordinate force control, we see each finger's impedance as a second-order sub-system to model an integrated coordinate dynamic system. Third level, single finger force control, execute the position and force command calculated in middle level by each finger, which has been presented in our early research. To verify the method, we set an experiment to grasp an apple assisted by a three fingers (thumb, index finger, and middle finger) exoskeleton. The results illustrate the effectiveness of the proposed method and also point out the direction for further research.

Keywords Grasp · Flexible · Coordinate · Multi-fingered · Exoskeleton

Q. Wei · C. Yang (✉) · Q. Bi · W. Yang
State Key Laboratory of Fluid Power Transmission and Control, Zhejiang University,
Hangzhou 310027, China
e-mail: ycj@zju.edu.cn

Q. Wei
e-mail: wqx@zju.edu.cn

Q. Bi
e-mail: biqianmyself@zju.edu.cn

W. Yang
e-mail: zjuaway@163.com

© Zhejiang University Press and Springer Science+Business Media Singapore 2017   225
C. Yang et al. (eds.), *Wearable Sensors and Robots*, Lecture Notes in Electrical
Engineering 399, DOI 10.1007/978-981-10-2404-7_18

1 Introduction

Exoskeleton robots have been developed to replace the traditional therapist-dependent post-stroke rehabilitation. Because the human hand is the most dexterous part of human body, including 5 fingers and 22 degrees of freedom (DOFs), development of hand exoskeletons has proceeded slowly in terms of both the mechanism and control method. Most of hand exoskeleton programs start from index finger exoskeleton then extend the method to other fingers. The common ultimate goal of them is recovering the function of hand. Thus the grasping with multi-fingered hand exoskeleton is studied widely. Here we classify the researches into three types, no interaction, interaction with virtual reality, and interaction with real object.

The first type hand exoskeletons emphasize the range of movement (ROM) of each joint. Nakagawara et al. (2005) developed an encounter-type multi-fingered exoskeleton to control a dexterous robot hand. When slave hand touches an object, master finger is controlled to produce a force that is equal to the force applied on the slave finger (force control). The slave finger is always controlled to take the same position as that of the master finger (position control). This bilateral control is the so-called force-reflecting servo method. Fang et al. (2009) developed another encounter-type exoskeleton with a different transmission mechanism for operation of DLR/HIT dexterous hand. They also used the similar force-reflecting servo control method. Schabowsky et al. (2010) presented a hand rehabilitation exoskeleton. Compensation algorithms were developed to improve the exoskeleton's backdrivability by counteracting gravity, stiction, and kinetic friction. Based on sliding mode position control, Wege et al. (2006) presented a force-based control mode for a hand exoskeleton. This method consisted of two parts: the force sensor value and previous motion. They were used to compute a desired motion. In the method, the finger impedance was neglected. Ho et al. (2011) presented an EMG-driven exoskeleton hand for task training in stroke rehabilitation. A threshold of 20 % of the maximum voluntary contraction (MVC) EMG signals was used to trigger the hand opening and hand closing motions. That is an open-loop EMG control method.

The second type hand exoskeletons look at the interaction between the exoskeleton and the human finger. Tzafestas (2003) presented a kinesthetic feedback methodology based on the solution of a generalized force distribution problem for the human hand during virtual manipulation tasks, which focused on the synthesis of whole-hand kinesthetic feedback and the experimental evaluation of the haptic feedback system. In general terms, an intuitive haptic interaction, integrating a large part of the human hand functionality within a VE, is oriented toward a more natural human–machine interaction, ultimately aiming at a more efficient human skill transfer. Ueki et al. (2011, 2012) presented a novel virtual reality (VR)-enhanced hand rehabilitation support system, which focused on recovering the ROM of each joint. With the system, they developed self-controlled rehabilitation therapy. That is open-loop control from healthy hand to impaired hand.

More than coordination, the third type hand exoskeletons focus on the interaction with the object. There are two important problems in this type. Each exoskeleton finger not only interacts with human finger directly but also interacts with other fingers indirectly. In other words, the fingers are not independent of each other, but are correlated with each other. Moreover, the human fingers are not rigid. Their mechanical impedance is nonlinear, which influence the stability of grasping seriously. It is necessary to take the impedance into consideration for the multi-fingered exoskeleton haptic force control. However, less researches involve these two problems, thus the method is lack of systematical investigation.

This paper is organized as follows: Sect. 1 introduces the motivation of these researches and the two problems to be solved here. Section 2 introduces the proposed method, including the principle and the derivation process. Section 3 is the experiment, including setting, result, and discussion. Section 4 concludes the method and experiment then points out the direction of further research.

2 Method

To solve the grasping task assisted with multi-fingered exoskeleton, we divide the method into three levels, as shown in Fig. 1.

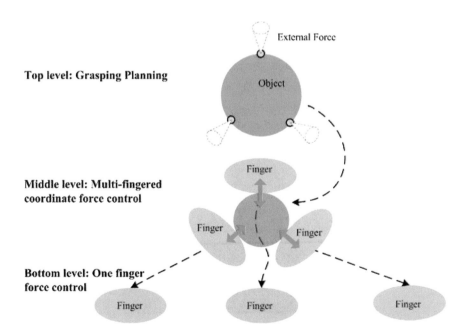

Fig. 1 Three levels in grasping with multi-fingered exoskeleton

Top level: Grasping Planning. According to the geometric and external force information of object, calculate the postures of each finger and the force applied on fingertip.

Middle level: Multi-fingered coordinate force control. Take the impedance of human finger into consideration and derive the coordinate dynamics.

Bottom level: One finger force control. Execute the position and force command calculated in middle level by each finger.

2.1 Grasping Planning

First, we need to find out the grasp plane for the object. As shown in Fig. 2, we consider four fingers in the grasping action and all the base joints are in the same plane, hand plane. The global coordinate is set at the base joint of thumb and X axis point to the base of middle finger, Y axis is in the same plane and perpendicular to the X axis. Z axis meets the right hand rule pointing to the vertical upward direction.

We use thumb, index finger, and middle to complete the basic grasping. The three fingertips construct the grasp plane. The manipulabilities of each finger under the grasp posture define hand manipulability.

Fig. 2 Coordinates in grasping task

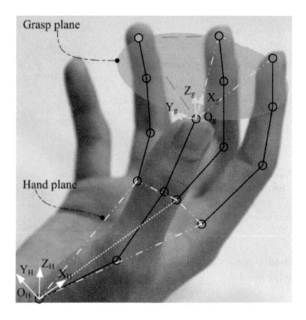

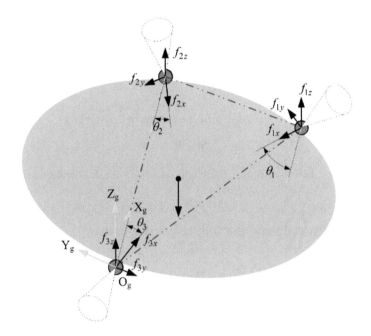

Fig. 3 Multi-fingered friction constraint

Figure 3 shows the force situation in grasp plane. We make two assumptions: (1) The object does not rotate when it is grasped. (2) All the contact points are hard finger frictional contact.

The contacting forces applied on each finger are

$$
\begin{aligned}
\mathbf{f}_1 &= [f_{1x} \quad f_{1y} \quad f_{1z}]^\mathrm{T} \\
\mathbf{f}_2 &= [f_{2x} \quad f_{2y} \quad f_{2z}]^\mathrm{T} \\
\mathbf{f}_3 &= [f_{3x} \quad f_{3y} \quad f_{3z}]^\mathrm{T}
\end{aligned}
\tag{1}
$$

where f_{1x} is the normal pressure, f_{2x} and f_{3x} are frictions. According to the friction cone theory, we have the stable condition of single finger contacting

$$
\mathbf{S}_i \mathbf{f}_i \geq 0;
\tag{2}
$$

where

$$
\mathbf{S}_i =
\begin{bmatrix}
1 & 0 & 0 \\
\mu_{iy} & 1 & 0 \\
\mu_{iy} & -1 & 0 \\
\mu_{iz} & 0 & 1 \\
\mu_{iz} & 0 & -1
\end{bmatrix}
; \quad
\begin{cases}
\mu_{iy} = \mu_i \cos \beta_i \\
\mu_{iz} = \mu_i \sin \beta_i \\
\beta_i = \arctan \frac{f_{iz}}{f_{iy}}
\end{cases}
\tag{3}
$$

We define

$$S = \begin{bmatrix} S_1 & & \\ & \ddots & \\ & & S_i \end{bmatrix}; \quad f = \begin{bmatrix} f_1^T & \cdots & f_i^T \end{bmatrix}^T \tag{4}$$

So we obtain the stable condition of multi-fingered contacting

$$Sf \geq 0 \tag{5}$$

To grasp the object stably, the force should satisfy the following equation:

$$Gf + w = 0 \tag{6}$$

where w is the external force, and G is the grasping matrix:

$$G = \begin{bmatrix} R_1 & \cdots & R_i \end{bmatrix}; \quad R_i = \begin{bmatrix} \cos\theta_i & -\sin\theta_i & 0 \\ \sin\theta_i & \cos\theta_i & 0 \\ 0 & 0 & 1 \end{bmatrix}$$

Unfolding Eq. (6), we obtain

$$\begin{bmatrix} (f_{1x}\cos\theta_1 - f_{1y}\sin\theta_1) + (f_{2x}\cos\theta_2 - f_{2y}\sin\theta_2) + (f_{3x}\cos\theta_3 - f_{3y}\sin\theta_3) \\ (f_{1x}\sin\theta_1 + f_{1y}\cos\theta_1) + (f_{2x}\sin\theta_2 + f_{2y}\cos\theta_2) + (f_{3x}\sin\theta_3 + f_{3y}\cos\theta_3) \\ f_{1z} + f_{2z} + f_{3z} \end{bmatrix} = w \tag{7}$$

Under stable situation, it's static friction at the contacting point is as follows:

$$\begin{aligned} f_{iy} &= \mu_{iy} f_{ix} \\ f_{iz} &= \mu_{iz} f_{ix} \end{aligned} \tag{8}$$

Substituting Eq. (8) into (7), we obtain

$$F_T = A^{-1} w \tag{9}$$

where

$$F_T = \begin{bmatrix} f_{1x} \\ f_{2x} \\ f_{3x} \end{bmatrix};$$

$$A = \begin{bmatrix} (\cos\theta_1 - \mu_{1y}\sin\theta_1) & (\cos\theta_2 - \mu_{2y}\sin\theta_2) & (\cos\theta_3 - \mu_{3y}\sin\theta_3) \\ (\sin\theta_1 + \mu_{1y}\cos\theta_1) & (\sin\theta_2 + \mu_{2y}\cos\theta_2) & (\sin\theta_3 + \mu_{2y}\cos\theta_3) \\ \mu_{1z} & \mu_{2z} & \mu_{3z} \end{bmatrix}; \tag{10}$$

As a result, we get the normal pressure applied on each fingertip.

2.2 *Multi-fingered Coordinate Force Control*

The results given by grasp planning are normal values, but the interaction processing is dynamic. Thus the mechanical impedances of fingers influence the grasp movement obviously. It's necessary to model the coordinate dynamics of grasping.

We see the mechanical impedance of each finger as a spring-damping system, as shown in Fig. 4.

Model the dynamics in X and Y direction, respectively as follows:

$$m\ddot{x} + \mathbf{b}(\dot{\mathbf{s}}\mathbf{C} - \dot{x}) + \mathbf{k}(\mathbf{s}\mathbf{C} - x) = \mathbf{F_T}\mathbf{C} \tag{11}$$

where

$$\mathbf{b} = \begin{bmatrix} b_1 & b_2 & b_3 \end{bmatrix}; \quad \dot{\mathbf{s}} = \begin{bmatrix} \dot{s}_1 & & \\ & \dot{s}_2 & \\ & & \dot{s}_3 \end{bmatrix}; \quad \mathbf{C} = \begin{bmatrix} \cos\theta_1 \\ \cos\theta_2 \\ \cos\theta_3 \end{bmatrix};$$

$$\mathbf{k} = \begin{bmatrix} k_1 & k_2 & k_3 \end{bmatrix}; \quad \mathbf{s} = \begin{bmatrix} s_1 & & \\ & s_2 & \\ & & s_3 \end{bmatrix}$$

$$m\ddot{y} + \mathbf{b}(\dot{\mathbf{s}}\mathbf{S} - \dot{y}) + \mathbf{k}(\mathbf{s}\mathbf{S} - y) = \mathbf{F_T}\mathbf{S} \tag{12}$$

Fig. 4 Coordinate dynamics of grasping

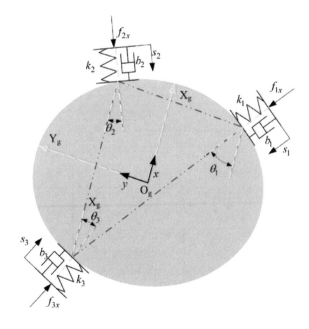

where

$$\mathbf{b} = \begin{bmatrix} b_1 & b_2 & b_3 \end{bmatrix}; \quad \dot{\mathbf{s}} = \begin{bmatrix} \dot{s}_1 & & \\ & \dot{s}_2 & \\ & & \dot{s}_3 \end{bmatrix}; \quad \mathbf{S} = \begin{bmatrix} \sin\theta_1 \\ \sin\theta_2 \\ \sin\theta_3 \end{bmatrix};$$

$$\mathbf{k} = \begin{bmatrix} k_1 & k_2 & k_3 \end{bmatrix}; \quad \mathbf{s} = \begin{bmatrix} s_1 & & \\ & s_2 & \\ & & s_3 \end{bmatrix}$$

Write Eq. (11) and (12) into an equation as follows:

$$m\ddot{\mathbf{X}} + \mathbf{b}\dot{\mathbf{s}}\varphi - \mathbf{b}\dot{\mathbf{X}} + \mathbf{k}\mathbf{s}\varphi - \mathbf{k}\mathbf{X} = \mathbf{F}_T\varphi \tag{13}$$

where

$$\mathbf{X} = \begin{bmatrix} x \\ y \end{bmatrix}; \quad \varphi = \begin{bmatrix} \mathbf{C} \\ \mathbf{S} \end{bmatrix}$$

2.3 Single-Finger Force Control

We have studied single-finger force control in our earlier research (Bi and Yang 2014). The mechanical impedance of finger is position-dependent. Any posture change or small displacement leads to change in parameters that cannot be formulated. Parameters identification online is an effective solution. Meanwhile prediction could eliminate the effect of time delay. Different from pure robot control, we care about the comfort of patients instead of speediness of exoskeleton in rehabilitation. Thus we want the system to perform as a reference model. In that paper, we present a model prediction-reference adaptive impedance control method (Fig. 5).

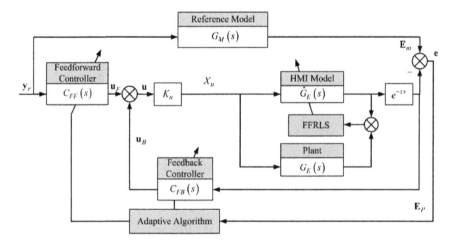

Fig. 5 Model prediction-reference adaptive impedance control diagram

Besides, the comparison of experiments using the MRAIC and the PID indicates that the MRAIC scheme performs better in terms of stabilization and speed. In other words, the MRAIC is effective for the nonlinear HMI problem in the hand exoskeleton control, where the PID or other linear method could not work well.

3 Test

3.1 Multi-fingered Hand Exoskeleton

A multi-fingered hand exoskeleton is developed in our research. There are three fingers in the exoskeleton, thumb has three active DOFs and one coupling DOF, while both index finger and middle finger have two active DOFs and one coupling DOF. The structures of them are similar, taking the index finger as an example, shown in Fig. 6. The two active degrees of freedom (DOFs) are at the metacarpophalangeal (MP) joint and proximal interphalangeal (PIP) joint, respectively, and one coupling DOF at the distal interphalangeal (DIP) joint. Two Maxon DC motors rotate screws, driving the multi-linkages to make the joints of the exoskeleton rotate. Both linkages and exoskeleton phalanxes are manufactured in nylon material through rapid prototyping. The force sensor is a pressure sensitive resistance between two metal gaskets.

Each finger has similar control diagram. They are controlled to exert the desired interface force on fingertips using the force control method mentioned in Sect. 2.3. As the three fingers are correlative, the commands for each motor are calculated in

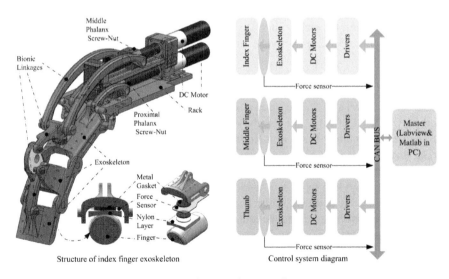

Structure of index finger exoskeleton   Control system diagram

Fig. 6 The structure of exoskeleton and its control system diagram

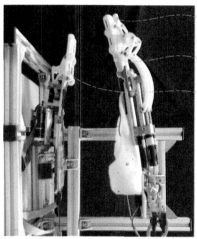

Middle Finger
Exoskeleton

Index Finger
Exoskeleton

Thumb
Exoskeleton

Hand Exoskeleton   Grasping an apple

Fig. 7 The multi-fingered hand exoskeleton

master PC by software Labview and MATLAB, then sent to motors via CAN bus.
Figure 7 shows the photos of the hand exoskeleton.

3.2 Test

To verify the method presented in this paper, an experiment to grasp an apple is set.
The apple is approximately 110 g weight and 70 mm diameter. According to
Eq. (10), we obtain the stable interface force applied on each fingertip, which are
0.92, 0.36, and 0.68 N, respectively. Figures 8, 9, 10 show the step responses of
three fingers.

After adjusting for 6 s, the apple is grasped steadily and all the fingers get into
steady state. The primary parameters of the experiment are listed in Table 1. The
percentages are relative to the desired values. These parameters indicate that our
method works effectively. The steady state errors are 0.08, 0.02, and 0.07 N and
they are all acceptable. But the maximum overshoot of index finger (147.2 %) is
larger than thumb (118.5 %) and middle finger (122.1 %), which means index
finger's damping coefficient is lower. Besides, the peak time of index finger (0.51 s)
is also shorter than thumb (0.98 s) and middle finger (0.83 s), which results from
the higher natural frequency. Similar phenomenon appears again when the external
disturbance is applied on the apple after 8 s.

In the experiment, the three fingers coordinate with each other. As the index
finger shows faster response, the other fingers' responses are correlative with index
finger. This gives us a hint that we could take one finger as the "master finger" and

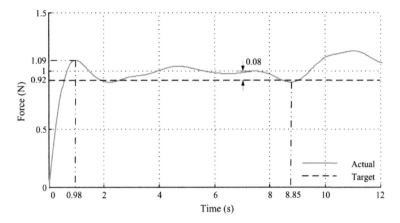

Fig. 8 Contacting force applied on thumb tip

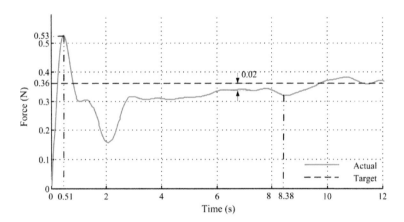

Fig. 9 Contacting force applied on index finger tip

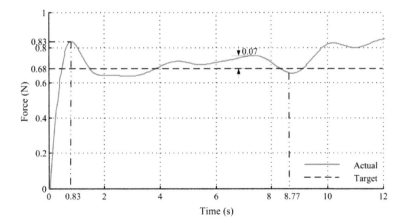

Fig. 10 Contacting force applied on middle finger tip

Table 1 The primary parameters of the experiment

Items	Thumb	Index finger	Middle finger
Steady-state error	0.08 N (8.7 %)	0.02 N (5.6 %)	0.07 N (10.3 %)
Maximum overshoot	1.09 N (118.5 %)	0.53 N (147.2 %)	0.83 N (122.1 %)
Peak time	0.98 s	0.51 s	0.83 s

take the thumb and index finger as "slave fingers." Thus an estimation could be calculated according to the state of master finger and the coordinate dynamics, then better control commands are sent to the slave fingers.

4   Conclusions

To solve the problems that the fingers are correlative in the movement the contacting part, the human finger, is flexible, this article presents a method, which takes not only all the fingers but also their mechanical impedance into a dynamic system. We divide the method into three levels: grasping planning, multi-fingered coordinate force control, and single-finger force control. The desired interface force of each fingertip is given, and an integrated coordinate dynamic system considering the finger impedance is derived. For the single-finger force control, we used the method presented in our early research. A three-fingered hand exoskeleton is developed here.

We set an experiment to grasp an apple to verify the method. The system gets stable after only once overshoot and the steady state errors are acceptable. What is more, the fingers show different mechanical impedance when they come into contacting with object, that leads to difficulty for the steady grasping.

Future work will involve improvements in stability and rapidity. Estimation according to the faster response finger and the coordinate dynamics is a potential feasible trend.

Acknowledgments This work was supported in part by Science Fund for Creative Research Groups of National Natural Science Foundation of China (No.: 51221004).

References

Wege A, Kondak K, Hommel G (2006) Force control strategy for a hand exoskeleton based on sliding mode position control. In: Proceedings of the 2006 IEEE/RSJ international conference on intelligent robots and systems, Beijing, China, 9–15 Oct 2006

Schabowsky CN, Godfrey SB, Holley RJ, Lum PS (2010) Development and pilot testing of HEXORR: hand EXOskeleton rehabilitation robot. J Neuroeng Rehabilitation 7:36

Tzafestas CS (2003) Whole-hand kinesthetic feedback and haptic perception in dextrous virtual manipulation. In: IEEE transactions on system, man, and cybernetics—part A: systems and humans, vol 33, no 1

Fang H, Xie Z, Liu H, Lan T, Xia J (2009) An exoskeleton force feedback master finger distinguishing contact and non-contact mode. In: 2009 IEEE/ASME international conference on advanced intelligent mechatronics, Suntec Convention and Exhibition Center Singapore, 14–17 July 2009

Ho NSK, Tong KY, Hu XL, Fung KL, Wei XJ, Rong W, Susanto EA (2011) An EMG-driven exoskeleton hand robotic training device on chronic stroke subjects. In: 2011 IEEE international conference on rehabilitation robotics rehab week Zurich, ETH Zurich Science City, Switzerland, June 29–July 1 2011

Bi Q, Yang C (2014) Human-machine interaction force control: using a model-referenced adaptive impedance device to control an index finger exoskeleton. J Zhejiang Univ Sci C 15(4):275–283. ISSN 1869-1951

Itoa S, Kawasakia H, Ishigureb Y, Natsumec M, Mouria T, Nishimoto Y (2011) A design of fine motion assist equipment for disabled hand in robotic rehabilitation system. J Franklin Inst 348:79–89

Ueki S, Kawasaki H, Ito S, Nishimoto Y, Abe M, Aoki T, Ishigure Y, Ojika T, Mouri T (2012) Development of a hand-assist robot with multi-degrees-of-freedom for rehabilitation therapy. In: IEEE/ASME Transaction on Mechantronics, vol 17, no 1

Nakagawara S, Kajimoto H, Kawakami N, Tachi S, Kawabuchi I (2005) An encounter-type multi-fingered master hand using circuitous joints. In: Proceedings of the 2005 IEEE international conference on robotics and automation barcelona, Spain, April 2005

Unscented Transform-Based Correlation Between Surrogate and Tumor Motion in Robotic Radiosurgery

Shu-mei Yu, Feng-feng Zhang, Meng Dou, Rong-chuan Sun and Li-ning Sun

Abstract Human respiration hinders real-time accurate radiation in robotic radiosurgery with most of the extracranial tumors. Respiratory tumor motion tracking is crucial for respiration compensation in radiosurgery treatments. This paper presents our work on correlating the surrogate motion with the tumor motion both caused by human respiration. Usual correlation models between surrogates and tumor motions are built by linear or polynomial fitting based on least square method. In those models, sensor data are regarded as accurate. Our work aims to solve the respiration tracking problem by Kalman filters. In this paper, sensor data along with noises are considered in correlation model. Moreover, uncertainty of the model itself is obtained by calculating covariance of the model parameters induced using Unscented Transform.

Keywords Robotic radiosurgery · Correlation model · Frontiers of information technology and electronic engineering

1 Introduction

Recently, image-guided stereotactic radiosurgery has grown as an important method for tumor treatment. Radiosurgery experienced several generations. Gantry-fixed radiosurgery is the earliest, mainly facing the intracranial tumors. As techniques developed, radiosurgery was able to deal with extracranial tumors, such as lung cancer, liver cancer, pancreas cancer, etc. However, most of the extracranial tumors are not static; their motions are mainly caused by human breathing. The disturbance is unfavorable for target's accurate location during radiation treatment. New generations of radiosurgery appeared to solve the problem.

S. Yu (✉) · F. Zhang · M. Dou · R. Sun · L. Sun
Robotics and Microsystems Center & Collaborative Innovation Center of Suzhou Nano Science and Technology, Soochow University, Suzhou 215006, China
e-mail: yushumei@suda.edu.cn

© Zhejiang University Press and Springer Science+Business Media Singapore 2017
C. Yang et al. (eds.), *Wearable Sensors and Robots*, Lecture Notes in Electrical Engineering 399, DOI 10.1007/978-981-10-2404-7_19

The simplest approach is extending the margins of the target, which harms the neighborhood normal tissues unavoidably. Another choice is breath-holding (Sarrut et al. 2005), which brings great pain to weak patients. Gating, as an improvement, allows patients to breathe normally (Berbeco et al. 2005). When the target moves into the predefined band, the beam is emitted. This increases the time for treatment, also the time delay for the beam opening is neglected.

Robotic radiosurgery solves the problem by respiration compensation. Cyberknife (Harold et al. 2007) by Accuray is the most successful technique in the robotic radiosurgery area worldwide. The respiration tracking task is accomplished by Synchrony of Cyberknife, which predicts tumor position through the movement of surrogates placed on patients' skin. As we all know, the internal tumor is invisible, so the tumor position is obtained by X-ray imaging. The movement of internal tumor is continuous while the X-ray must be controlled in a limited frequency. Therefore, surrogates, such as markers on patients' skin (McClelland et al. 2011) or inspiration/expiration volume (Hoisak et al. 2004) are chosen to correlate with the internal tumor's movement. Skin markers are the most often choice.

Traditional method is capturing the relationship between the surrogates and target using linear or polynomial fitting based on least square standard. The surrogates data used to do the fitting can be raw data, which are the 3D coordinates (Ernst et al. 2009), or data treated by principal component analysis (Yvette et al. 2007). Another choice is to build a deformation model for the surrogate and internal movements, with which internal position can be induced through external surrogates by equal proportion with the two courses (Schweikard 2000). In these situations, correlation models were made to obtain the current relationship between external markers and internal tumor. Prediction with the internal target's motion was needed separately. The following studies solve the correlation and prediction in one model. Artificial neural networks were used to estimate the internal position from external markers' position (Seregni et al. 2013). Adapted fuzzy logic models were developed for prediction of the internal tumor position directly for the surrogate data (Esmaili et al. 2013). Support vector regression was used to correlate the external signal to internal motion using data from porcine (Ernst et al. 2012). Kalman filtering (Sharp et al. 2004) and particle filtering (Rahni et al. 2011) were used in the correlation study between the external signals and internal tumor motions.

The method of linear or polynomial fitting, neural networks, and support vector regression treats the external data as accurate. Nevertheless, external mark positions are obtained by sensors that have errors, and the errors should be considered in model building. Kalman filters handle the problem by adding Gaussian noises in the observation model (Sharp et al. 2004). Our method includes the uncertainty of external mark data in the correlation model, heading for a more accurate modeling of the respiration compensation.

2   Methods and Materials

2.1   *Basic Method of Linear Correlation Model*

In respiration compensation for robotic stereotactic radiosurgery, surrogates such as LED markers are used to measure patients' chest and abdominal movements continuously caused by respiration. The internal movements are derived by X-ray image processing at intervals. Also, the internal motion should be predicted so as to plan the robotic arm's trajectory in advance. However, the internal data are too sparse and not reliable. Since the surrogate data can be obtained continuously, the correlation model between the surrogate motion and internal motion is essential for respiration compensation.

Our work aims to validate the accuracy of the new correlation model, so we used the open dataset from Institute for Robotics of Lubeck University (Ernst 2011). The dataset contains 7 male adults' respiratory motion traces during 5–6 min. The surrogate data are from optical tracking results of one infrared LED placed on the patient chest. The internal data are from a vessel bifurcation's movements of human liver measured by GE Vivid1 Dimension, which is a 4D ultrasound equipment. Both the surrogate data and internal data are obtained simultaneous with the frequency from 17.5 to 21.3 Hz.

A typical correlation model between surrogate and internal target is the linear model based on least mean squares (Hoogeman et al. 2009). Traditionally, we minimize the following second-order function to obtain the linear model $(y = ax + b)$:

$$\begin{cases} \chi_1^2(a, b, \bar{y}_1, \ldots, \bar{y}_n) = \sum_{i=1}^{n} (\bar{y}_i - y_i)^2 \\ \text{s.t. } \bar{y} = ax_i + b \\ <\hat{a}, \hat{b}> = \underset{\forall \bar{y}_1, \ldots, \bar{y}_n}{\text{argmin}} \chi_1^2(a, b, \bar{y}_1, \ldots, \bar{y}_n) \end{cases} \tag{1}$$

where x_i $(i = 1, \ldots, n)$ are the principal components of the marker position's measurements; y_i $(i = 1, \ldots, n)$ are the measurements of the tumor's position in x, y, or z direction; $\bar{y}_i$ $(i = 1, \ldots, n)$ are the optimized value corresponding to y_i; a and b are the parameters of the fitted linear model. The method is widely used in current robotic radiosurgery such as Cyberknife. It is considered as conservative but reliable. Equation (1) assumes that the marker position, as x are accurate and errorless. In the precondition, $<\hat{a}, \hat{b}>$ is calculated by minimizing the error between the true y and estimated y. In fact, we all know that error exists in the marker positions introduced by sensor noises, so errors of x should be considered in the correlation model building.

2.2 Improved Method of Linear Correlation Model

In our method, errors of the external marker coordinates (or the principal components of the external marker movements) are considered in the model's derivation. The linear correlation model can be derived by optimization of the following target function:

$$
\begin{cases}
\chi_2^2(a,b,\bar{x}_1,\ldots,\bar{x}_n,\bar{y}_1,\ldots,\bar{y}_n) = \sum_{i=1}^{n}\left((\bar{x}-x_i)^2 + (\bar{y}-y_i)^2\right) \\
\text{s.t. } \bar{y} = a\bar{x}_i + b \\
<\hat{a},\hat{b}> = \underset{\forall \bar{x}_1,\ldots,\bar{x}_n,\bar{y}_1,\ldots,\bar{y}_n}{\text{argmin}} \chi_2^2(a,b,\bar{x}_1,\ldots,\bar{x}_n,\bar{y}_1,\ldots,\bar{y}_n)
\end{cases}
\tag{2}
$$

Replacing $\bar{y}_i$ using Eq. (1) in the above constraints, we obtained

$$
\begin{cases}
\chi_3^2(a,b,\bar{x}_1,\ldots,\bar{x}_n) = \sum_{i=1}^{n}\left((\bar{x}_i-x_i)^2 + (a\bar{x}_i+b-y_i)^2\right) \\
<\hat{a},\hat{b}> = \underset{\forall \bar{x}_1,\ldots,\bar{x}_n}{\text{argmin}} \chi_3^2(a,b,\bar{x}_1,\ldots,\bar{x}_n)
\end{cases}
\tag{3}
$$

Equation (3) is a non-concave second-order function, so the optimal estimation of (a, b) cannot be calculated as analytic solutions. We solved the function in a numerical approach.

For Eq. (3), given that $a = \check{a}$, then $\chi_3^2(a = \check{a},b,\bar{x}_1,\ldots,\bar{x}_n)$ would be a concave second-order function of variable b and $\bar{x}_i$ $(i = 1,\ldots,n)$:

$$
\chi_4^2(b,\bar{x}_1,\ldots,\bar{x}_n) = \sum_{i=1}^{n}\left((\bar{x}_i-x_i)^2 + \check{a}\bar{x}_i+b-y_i)^2\right)
\tag{4}
$$

Minimum values of $\check{b},\bar{x}_1,\ldots,\bar{x}_n)$ of the object function χ_4^2 $(b,\bar{x}_1,\ldots,\bar{x}_n)$ can be obtained by least square method.

Set $X = [b,x_1,\ldots,x_n]^T$, Eq. (4) can be rewritten in the form of matrix as follows:

$$
\chi_4^2(X) = \chi_4^2(b,\bar{x}_1,\ldots,\bar{x}_n) = X^T A X + X^T B + C
\tag{5}
$$

where

$$
A = \begin{bmatrix}
1+a^2 & 0 & \cdots & \cdots & 0 & a \\
0 & \ddots & & & \vdots & \vdots \\
\vdots & & \ddots & & \vdots & \vdots \\
\vdots & & & \ddots & 0 & \vdots \\
0 & \cdots & \cdots & 0 & 1+a^2 & a \\
a & \cdots & \cdots & \cdots & a & n
\end{bmatrix}
\tag{6}
$$

$$B = \begin{bmatrix} -2x_1 - 2ay_1 \\ \vdots \\ -2x_n - 2ay_n \\ -2(y_1 + \cdots + y_n) \end{bmatrix} \tag{7}$$

$$C = \sum_{i=1}^{n} x_i^2 + y_i^2 \tag{8}$$

The minimum value of $\chi_4^2(X)$, which is $\check{X} = [\check{b}, \check{x}_1, \ldots, \check{x}_n]^{\mathrm{T}}$ can be calculated by least square method as

$$\check{X} = -A^{-1}.B/2 \tag{9}$$

The optimum estimation of $\hat{a}$ and $\hat{b}$ by numerical solution follows the steps as

(1) Obtain $\tilde{a}$ as the approximate optimal estimation of a through Eq. (1);
(2) Set a region $[a_1 \quad a_2]$ as the optimal estimation area in the neighborhood of $\tilde{a}$;
(3) Partition the region $[a_1 \quad a_2]$ equally, for each probable value of a in the region, use Eq. (9) to obtain the optimal estimation of b, replacing it into Eq. (5) to calculate the objective function $\chi_4^2(X)$. Store the values of $\chi_4^2(X)$ in a table $\mathcal{L}$;
(4) Find the minimum value in $\mathcal{L}$, the corresponding a and b are the optimal estimation of $(\hat{a}, \hat{b})$ in Eq. (3).

2.3 Unscented Transform-Based Correlation Model Building with Probability

2.3.1 The Unscented Transform-Based Modeling

To predict and track a tumor, researchers have proposed many methods based on state estimation, such as extended Kalman filter, unscented Kalman filter, and particle filter. To propagate the information among the states, these methods need to use the correlation model or its variants as their measurement model, and then an error propagation formula performed to calculate the state covariance. These classes of methods only consider the measurements as stochastic variables. While at the same time, the correlation model in the traditional tumor tracking methods is considered as an accurate and determinate model, and it will not change during the tracking progress. However, because the correlation model is achieved by fitting measurement data which is polluted by noise, the correlation model should not be considered as determinate. The correlation model's stochastic properties need to be considered in the progress of state estimation.

In this paper, the stochastic properties of the correlation model are analyzed based on Eq. (10). In the domain of state estimation, the stochastic properties (the most common representatives are variable mean and its covariance) are calculated by error propagation formula. If the stochastic properties of the independent variable are known, then the stochastic properties of the dependent variable can be calculated analytically by the formula. However, in the area of radiation therapy, the correlation model is achieved by fitting methods, and does not have an analytical formula on the measurements. Then the stochastic properties of the model cannot be calculated by error propagation formula. To solve this problem, this paper proposes to use the Unscented Transformation method to achieve the model's stochastic properties. The Unscented Transformation (UT) is a method that calculates the statistics of a random variable which undergoes a nonlinear transformation (Simon et al. 1997). More importantly, it can deal with a transformation without analytical formulas. For example, the optimal estimates of a and b cannot be described with an analytical formula respect to the variables X.

Suppose the correlation model fitted by Eq. (3) is described as

$$Y = \begin{bmatrix} a \\ b \end{bmatrix} = f(S) \tag{10}$$

Here, S is an augmented vector composed by all the measurements:

$$S = [x_1, \ldots, x_n, y_1, \ldots, y_n]^{\mathrm{T}} \tag{11}$$

Its covariance matrix is denoted as P.

First, a set of sigma points consisting of 2L + 1 vectors and their associated weights W are calculated (Here, L is the dimension of the vector variable S):

$$\chi_0 = S \tag{12}$$

$$\chi_i = S + \left(\sqrt{(L+\lambda)P} \right)_i \quad i = 1, \ldots, L \tag{13}$$

$$\chi_i = S - \left(\sqrt{(L+\lambda)P} \right)_{i-L} \quad i = L+1, \ldots, 2L \tag{14}$$

$$W_0^{(m)} = \lambda/(L+\lambda) \tag{15}$$

$$W_0^{(c)} = \frac{\lambda}{L+\lambda} + (1 - \alpha^2 + \beta) \tag{16}$$

$$W_i^{(m)} = W_i^{(c)} = \frac{1}{2L+2\lambda} \quad i = 1, \ldots, 2L \tag{17}$$

where $\lambda = \alpha^2 (L + \kappa) - L$ is a scaling parameter. α is a scaling factor, and β is a function of the kurtosis. κ is a secondary scaling parameter which is usually set to $0.\left(\sqrt{(L+\lambda)P}\right)_i$ is the ith row of the matrix square root of $(L+\lambda)P$.

After the sigma points have been calculated, they are propagated through the nonanalytical transformation described by Eq. (10):

$$\mathcal{Y}_i = f(\chi_i) \quad i = 0, \ldots, 2L \tag{18}$$

Lastly, the optimal estimate of Y and its covariance are approximated using a weighted sample mean and covariance of the sigma points:

$$\hat{Y} = \sum_{i=0}^{2L} W_i^{(m)} \mathcal{Y}_i \tag{19}$$

$$P_y = \sum_{i=0}^{2L} W_i^{(c)} \left(\mathcal{Y}_i - \hat{Y}\right)\left(\mathcal{Y}_i - \hat{Y}\right)^{\mathrm{T}} \tag{20}$$

2.3.2 The Compatibility Test

.The Chi-squared test method is used to validate the effectiveness of the correlation model. Suppose the correlation model is described as $y = h(Y, x) = ax + b \ (Y = \begin{bmatrix} a & b \end{bmatrix}^{\mathrm{T}})$. Then for each set of measurements (x_i, y_i), the Mahalanobis distance D can be calculated as follows:

$$v = y_i - h(a, b, x_i) \tag{21}$$

$$C = \frac{\partial h}{\partial Y} Q \left(\frac{\partial h}{\partial Y}\right)^{\mathrm{T}} + \frac{\partial h}{\partial x_i} q_x \left(\frac{\partial h}{\partial x_i}\right)^{\mathrm{T}} + q_y \tag{22}$$

$$D = v^{\mathrm{T}} C^{-1} v \tag{23}$$

Here, Q is the covariance matrix calculated by Eq. (19). q_x and q_y are covariance of x_i and y_i, respectively.

The Chi-squared test then uses the Mahalanobis distance to judge whether the measurements (x_i, y_i) is compatible with the correlation model:

$$D < \chi^2_{d,1-\alpha} \tag{24}$$

where $\chi^2_{d,1-\alpha}$ denotes a Chi-square inverse cumulative distribution function. d is the dimension of y. $1 - \alpha$ is the confidence degree, which is usually set as 95 %.

3   Experiments and Results

As mentioned in Sect. 2, the dataset we used to validate the Unscented Transform-based correlation model is by Institute for Robotics, Lubeck University, Germany.

The first subject's data among the seven subjects were chosen for the experiments. The data we used contains 7565 arrays of data collected in 360 s. Before the validation, we used PCA (principal component analysis) to calculate the principal vector of the external surrogate's motion. Inhalation and exhalation stage partition was made to the principal component of surrogate's respiration motion, with the results shown by Fig. 1.

Based on the results of respiratory phase partition, we used Eq. (1) and (3) in Sect. 2 separately to accomplish the correlation model fitting. The fitting errors of the first 250th arrays of dataset are shown in Fig. 2. We can see from the figure that the improved fitting method owns the results approaching that of polynomial fitting. At the peaks of the errors, the improved method reduced the fitting error slightly. In the future work, we will consider the covariance's influence on the objective functions. The dataset collected by sensors will be weighted using covariance in order to obtain better results.

Figure 3 shows the examination results of the Unscented Transform-based correlation model's effectiveness. As we can see from Fig. 3, most of the data are probabilistically consistent with the correlation model, and merely several arrays of data display as inconsistent. These are due to the error of model building at the boundaries of inhalation and exhalation phases.

As a comparison, Fig. 4 shows the results of chi-square test on the correlation model that is not built with model error consideration. We can see that in Fig. 4 there are much more data that cannot pass the consistency test than those in Fig. 3. This is due to the error of the correlation model itself, the results of the estimated

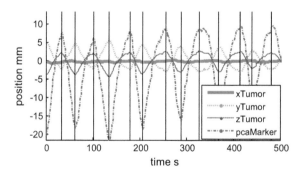

Fig. 1 Results of the dataset partition

Fig. 2 Improved fitting method compared to polynomial fitting

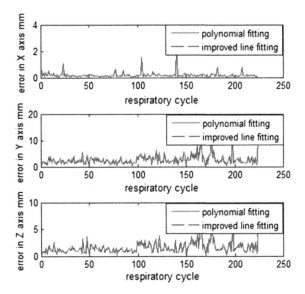

Fig. 3 Chi-square test results of UT-based correlation model

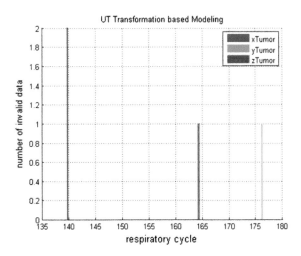

data are overconfident. In the following prediction work for the tumor tracking, failure of data consistency test will lead to singularity of corresponding EKF/UKF's states, or singularity of the corresponding particle filters' samples set. The singularity mentioned above will reduce the accuracy of the tumor tracking.

Fig. 4 Chi-square test results
of the model lacking error
consideration

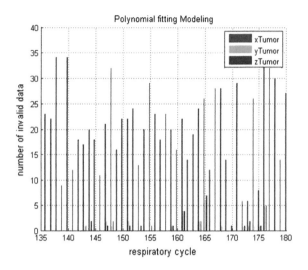

4 Conclusions and Future Work

We proposed a probability-based frame of tumor tracking method in robotic radiosurgery in this paper. Compared to the traditional correlation model building, we
considered the error of the sensors that measure the external surrogate's motion.
The improved correlation model building used a numerical solution to obtain the
optimal estimation of the model parameters. In previous research, covariance of the
correlation model itself has not been considered. Because the model parameters in
the basic methods are derived by least mean squares, the uncertainty of the model's
error cannot be propagated using analytical form. For probabilistic accuracy, we
used Unscented Transform to calculate the covariance of the model parameters.
Experiments with open dataset validate the efficiency of our method.

The work presented in this paper is a preparation and foundation part for the
whole probability-based respiration tracking frame. In future, the improved correlation model building method and covariance of the model parameter will be used
in Kalman filtering process. With sensor error considered, the accuracy of tumor
tracking algorithm would be increased.

Acknowledgments The authors would like to thank Dr. Floris Ernst from Institute for Robotics
of Lubeck University. Dr. Ernst has been on the research of human respiratory motion signal for
nearly ten years. The open dataset we used in this paper are found from https://signals.rob.uni-
luebeck.de/index.php/Main_Page, on which Dr. Ernst and his colleagues share their data for
scientific research.

Project supported by the National Natural Science Foundation of China (No. 61305108),
Natural Science Foundation of Jiangsu Province (No. BK20130323), and High School Research
foundation of Jiangsu Province (No. 13KJB520033).

References

Berbeco RI, Nishioka S, Shirato H et al (2005) Residual motion of lung tumours in gated radiotherapy with external respiratory surrogates. Phys Med Biol 50:3655–3667

Ernst F (2011) Compensating for quasi-periodic motion in robotic radiosurgery. Springer. doi:10. 1007/978-1-4614-1912-9

Ernst F, Martens V, Schlichting S et al (2009) Correlating chest surface motion to motion of the liver using epsilon-SVR–a porcine study. Med Image Comput Comput Assist Interv 12:356–364

Ernst F, Bruder R, Schlaefer A, Schweikard A (2012) Correlation between external and internal respiratory motion: a validation study. Int J Comput Assist Radiol Surg 7(3):483–492

Esmaili TA, Riboldi M, Imani Fooladi AA, Modarres Mosalla SM, Baroni G (2013) An adaptive fuzzy prediction model for real time tumor tracking in radiotherapy via external surrogates. J Appl Clin Med Phys 14(1):102–114

Hoisak JD, Sixel KE, Tirona R, Cheung PC, Pignol JP (2004) Correlation of lung tumor motion with external surrogate indicators of respiration. Int J Radiat Oncol Biol Phys 60(4):1298–1306

Hoogeman M, Prévost JB, Nuyttens J, Pöll J, Levendag P, Heijmen B (2009) Clinical accuracy of the respiratory tumor tracking system of the cyberknife: assessment by analysis of log files. Int J Radiat Oncol Biol Phys 74(1):297–303

McClelland JR, Hughes S, Modat M, Qureshi A, Ahmad S, Landau DB, Ourselin S, Hawkes DJ (2011) Inter-fraction variations in respiratory motion models. Phys Med Biol 56:251–272

Rahni AA, Lewis E, Guy MJ, Goswami B, Wells K (2011) A particle filter approach to respiratory motion estimation in nuclear medicine imaging. IEEE Trans Nucl Sci 58(5):2276–2285

Sarrut D, Boldea V, Ayadi M et al (2005) Nonrigid registration method to assess reproducibility of breath-holding with ABC in lung cancer. Int J Radiat Oncol Biol Phys 61:594–607

Schweikard A (2000) Robotic motion compensation for respiratory movement during radiosurgery. Comput Aided Surg Off J Int Soc Comput Aided Surg 5(4):263–277

Seregni M, Pella A, Riboldi M, Orecchia R, Cerveri P, Baroni G (2013) Real-time tumor tracking with an artificial neural networks-based method: a feasibility study. Phys Med 29(1):48–59

Sharp GC, Jiang SB (2004) Predict ion of respiratory tumour motion for real time image guided radiotherapy. Phys Med Biol 49:425–440

Simon JJ, Jeffrey KU (1997) A new extension of the kalman filter to nonlinear systems. In: Proceedings of SPIE—the international society for optical engineering, pp 182–193

Urschel HC, John JK, James DL, Lech P, Robert DT, Raymond AS et al (2007) Treating tumors that move with respiration. Springer

Yvette S, Ross IB, Seiko N, Hiroki S, Ben H (2007) Accuracy of tumor motion compensation algorithm from a robotic respiratory tracking system: a simulation study. Med Phys 34 (7):2774–2784

A Pulmonary Rehabilitation Training Robot for Chronic Obstructive Pulmonary Disease Patient

Zhi-hua Zhu, Tao Liu, Bo Cong and Fengping Liu

Abstract As a kind of progressive obstructive pulmonary disease, chronic obstructive pulmonary disease (COPD) is seriously harmful to people's health, especially for the elder. People with COPD usually have symptoms of respiratory discomfort, leg tiredness, and even dyspnea. Traditionally, pulmonary rehabilitation is considered as an important management and treatment for COPD and is widely applied in clinics. The pulmonary rehabilitation training robot proposed in this paper is designed based on the pulmonary rehabilitation principle. Additionally, this respiratory robot can effectively assist COPD patients with respiratory rehabilitation training exercise, relief the burden of respiratory muscle and help them adjust the respiratory strategy. In the experimental study, we used the wireless EMG measurement technique to study the activity of respiratory muscle during training exercise and found that the activity of the diaphragm muscle decreases under training state comparing with that the nonassisted state. In addition, with the increase of the motor's speed, the diaphragm muscle becomes less active. We also studied the motion changes of the subject's chest and diaphragm muscle during training through motion capture system.

Keywords COPD · Pulmonary rehabilitation · Robot · Respiratory muscle · EMG

1 Introduction

COPD has traditionally been considered as an inexorably multi-factorial progressive chronic lung disease, characterized by progressive airflow obstruction that is irreversible, inflammation in the airways, and several extra-pulmonary consequences

Z. Zhu · T. Liu (✉) · B. Cong
The State Key Laboratory of Fluid Power Transmission and Control, College of Mechanical Engineering, Zhejiang University, Hangzhou 310027, China
e-mail: liutao@zju.edu.cn

F. Liu
College of Medicine, Zhejiang University, Hangzhou 310027, China

© Zhejiang University Press and Springer Science+Business Media Singapore 2017   251
C. Yang et al. (eds.), *Wearable Sensors and Robots*, Lecture Notes in Electrical Engineering 399, DOI 10.1007/978-981-10-2404-7_20

(such as accelerated muscle protein degradation, skeletal muscle weakness, and osteoporosis) or comorbidities (such as ischemic heart disease, diabetes, and lung cancer) (Lopez-Campos et al. 2013; Cielen et al. 2014; Troosters et al. 2005). Symptoms of COPD include breathlessness, coughing, fatigue, and frequent chest infection. According to WHO, COPD has become the third leading cause of death, killing over 3 million people throughout the world in 2012. Moreover, COPD mortality rates are likely underestimated because of underdiagnosis and omission when comorbidities are commonly recorded as causes of death (Sin et al. 2006). Cigarette smoking, air pollution, occupational dusts and chemicals, genetic factors, and recurrent bronchopulmonary infections are the main causes of COPD (William 2003).

Currently, COPD, with progressive disease severity, is an incurable condition that often leads to frequent hospital admissions and readmissions and definitely associated with significant economic costs. But smoking cessation, effective inhaled therapy, noninvasive ventilation, surgical treatment, and pulmonary rehabilitation are encouraged for patients to treat symptoms and delay its progression (McCarthy et al. 2015; Scullion and Holmes 2010). Pulmonary rehabilitation (also known as respiratory rehabilitation) training is considered as an important management and treatment for COPD in clinical as it can increase endurance time, decrease respiratory effort and much more arm effort without affecting chest wall dynamic hyperinflation or configuration, which greatly improves health-related life quality and exercise capacity (Romagnoli et al. 2013). However, owing to the uncomfortable sensations of COPD patients such as dyspnea, leg tiredness and respiratory discomfort, patients with moderate or severe COPD always have trouble performing training exercise and any other normal daily tasks. The development of dyspnea is related to dynamic hyperinflation and mechanical restriction, such as expanding tidal volume and minimal inspiratory reserve volume (Marc 2009). Accordingly, though best evidence and all current international guidelines ratify the central role of pulmonary rehabilitation training exercise in the treatment of people with COPD (GOLD 2014, The Global Initiative for Chronic Obstructive Lung Disease), clinical studies have shown that only 1 % (Ambrosino et al. 2008) of COPD patients can keep on training and most patients complain of dyspnea during rehabilitation exercise (such as six-minute walk, upper-limb activities and lower-limb activities) and refuse to keep training.

To solve this problem and find an innovative and convenient way for pulmonary rehabilitation training, we develop a respiratory training robot for patients with COPD. In the literatures, we have not found any applications of the rehabilitation robots for respiratory rehabilitation. Nearly all of the respiratory rehabilitation programs rely on patients to carry out unsupported exercises at regular intervals under the guidance of a physician or nursing staff (McCarthy et al. 2015; Marc 2009). However, patients (especially for the severe COPD patients) are easy to give up training because of the exercise limitations resulting from dyspnea, ventilatory limitation, and skeletal muscle dysfunction (Rochester 2003). Additionally, these rehabilitation programs require the nursing staff to accompany with the patient along the training process in order to adjust the training strategy according to the physical and psychological conditions of patients, which needs a lot of manpower and can hardly achieve the best rehabilitation result.

This paper concentrates on a novel design of respiratory rehabilitation training robot which can assist COPD patients with training exercise. It consists of three related sections. In the first part, we introduce the basic design principles and details of the robot. Through reducing the weight of the upper body, the robot can relief the rehabilitation training burdens of patients during exercise. In addition, through alleviating pain and dyspnea and improving the performance of respiratory muscles, the robot can effectively promote long-term adherence to the rehabilitation exercise. Second, some experimental studies are carried out to verify the rehabilitation training effects of the robot. Through utilizing wireless surface EMG measurement technique, we found that in the static state, the respiratory muscle is much more active than that in the training state, which indicates the robot relieves the load of the respiratory muscle. We also study the motion details of the chest and respiratory muscles by the motion capture system and analyze the difference between the static state and the training state, which provide an apparent evidence of the improvement acquired from the robot. In the third part, we summarize and discuss the experimental results and conclude the advantages of the robot. The improvement of the robot in the future design is also discussed in the last section.

2 Design Details of the Respiratory Rehabilitation Robot

2.1 Mechanical Design

Most COPD patients are over the age of 45, and they always tolerate the discomfort of dyspnea, limb weakness, and depression. Therefore, the design of the respiratory training robot for patients with COPD should meet the following requirements with consideration of both user aspects and technical functions:

- Effectively assist patients in pulmonary rehabilitation training, improve respiratory muscle endurance, and alleviate the respiratory muscle burden.
- Be able to work continuously and be controlled according to the training requirements.
- Be nonintrusive, have esthetic appearance, and avoid adding discomfort and disturbance to the patient.
- Be able to make any appropriate or reasonable adjustments timely according to patient's condition during training.
- Be safe and reliable for its use in the ward or in the bed.

To satisfy the above design requirements, we develop a respiratory rehabilitation training robot for patients with COPD. As described in Fig. 1, two curved support plates made of rubber are placed under the armpit of patient to support the patient's

Fig. 1 Mechanical design of
the respiratory rehabilitation
training robot

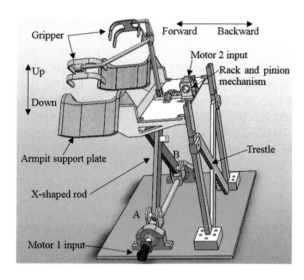

Fig. 1 Mechanical design of the respiratory rehabilitation training robot

upper body weight while respiratory training, which helps the patient to lift up and put down his chest in the vertical direction. In the bottom of the robot, the screw has two screw threads with different direction of rotation in each side. The X-shaped support rod joint with the screw through a threaded connection. The X-shaped rod under the curved plates supports the patient and when side A and side B move toward each other with the same speed driven by the motor, the curved plates go up and the patient will be raised, and when they move far from each other, the patient goes down. This motion of robot imitates the movement of the chest in the vertical direction during respiratory process, which will alleviate the burden of respiratory muscles by relieving the weight of upper body when patients inhale and exhale. Above the armpit support plates, two arcuate grippers are used to grasp patient's arm. Driven by the motor, the two grippers which connect with a rack and pinion mechanism will move forward and backward with the arms to expand the volume of the chest cavity. The combination of the above robot movements constitutes the two-dimensional movement of the chest. The vertical fixed trestle behind the robot contains two parallel rails, which ensure the robot move in the vertical direction and enhance the stability of the system. The training robot assists the patient with the motion of the chest during respiratory training, which will improve the performance of respiratory muscles.

The safety of the robot is the most important thing in the clinical application. The robot's range of motion in the vertical direction is about 0–1.5 cm, which is safe enough to the application. The speed and the direction of the motor can be controlled according to the requirements of the patient. Additionally, though the working voltage of the motor is about 110 V, the motor and the external circuits can be isolated from the patient through some insulated protective device. The robot is also can be covered by some soft materials to ensure the comfort of the patient, which will be improved in the future design.

2.2 Control Strategy and Workflow

Another innovation of this robot is that the control strategy utilizes respiratory information from patients to feedback control the motor's output parameters, such as torque, speed, and sense of rotation, which aims to achieve good rehabilitation result. The wearable sensors (such as EMG sensors, tactile sensors, force sensors, CO_2 sensors (Zhu et al. 2015a), and inertial sensors) on the COPD patient can measure the respiratory information inside the body, including chest motions, respiratory rate, and activity of respiratory muscles. The degree of dyspnea detected by the wearable sensors can be measured as the feedback signal to the motor control.

When the wearable sensors are put on the patient's body and all devices are connected correctly, the respiratory information are sent to the CPU and are processed by the software. Then the CPU sends control commands according to the degree of dyspnea of the patient to the AC servo drive to control the movements of the motor. The armpits of the patient are put on the armpit support plates and when the motor starts, the X-shaped rod moves with the motor to raise up and put down the chest of the patient according to the respiratory rhythm and the degree of dyspnea. The movement of the motor can be controlled by the AC servo drive, which is connected with the PC. Meanwhile, the gripper and the rack and pinion mechanism driven by the motor 2 grasp the arms of the patient and move forward and backward to expand the volume of the chest cavity. In the process of respiratory rehabilitation training, the chest is lifted to relief the respiratory load and patients will feel better to do breathing exercise and chest exercise. As a result, the robot can effectively promote long-term adherence to the rehabilitation exercise.

3 Experimental Verification

Experiments were carried out to demonstrate the robot's effects on respiratory muscles. We utilized the wireless surface EMG technique (KinePro 3.2) to measure the activities of respiratory muscles, which could verify the effectiveness of rehabilitation training of the robot. Additionally, we also use the motion capture system to study the movement of the chest and respiratory muscles under the static state and training state, respectively.

All the subjects have been fully informed about the experiments and all the experiments are agreed by the ethic committee of medicine of Zhejiang University.

3.1   Experiments on the Robot Utilizing Wireless Surface EMG Technique

The experimental setup is shown in Fig. 2. Two wireless disposable surface electrodes are placed over the diaphragm muscle, which is the most important respiratory muscle characterizing the symptom of the severity of COPD, for instance sarcomere length, mitochondrial density, and enzyme activity which always occur within diaphragm muscle fibers of COPD patients (Ottenheijm et al. 2005). Electrode placement for this study is determined based on the anatomy of the diaphragm muscle (Duiverman et al. 2004). The electrode measures the EMG of the subject and sends the signal through the antenna to the Kine Measuring Units (KMU), which transmits the data to the computer for data processing. The sampling frequency is 1600 Hz and the sampling time is 20 s in every measurement. All the raw EMG datum is processed on the PC and can be presented on the screen.

Through controlling the speed of the motor, we can change the speed of the movement of the robot and assist the subject in adjusting the respiratory strategy. In the study, the speed of the motor is set to 600, 1200, and 1800 rpm, respectively. Assisted with the reducer, the robot takes 4, 2, and 1 s moving up and down in a cycle, correspondingly. With different training speeds, the subject has different respiratory rates. We measure the EMG of the subject under training state with these three different respiratory rates, including slow speed (motor speed is 600 rpm), moderate speed (motor speed is 1200 rpm), and fast speed (motor speed is 1800 rpm). In addition, we also measure the EMG of the subject with the same respiratory rate under nonassisted state. Through comparing the EMG data of the

Fig. 2 The experimental setup for the robot with wireless surface EMG measurement

subject under the two different kinds of states, we can find the relationship between the training speed and the activity of the respiratory muscles.

3.2 Results and Discussions

In our previous study, we have verified that under static state the respiratory muscle is more active than that of training state. As described in Fig. 3, the EMG of the two diaphragm muscles (the left diaphragm and the right diaphragm) of the subject with slow respiratory rate under nonassisted state on the left is more active than that under training state on the right. This analysis result has proved that the respiratory rehabilitation training robot is feasible.

The EMG of the subject of the other two kinds of respiratory rates under the static state and the raining state are described in Fig. 4. We can find that with the increase of the motor's speed, the diaphragm becomes less active. Table 1 shows the mean EMG of the diaphragm under different states, which quantitatively proves this result. Under the training state, the mean EMG of the right diaphragm is 5.95×10^{-6} V, 4.61×10^{-6} V, and 4.46×10^{-6} V with the motor speed of 600, 1200, and 1800 rpm. The experiment result indicates that the activity of the respiratory muscle will decline with the increase speed of the motor. This means that, through controlling the speed of the motor, we can assist the patient in adjusting the respiratory strategy, such as relieving the burden of the respiratory muscle and inducing patients to abandon shrink lip breathing or chest breathing to perform abdominal breathing.

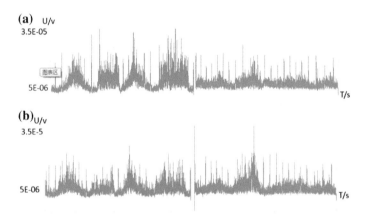

Fig. 3 The diaphragm under nonassisted and training state (motor speed is 600 rpm). **a** The left diaphragm under nonassisted (*left*) and training state (*right*), **b** The right diaphragm under nonassisted (*left*) and training state (*right*)

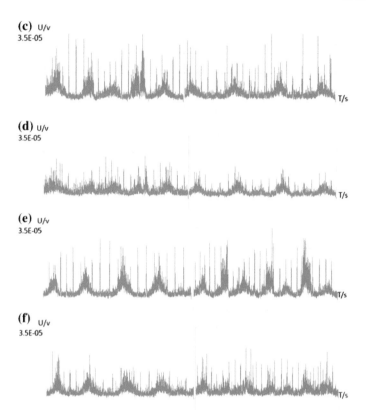

Fig. 4 The diaphragm under nonassisted state and training state [motor speed is 1200 rpm in (**c**) and (**d**), and 1800 rpm in (**e**) and (**f**)]. **c** The left diaphragm of moderate respiratory rate under nonassisted state (*left*) and training state (*right*). **d** The right diaphragm of moderate respiratory rate under nonassisted state (*left*) and training state (*right*). **e** The left diaphragm of fast respiratory rate under nonassisted state (*left*) and training state (*right*). **f** The right diaphragm of fast respiratory rate under nonassisted state (*left*) and training state (*right*)

	Motor speed (rpm)	Left ($\times 10^{-6}$ V)	Right ($\times 10^{-6}$ V)
Table 1 The mean EMG of the diaphragm muscle under training state	600	5.96	5.95
	1200	5.34	4.61
	1800	5.20	4.46

3.3 Experiments on the Robot Utilizing the Motion Capture System

Utilizing the Vicon motion capture system, we want to study the motion of the chest and the respiratory muscle when the subject trains on the robot. We have six cameras symmetrically set in the laboratory of different positions, and the subject sit

Fig. 5 Experimental setup with motion capture system. **a** The subject have five markers pasted on the chest. **b** Markers and cameras displayed on the screen

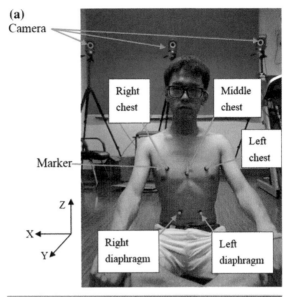

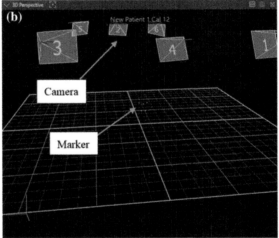

on the ground with five makers pasted on the chest, as shown in Fig. 5a. In Fig. 5b, five markers are shown on the screen and all the markers have their own coordinates in the room. These cameras are connected with the PC, and the motion data (co-ordinates of markers) captured by the camera will be sent to the PC in real time and be saved by the software named Vicon Nexus 2.1.

The main motion of the chest and diaphragm is along with the Y-axis and Z-axis, which changes cyclically along with breathing. We compared the nonassisted state and the training state of the chest motion of the subject, as an example shown in Fig. 6. The specific amplitude of the chest and diaphragm is compared in Table 2.

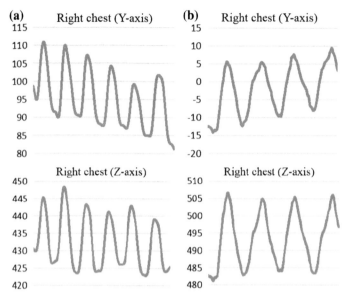

Fig. 6 Examples of the motion curves of the markers (**a** is under nonassisted state, and **b** is under training state)

Table 2 The amplitude of the chest and diaphragm in Y-axis

Y-axis	Nonassisted state (mm)		Training state (mm)		Amplitude (mm)	
	Maximal	Minimal	Maximal	Minimal	SS	TS
Right chest	110.965	89.9404	5.6937	−12.357	21.0246	18.0507
	110.007	90.2561	5.4322	−9.6608	19.7509	15.093
	107.25	87.5611	7.6841	−8.0629	19.6889	15.747
Middle chest	98.3136	82.8881	6.1703	−9.0455	15.4255	15.2158
	96.0132	82.9197	5.6363	−6.2265	13.0935	11.8628
	95.1681	79.3735	6.8647	−5.2245	15.7946	12.0892
Left diaphragm	62.1075	35.6264	13.3723	−2.0420	26.4811	15.4143
	59.1663	27.986	13.0486	−1.1763	31.1803	14.2249
	56.667	22.9737	13.3396	−1.9736	33.6933	15.3132

4   Conclusions

In our previous study, we have verified the effectiveness and feasibility of the training robot by analyzing the different reaction of the diaphragm in four different degrees of dyspnea and comparing the respiratory muscles of subjects in static state and in training state (Zhu et al. 2015b). In this paper, our experiments are conducted to study the influences and changes brought by the respiratory rehabilitation robot to the respiratory muscles. Through comparing the different reactions of the diaphragm under different training speeds, we conclude that the activity of the respiratory decline with the increase of the motor's speed. Though now it is hard to determine the accurate mathematical relationship between the activity of the respiratory muscle and the motor's speed, we are convinced that we can improve the control algorithm of the motor to achieve real-time control and feedback control according to the relationship in the future. With the help of wearable biosensors, such as the EMG sensor, inertial sensor, and CO_2 sensor, it is easy to acquire the respiratory information of the patient to guide the motion of the robot.

In the future design, we need to improve the control strategy to achieve more accurate feedback control. The mathematical relationship between the activity of the respiratory muscle and the motor's speed and output torque need to be found to give guidance to the motor control. In addition, the appearance of the robot needs to be improved, and combined with the conception of the soft robot, the respiratory robot needs to become more safe and adorable.

Acknowledgments Project supported by the National Natural Science Foundation of China (NSFC) under grants 61428304, Zhejiang Provincial Natural Science Foundation of China under Grant No. LR15E050002, and the China State Key Laboratory of Robotics and System (HIT) SKLRS-2014-ZD-04.

References

Ambrosino N, Casaburi R, Ford G et al (2008) Developing concepts in the pulmonary rehabilitation of COPD. J Respir Med 102(1):17–26. doi:10.1016/S0954-6111(08)70002-7

Cielen N, Maes K, Gayan-Ramirez G (2014) Musculoskeletal disorders in chronic obstructive pulmonary disease. BioMed Res Int, 2014:1–17. [http://dx.doi.org/10.1155/2014/965764]

Duiverman ML, Eykern LAV, Vennik PW et al (2004) Reproducibility and responsiveness of a noninvasive EMG technique of the respiratory muscles in COPD patients and in healthy subjects. J Appl Physiol 96(5):1723–1729. doi:10.1152/japplphysiol.00914.2003

Lopez-Campos JL, Calero C, Quintana-Gallego E (2013) Symptom variability in COPD: a narrative review. Int J COPD 8:231–238. doi:10.2147/COPD.S42866

Marc D (2009) Response of the respiratory muscles to rehabilitation in COPD. J Appl Physiol 107 (3):971–976. doi:10.1152/japplph-ysiol.91459.2008

McCarthy B, Casey D, Devane D et al (2015) Pulmonary rehabilitation for chronic obstructive pulmonary disease (Review). Cochrane Database Syst Rev 2(2):CD003793. [10.1002/14651858.CD003793.pub3]

Ottenheijm CAC, Heunks LMA, Sieck GC et al (2005) Diaphragm dysfunction in chronic obstructive pulmonary disease. Am J Respir Crit Care Med 172(2):200–205. doi:10.1164/rccm.200502-26OC2OC

Rochester CL (2003) Exercise training in chronic obstructive pulmonary disease. J Rehabil Res Dev 40(5):59–80

Romagnoli I, Scano G, Binazzi B et al (2013) Effects of unsupported arm training on arm exercise-related perception in COPD patients. J Respir Physiol Neurobiol 186(1):95–102. doi:10.1016/j.resp.2013.01.005

Scullion J, Holmes S (2010) Chronic obstructive pulmonary disease (COPD): updated guidelines. Prim Health Care 20(8):33

Sin DD, Anthonisen NR, Soriano JB et al (2006) Mortality in COPD: role of comorbidities. Eur Respir J 28(6):1245–1257. doi:10.1183/09031936.00133805

Troosters T, Casaburi R, Gosselink R et al (2005) Pulmonary rehabilitation in chronic obstructive pulmonary disease. Am J Respir Crit Care Med 172(1):19–38. doi:10.1164/rccm.200408-1109SO

William M (2003) COPD: causes and pathology. Medicine 31(12):71–75. doi:10.1383/medc.31.12.71.27170

Zhu ZH, Liu T, Li GY et al (2015a) Wearable sensor systems for infants. Sensors 15(2):3721–3749. doi:10.3390/s150203721

Zhu ZH, Cong B, Liu FP et al (2015b). Design of respiratory training robot in rehabilitation of chronic obstructive pulmonary disease. In: 2015 IEEE international conference on advanced intelligent mechatronics (AIM), p 866–870

Research and Development for Upper Limb Amputee Training System Based on EEG and VR

Jian Li, Hui-qun Fu, Xiu-feng Zhang, Feng-ling Ma, Teng-yu Zhang, Guo-xin Pan and Jing Tao

Abstract It is necessary to have electromyogram (EMG) training before installing an EMG prosthetic for upper extremity amputees. Aiming to improve the training effect, in this paper a training system based on EMG and virtual reality (VR) is designed. The hardware and software of the training system were designed. And based on the VR technology and EMG technology, in this paper an interesting game in the software has been developed. Meanwhile some experiments were done in the hospital. After actual upper limb amputee experiment, the feasibility and rationality of the system is proved. This paper develops an upper limb amputees training system which has a lively and interesting game, can actively mobilize the amputees' subjective training initiative, which has good effects and positive meanings for clinical prosthetics installation and usage.

Keywords EMG · Virtual reality · Upper limb · Amputation · Rehabilitation training

1 Introduction

For the upper limb amputees, myoelectric prosthetic hand is very useful. It can compensate some or most of their daily hand functions, and improve their self-care ability and quality of life greatly. In order to control the myoelectric prosthetic hand

J. Li (✉) · X. Zhang · F. Ma · T. Zhang · G. Pan · J. Tao
National Research Center for Rehabilitation Technical Aids,
Beijing 100176, China

J. Li
Robotic Institute, Beihang University, Beijing 100191, China

H. Fu (✉)
101 Institute of the Ministry of Civil Affairs, Beijing 100070, China
e-mail: redbomb628@163.com

© Zhejiang University Press and Springer Science+Business Media Singapore 2017
C. Yang et al. (eds.), *Wearable Sensors and Robots*, Lecture Notes in Electrical
Engineering 399, DOI 10.1007/978-981-10-2404-7_21

flexibly, the amputees usually need a long-term training to study how to control. The training is very important as it can strengthen the electromyogram (EMG) intensity for specific parts of the brain. Aiming to this, some specialized training systems were developed, such as Ottobock company which has developed a system for the training. But the system is so expensive that a small limb-fitting center cannot afford it. And also the aid-disability career is a charity cause, and has minimal profits. Then many domestic companies do not want to do. So we need a lower price similar training system urgently in China. Based on the above, our research builds a three-dimensional model of the hand and a virtual recovery environment (NRRA 2011) according to the physiological mechanism and rehabilitation mechanism of hand movement for patients. And a training system was developed with the upper limb surface EMG. By the detection of the stumps' EMG signal, the amputees can control the hand mode in the virtual reality environment (Guberek et al. 2008). This provides a scientific and effective way for training and assessment. The whole system includes hardware and software as follows.

2   Hardware Design

The hardware devices includes signal extraction circuits, signal acquisition circuits, computer interface, data storage units, power supply modules and so on. As shown in Fig. 1, the signal extraction circuit is used for extracting surface EMG of stumps. It can remove industrial frequency, electrocardiograph (ECG) and other noise, and get the signal which can be used to control the training system by gain amplifier, filter, and weak signal processing. The acquisition circuits use a microprocessor to A/D conversion and acquisition for multiple filtered EMG signals. The data storage units use a CF or SD card as a storage device, can deposit the EMG signal independently in the case of no computer intervention. And the power module includes voltage converter and regulator circuit, providing reliable power for amplification circuits and microprocessors of EMG.

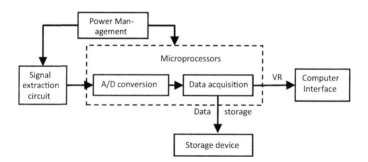

Fig. 1 Hardware architecture of the training system

2.1 Design of the Signal Extraction Circuit

Signal extraction circuit is a key issue to EMG test system, so the quality of it directly affects the quality of the extracted EMG signals. And it also has a certain influence on the stability of consecutive control. Therefore, in the design of EMG amplification extraction circuit, following aspects were mainly considered:

- EMG is a kind of weak electrical signal, and its amplitude is in the range of µv level. Thus higher gain of the amplifier needs to be introduced. And taking into acquiring of surface EMG is noninvasive, the electromyographic signals may be much weaker. Therefor, 6800 times adjustable magnification was taken as the maximum in the design.
- The frequency range of EMG signal is very wide, about between 2 Hz and 10 kHz. And the main signal energy is concentrated between 10 and 1000 Hz. Since the frequency range of surface EMG which is detected by electrodes is between 10 and 500 Hz. So this design uses the EMG signal between 5 and 600 Hz.
- Usually the contact resistance between surface electrode and skin is relatively high, is about several thousand ohms to hundreds of thousands of ohms. And the influence of tightness between electrode and skin, cleanliness of the skin, skin moisture, season changes, and many other factors are very sensitive to the contact resistance. Therefore, a high input impedance need to balance and reduce the impact when we designed the amplifier.
- Due to ECG, EEG, and other signals interference of human and external high-frequency electromagnetic signals, we design a high rejection ratio to suppress these effects.
- EMG is very weak, so low-noise and low drift must be considered in the design of amplifier.

Based on the above analysis, our study designs three kinds of signal extraction circuits and amplification circuits through analysis of EMG amplifier's working principle and compare of the performance deficiencies of domestic amplifier in order to improve extraction capacity of signal, avoid all kinds of interference and noise, shrink board size as much as possible, and facilitate patient to fixed on the stump. Then the magnification of circuit, suppression of frequency interference, suppression of common mode interference, anti-high-frequency interference, and others were tested. And the testing found that all performance of our EMG electrodes have been improved greatly than domestic. Figure 2 is the circuit design of EMG amplifier. Figure 3 is the measured EMG signals of the above design. And also the EMG signals were converted by root mean square, just as Fig. 4.

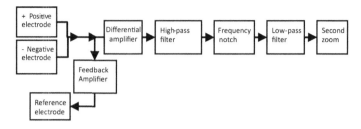

Fig. 2 Design principles of the EMG amplifier

Fig. 3 Measured EMG signals

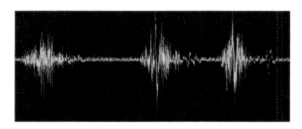

Fig. 4 EMG signal which converted by root mean square (RMS)

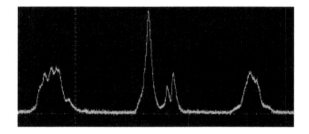

2.2 Material Study of the Electrode Sheet

Currently, nickel silver is the common material which is used to make electrode for EMG amplifier in China. However, in the process of clinical use we found that some patients' stump skins will be allergic when it contacted with the electrode sheet in long time. For the stump skin, allergies can lead to skin ulcers, and cause secondary damage to the patients easily. Therefore, our study consulted large numbers of documents, and make varieties of chemical analysis of EMG amplifier electrode sheet on the market. And the study find silver chloride electrode which has good conductivity, but the price is higher. In addition, the metal Ti12 has excellent conductivity and stability. It is also a good material for electrode sheet. Based on the above materials, our study produced different electrode sheets and carried out related clinical validation. After a long time wearing, patients generally reflected in the course of a silver electrode less affected by different skin, the signal is more stable and does not produce allergy when contact with the skins.

Fig. 5 Prototype system of EMG

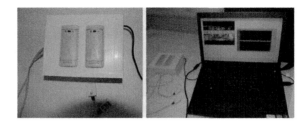

2.3 Integration of the EMG System

AVR microcontroller is used as control chip for the EMG acquisition circuit. The signals which are extracted by EMG amplifier are input to the microcontroller AD and get electric-mode conversion. And the microcontroller communication with computer via USB interface module, it can collect EMG signals and send them to the host computer. Figure 5 is the EMG prototype system after integration and packaging.

3 Software Design

Combined with EMG hardware system, our study has developed corresponding training software. By extraction, treatment, and identification of EMG, our study established a realistic three-dimensional model of the hand and upper limb, and create a virtual reality game training scenarios. These can achieve finger opening and closing, wrist rotating and flexion, elbow flexion and other movements for single degree of freedom myoelectric hand. However, the whole software has scientific training methods and evaluation systems.

3.1 Mode and Animation of the Hand

As shown in the Figs. 6 and 7, the virtual reality training scenarios use skeleton model and grid model for the hand (Novak et al. 2011). These realized 3D models and animations for the upper limb amputees' hand (Chu et al. 2006). Among them key-frame is the key animation techniques. All of the bone mode and animation of hand was designed in the Maya software (Li et al. 2009). The total animation includes 100 frames. The former 50 frames are the fisting movement, and the later 50 frames are the opening movement. The model of palm bones is segment model, which are separated in each bone sections (Shi et al. 2010).

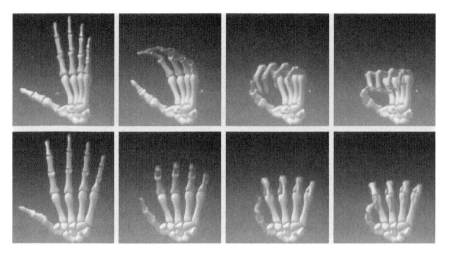

Fig. 6 Model of palm bone

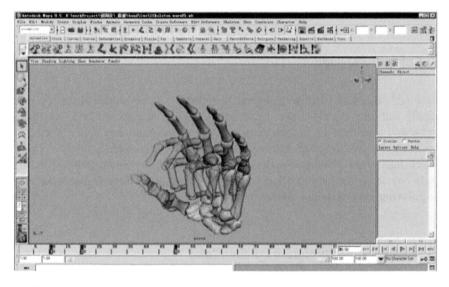

Fig. 7 Animation of palm bone

3.2 Design of Game

Based on the above, our study developed an eating sushi game, which can be controlled through fisting and opening with hand by upper limb amputees. If the amputees eat sushi correctly in the mind, he will receive corresponding scores. If the role of upper limb amputees eat sushi which is not in his thoughts, he will be deducted some scores. It can be controlled by up and down keys and left and right keys before the game. Even after connecting the EMG hardware (Wu et al. 2009), it can be controlled by the fisting and opening of palm to operate (Fig. 8).

The whole game time is 120 s. The final score will display on the summary screen with the best results in the past at the same time (Zheng et al. 2009). If the current score is higher than the previous best score, a congratulations text will appear on the screen. And animation on the lower right corner tips the upper limb amputees need fisting operation to take a game again. In order to make the game more lively and exciting, the game also has a corresponding music and sound effects, such as angry grumbling, clapping, and humming.

If the difficulty of game exceeds the level of upper limb amputees, he or she will feel anxiety (Cai et al. 2012). Obviously if the game is too simple, players will get bored. Both of which may cause upper limb amputees give up the game. So let us upper limb amputees immerses in the game is the ability to maintain homeostasis between upper extremity amputation and difficulty of the game. The aim of the software is to let all of the amputees find their own level of game currently. Thus, as shown in Fig. 9 technology of multidimensional role and expression are used in the software. Such as the default role is a wait and considerations state, the eyes are open and toward when in the current role, coming to a satisfactory state when eating the correct sushi, getting angry when eating wrong sushi.

3.3 Evaluation of Training System

The system can record the intensity and flexibility of control muscle every training time, and compare control effect of each muscle group for different people (Davis et al. 1989). It provides guidance for the next rehabilitation training. Specific performances are as follows:

- Record the patient's basic information, including name, age, amputation part, and amputation time.
- Each training session, including training time, muscle group, strength and accuracy to control muscles;
- Accessing to training records at any time, and generating comparison chart for every time's training.
- Exporting training data with a variety of table type (e.g., Excel and Word), just as rehabilitation medicine data.

(a)

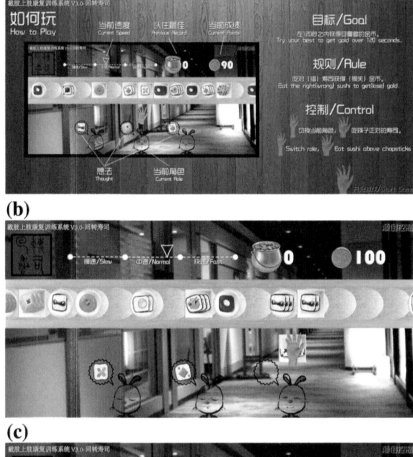

(b)

(c)

Fig. 8 Game interface. **a** Guide interface. **b** Start interface. **c** Eating sushi interface

| head1.1.jpg | head1.2.jpg | head2.1.jpg | head2.2.jpg | head3.1.jpg | head3.2.jpg |

Fig. 9 Various facial expressions of role

4 Patient Experiment

In order to verify the usability of the system, an experiment was done in the hospital which is affiliated to National Research Center for Rehabilitation Technical Aids. Experiment strictly follows relevant provisions of ethics. And all patients

(a)

(b)

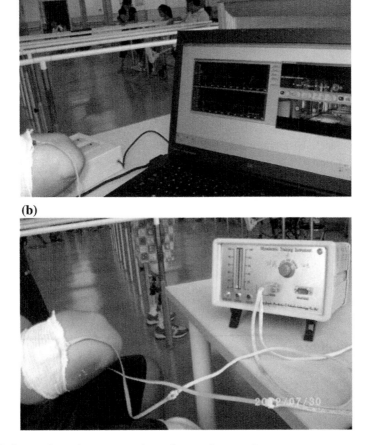

Fig. 10 Scene of experiment. **a** experimental group. **b** comparison group

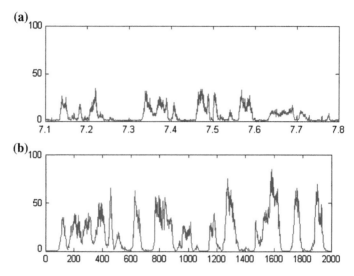

Fig. 11 EMG signals of a male patient with new virtual reality training system. **a** EMG signals before the training. **b** EMG signals after 7-days training

consensually participated and knew that the system is a prototype. After signing an agreement, about 22 upper extremity amputees with different ages were selected for the experiment. All of the selected patients were divided into comparison group and experimental group, who got an EMG strength training with traditional EMG training instrument and new virtual reality training system respectively. Before the training, patient's basic information is recorded in detail. And every period of training is two hours a day. The stump's EMG amplitude, selected parts' girth and weight-bearing capacity were tested after training for three days, seven days, and ten days respectively (Fig. 10).

The study finds: the comparison group with conventional training equipment have no improvement after 3 days and just have a little bit after 7 days. The magnitude of EMG signal is not very high. But the experimental group with virtual reality training system is generally 5–15 µv after 3 days and about 35 µv after 7 days (Fig. 11). After 10 days, muscle strength is increased significantly, its magnitude is about 40 µv which is the basic conditions to installing prosthetic hand.

5 Conclusion

The availability of the system is tested by the preliminary experiment. And since the training system with virtual reality has an interesting game which can make the upper limb amputees focused on the training, so it has a good effect compared to the traditional training instrument. Meanwhile, the VR game increases interesting and endurance training time significantly for each patient, can give patients right psychological suggestion and encouragement continuously, which combining physical therapy with psychotherapy improve the training effect greatly and reduce the training cycle. All of these are very important for the return of patients to society.

Acknowledgments This research is supported in part by the National Natural Science Foundation of China (Grants 51275101, 51335004), and the National Science & Technology Pillar Program of China (Grants 2009BAI71B07). Thank you for the help of Yang-sheng Wang, Guo-qing Xu and Can-jun Yang.

Project supported by the National Natural Science Foundation of China (Grants 51275101, 51335004), and the National Science & Technology Pillar Program of China (Grant 2009BAI71B07).

References

Cai H, Wang X, Wang Y (2012) Entropy-based maximally stable extremal regions for robust feature detection. Math Probl Eng 2012(7):1627–1632. doi:10.1155/2012/857210

Chu J-U, Moon I, Mun M-S (2006) A real-time EMG pattern recognition system based on linear-nonlinear feature projection for a multifunction myoelectric hand. IEEE Trans Biomed Eng 53(11):2232–2239. doi:10.1109/TBME.2006.883695

Davis FD et al (1989) User acceptance of computer technology: a comparison of two theoretical models. Int J Man–Mach Stud 35(8):982–1001. doi:10.1287/mnsc.35.8.982

Guberek R, Schneiberg S et al (2008) Application of virtual reality in upper limb rehabilitation. In: Virtual rehabilitation, p 65. doi:10.1109/ICVR.2008.4625128

Li J, Bai L, Wang Y, Tosas M (2009). Hand tracking and animation. In: IEEE international conference on computer-aided, pp 318–323. doi:10.1109/CADCG.2009.5246885

National Research Center for Rehabilitation Technical Aids (NRRA) (2011) Upper limb amputation based on virtual reality rehabilitation system. CN Patent, 200910093340.2

Novak D, Mihelj M et al (2011) Psychophysiological responses to different levels of cognitive and physical work load in haptic interaction. Robotica 29(3):367–374. doi:10.1017/S0263574710000184

Shi L, Wang Y, Li J (2010) A real time vision-based hand gestures recognition system. In: The 5th international symposium on intelligence computation and applications, pp 349–358. doi:10.1007/978-3-642-16493-4_36

Wu X, Wang Y, Li J (2009) Video background segmentation using adaptive background models. In: Image analysis and processing – ICIAP 2009, pp 623–632. doi:10.1007/978-3-642-04146-4_67

Zheng X, Wang Y et al (2009) A comprehensive approach for fingerprint alignment. In: Intelligent networks and intelligent systems, 2009. ICINIS'09, pp 382–385. doi:10.1109/ICINIS.2009.104

Evaluation on Measurement Uncertainty of Sensor Plating Thickness

Xu-cheng Rong and Jian-jun Yu

Abstract The thickness of the sensor plating is a very important parameter, but it is difficult to measure it accurately. The sensor plating thickness is studied, and it is measured by metallographic method. From the aspects of repeatability of measurement, verticality of inlaying specimen, calibration of microscopic ruler, and distinguishability of optical microscope; the influences are analyzed and the measurement is done. Then the uncertainty components are combined. Finally the value of uncertainty of plating thickness is calculated, the method of uncertainty measurement and the notes during the measurement are illuminated.

Keywords Sensor · Plating thickness · Metallographic method · Uncertainty evaluation

1 Introduction

In recent years, people pay more attention to the uncertainty analysis of experimental data because there must be uncertainty during the testing or experiment, no matter how perfect are the test methods or the test equipments. For a long time, this uncertainty was described with error but now, describing that by uncertainty of measurement evaluation is rather reasonable than error because error is only of a single data. Uncertainty of measurement is an index, which can be calculated and can describe the result of measurement, and it is going to be unified in worldwide scale (Wang 2012). Metal plating is used for enhancing the corrosion resistance of the base material, and makes the material more beautiful. So this craft is widely used on sensor. The common plating materials are copper, nickel, and chrome. The thickness of the plating is a very important parameter, which ensures the corrosion resistance of the base material. Therefore, the thickness of the plating is examined

X. Rong · J. Yu (✉)
Department of Industrial Engineering, South China University of Technology,
Guangzhou 510640, China
e-mail: yujj@scut.edu.cn

© Zhejiang University Press and Springer Science+Business Media Singapore 2017   275
C. Yang et al. (eds.), *Wearable Sensors and Robots*, Lecture Notes in Electrical
Engineering 399, DOI 10.1007/978-981-10-2404-7_22

frequently. Generally, the magnification of the optical microscope is 50–1000 times, and the thickness of plating which is greater than 1 μm can be examined.

In this article, the examination of the sensor plating thickness will be an example of case, and the process of uncertainty evaluation will be illuminated.

2 Specimen Preparation and Test Method

The test specimen was a sensor part whose base material was plastic. One cross section of the sensor part was cut from the original sample to be a test specimen before testing. Then the specimen should be inlaid, grinded, and polished. After these processes, the test surface of the specimen meets the testing requirement. In the process of specimen preparation, two aspects should be noticed. First, cold curing inlaying material should be used because the base material of the trim part is plastic, which cannot bear the high temperature. Second, pay attention to the verticality between the test surface and the inlaying material, otherwise the testing result will diverge seriously. For the large size specimen, pre-grinding should be done first, and after obtaining a flat cross section, the process of inlaying can be done. For the tiny specimen, the specimen should be fixed first, and then to be inlaid. The specimen will move in the process of inlaying if there is no jig to fix it. In this article, Carl Zeiss microscope and the assorted measuring software were selected. The magnification is 500 times and the detected object is the copper layer.

3 Building of the Mathematical Model

Because the thickness was read directly from the microscope without other mathematic operation, the mathematical model was as follow:

$$y = x \tag{1}$$

In Eq. (1), y is the output quantity, it is the measurement result of the copper layer thickness and its unit is μm. x is the input quantity, it is the thickness value, which was read from the microscope and its unit is μm.

Equation (1) above is the mathematical model of comprehensive assessment method of the uncertainty evaluation. This model meets the requirement of the uncertainty evaluation, which was finished by the sensor instruments, and it is suitable for the test item, which is difficult to evaluate directly, such as:metallographic method, spectrograph, energy spectrum analysis, and so on.

4 Analysis of Uncertainty Source

By measuring the plating thickness with metallographic microscope, there are four uncertainty sources. The first uncertainty component was from the measurement repeatability, which is named $u_1(x)$. The second uncertainty component was derived from the inlaying verticality, which is named $u_2(x)$. The third uncertainty component derived from the calibration of the microscopic ruler, which is named $u_3(x)$. The fourth uncertainty component derived from the distinguishability of the microscopic measuring system, which is named $u_4(x)$.

5 Evaluation of Standard Uncertainty Component

5.1 Evaluation of Uncertainty Component $u_1(x)$

By mean of measuring continuously to obtain the observation column, type A evaluation method should be used. According to the test method of GBT 6462-2005 (Metallic and oxide coatings—Measurement of coating thickness—Microscopical method), five testing operator should use the same calibrated microscope to measure the plating thickness on the specimen within ten random field of view. Therefore, five groups of data were obtained. The plating thicknesses of copper layers were shown in Table 1 and the metallographic photo was as Fig. 1 (The thicker layer is the copper layer).

The mean value of the plating thickness of group j was as shown in Eq. (2):

$$\bar{x}_j = \frac{1}{n} \sum_{i=1}^{n} x_i \tag{2}$$

In Eq. (2), j was the value of the number of groups, and $j = 1, 2, \ldots, m$ ($m = 5$), i was the value of the times of measurement, $i = 1, 2, \ldots, n$ ($n = 10$), x_i was the value of the plating thickness.

The standard deviation of each group of sample was as follow:

$$s_j = \sqrt{\frac{\sum_{i=1}^{n} (x_i - \bar{x}_j)^2}{n - 1}} \tag{3}$$

Combined sample standard deviation:

$$s_p = \sqrt{\frac{1}{m} \sum_{j=1}^{m} s_j^2} = 0.265$$

Table 1 Repetitive measurement data and the calculation results (unit: μm)

Group	1st	2nd	3rd	4th	5th	6th	7th	8th	9th	10th	Mean value	Standard deviation
1	34.50	34.07	34.07	34.72	34.29	34.01	34.01	34.01	34.01	34.01	34.31	0.28
2	34.34	34.34	34.34	34.34	34.34	34.34	34.34	34.34	34.34	34.34	34.51	0.29
3	34.39	34.39	34.39	34.39	34.39	34.39	34.39	34.39	34.39	34.39	34.29	0.20
4	34.78	34.78	34.78	34.78	34.78	34.78	34.78	34.78	34.78	34.78	34.50	0.28
5	34.60	34.60	34.60	34.60	34.60	34.60	34.60	34.60	34.60	34.60	34.52	0.26

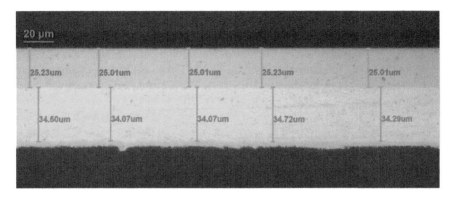

Fig. 1 Metallographic photo of measurement of plating thickness of copper layer (local)

The standard deviation of S_j: 0.036.

The estimated value of $\hat{\sigma}Y_sY = 0.062$.

Because of $\hat{\sigma}Y_sY, \hat{\sigma}_{ev}Y_sY$ it means that the measuring condition is stable. The combined sample standard deviation (s_p) could be recognized as the uncertainty component. In the actual measurement, the mean value of five time repetitive measurement was as the measuring result. Therefore, the uncertainty component form measurement repeatability would be:

$$u(x_1) = 0.119$$

5.2 Evaluation of Uncertainty Component $u_2(x)$

According to the standard, the relation between inclination of specimen cross section and actual plating thickness was shown in the following equation: (Hu 2010);

$$d = d' \cos \alpha \tag{4}$$

In Eq. (4), α was the degree of deviation between the cross section and the vertical surface of the plating, the unit is degree. When $\alpha = 0$, the value of d equaled to the plating thickness, when, the measured value of the plating thickness equaled to d'.

From the Eq. (4), the verticality problem which is introduced by inlaying can be indicated with $\cos \alpha$, and the minimum distinguishability of optical microscope was 1 μm. The second digit behind the decimal point was an uncertain value, hence the deviation of $\pm 5°$ should be the acceptable scope, and this deviation belonged to uniform distribution. So, the uncertainty component which is derived from inlaying would be $u(x_2) = 0.575$.

5.3   *Evaluation of Uncertainty Component* $u_3(x)$

According to the calibration certificates of the equipment, the uncertainty of the optical microscopic ruler should be $U = 1.3$; $k = 2$. The calibration condition was $23 \pm 2 \,°C$, 50 %RH. According to JJF 1059.1-2012, the uncertainty derived from optical microscopic ruler should be $u(x_3) = 0.650$.

5.4   *Evaluation of Uncertainty Component* $u_4(x)$

The formula of distinguishability of optical microscope was as follow:

$$\frac{0.61\lambda}{NA} \tag{5}$$

In the equation, λ equaled to the wave length of the light irradiated from the equipment, the unit was nm, NA equaled to the numerical aperture of the objective lens. The wave length of visible light is 400–700 nm, the mean value should be 550 nm. In general, larger the magnification, larger is the numerical aperture. In this article, the magnification was 500 times, the magnification of objective lens was 50 times, d equaled to 0.75 (this parameter decided by the equipment). Therefore, the distinguishability should be 0.45 μm, and this deviation belonged to uniform distribution. So the uncertainty component, which is derived from distinguishability should be:

$$u(x_4) = 0.130$$

6   Evaluation of Combined Standard Uncertainty

Because the uncertainty components of repeatability of measurement, verticality of inlaying specimen, calibration of microscopic ruler, and distinguishability of optical microscope were mutually independent, and the sensitivity coefficient equaled to $c_i = \frac{\partial y}{\partial x_i} = 1$. Since the mathematic model of plating thickness   measuring was $y = x$, the combined standard uncertainty should be

$$u_c(y) = \sqrt{\sum_{i=1}^{n} u^2(x_i)} = 0.886 \, \mu m$$

7 Evaluation of Expanded Uncertainty

Expanded uncertainty U is the product of combined standard uncertainty $u_c(y)$ and the coverage factor. The value of k should be 2 in general, and the confidence probability would be 95 %. Therefore, the expanded uncertainty equaled to

$$U = kuc = 1.772$$

After rounding, it would be 1.77.

8 Uncertainty Budget

The final result should be the mean value of the measured value of five groups, and the result was 34.43. Finally, the plating thickness of the trim part was 34.43, $U = 1.77$, $k = 2$, confidence probability was 95 %.

9 Conclusion

1. By a series of analysis and calculation, the uncertainty of the trim part copper plating thickness was obtained. Besides the copper layer, the other uncertainty of plating layer on the sensor could be calculated and analyzed by the same method.
2. From the analysis and calculation process above, we found that the uncertainty of the plating thickness mainly derived from specimen preparation, repeatability of measurement, and the condition of the equipment, if the environment is relatively stable. Therefore, when the similar measurement is proceeded, the operators should pay attention to the operation specification and the calibrated equipment should be used, then the result of measurement might be valid and accurate.

Acknowledgments Project supported by the national natural science foundation of China (No. 71071059, 71301054), training plan of Guangdong province outstanding young teachers in higher education institutions (Yq2013009), open project of laboratory of innovation method and decision management system of Guangdong province (2011A060901001-03B).

References

Hu YX (2010) Practical technology of metallographic examination. China Machine Press, Beijing, pp 2–8
Wang CZh (2012) Material physical and chemical inspection measurement uncertainty evaluation guidelines and examples. China Metrology Press, Beijing, pp 3–34

Research on the Stability of Needle Insertion Force

Qiang Li and De-dong Gao

Abstract The unstable needle insertion, contributing to imprecise insertion, can be reflected from insertion force stability. To improve the accuracy of needle-based intervention procedures and guide the investigation on needle steering technologies, a series of needle insertion experiments were performed on different soft tissues including single-layer PVA (Polyvinyl Alcohol) phantoms, multi-layer PVA phantoms and porcine livers. The effects of insertion velocities, tissue properties and tissue structures on insertion force stability were investigated. For mechanical noises in force data vary with interventional equipment, they were filtered before quantitative analysis of insertion force stability. The unit amplitude of insertion force was directly used to reflect the insertion stability. The results from both the single-layer PVA phantoms and porcine livers show that there is a critical velocity, under which the unit amplitude sharply decreases with the increase of velocity and above which it almost does not vary with velocity. In the actual application, insertion velocity above this critical value can be adopted to improve the insertion stability. The multi-layer PVA phantom tests show that the unit amplitude increases firstly and then decreases with the increase of PVA composition. By changing the direction of insertion into the same multi-layer PVA phantom, results indicate that both friction and cutting force can lead to unstable insertion.

Keywords Needle insertion · Stability · Insertion force · Multi-layer phantom

Q. Li · D. Gao (✉)
School of Mechanical Engineering, Qinghai University, Xining 810016, China
e-mail: gaodd@qhu.edu.cn

Q. Li
e-mail: clyz4liqiang@163.com

D. Gao
The State Key Lab of Fluid Power Transmission and Control, Zhejiang University, Hangzhou 310027, China

© Zhejiang University Press and Springer Science+Business Media Singapore 2017   283
C. Yang et al. (eds.), *Wearable Sensors and Robots*, Lecture Notes in Electrical Engineering 399, DOI 10.1007/978-981-10-2404-7_23

1   Introduction

In recent years, robot-assisted needle interventions have become a research focus (Mallapragada et al. 2009; Misra et al. 2010a; Lorenzo et al. 2011; Fukushima and Naemura 2014). As a kind of minimally invasive procedures, needle insertion procedures include vaccinations, blood/fluid sampling, tissue biopsy, brachyther-apy, regional anesthesia, abscess drainage, catheter insertion, cryogenic ablation, electrolytic ablation, neurosurgery and deep brain stimulation (DiMaio and Salcudean 2005a). Some general procedures require accuracy of millimeters while accuracy of micro-millimeters is desired in procedures involving fetus, eye and ear (Abolhassani et al. 2007a). Although robot-assisted needle interventions are most potential to realize precise placement of a needle tip, due to inhomogeneity of biological tissues, needle bending and tissue deformations happening in needle insertion, many researches have to be done on such as force modeling, kinematics of needles, path planning, control algorithm, etc.

Several studies have been dealing with needle steering strategies by using the needle–tissue interaction models. Both needle deflection and soft tissue deformation models were used to establish a needle manipulation Jacobian for needle steering based on a potential-field-based path planning technique (DiMaio and Salcudean 2005b). The tissue deformations simulator using finite element model has been adopted to design a planer for path planning (Alterovitz et al. 2005). The virtual spring's model of insertion force was used in trajectory planning (Glozman and Shoham 2007). Euler–Bernoulli beam bending equations were also used to establish a needle deflection model which can predict the needle tip position with the help of real-time force/moment feedback (Abolhassani et al. 2007b). An analytical force model based on bevel edge geometry and soft-tissue material properties was used for predicting needle curvature produced in the process of needle insertion, which is useful for mechanics-based path planning and needle steering (Misra et al. 2010b).

Accurately modeling the needle–tissue interaction is very important for precise insertion. These models usually predict the deflection of needle based on force/torque readings at the needle base. A model based on cantilever beam for estimating the deflection of needle in soft tissues was presented (AbolhassanI and Patel 2006; Lee and Kim 2014). Needle insertion can be controlled by axially rotating needle at certain depths (Webster et al. 2006; Alterovitz et al. 2008; Duindam et al. 2010; Park et al. 2010). Flexible beam model for needle was used to estimate needle deflection to find needle axially rotating depths (Abolhassani et al. 2007d). The method of obtaining dynamic friction coefficient and cutting force was ever presented in literature (Fukushimaa et al. 2013). The model of friction force was established to study friction's effects on needle insertion (O'Leary et al. 2003; Asadian et al. 2014). The FE method was also adopted to determine needle–tissue interaction forces (Misra et al. 2008).

Today, China is seriously affected by liver cancer and most cancer deaths are related to liver cancer (Chena and Zhang 2011), so porcine liver was used for experiments. The method of preparing the multi-layer PVA phantoms were detailedly presented (Lee and Kim 2014). The effects of insertion velocity and tissue property were quantitatively studied. The results can guide selecting an appropriate insertion velocity and preparing an optimal phantom for needle insertion studying. The method of filtering mechanical noise has first been adopted in this paper, which can make insertion force modeled more accurately. Compared with the stability analysis method shown in literature (Alja'afreh 2010), the method presented in this paper is more direct.

The rest of this paper is organized as follows: Sect. 2 presents materials and methods used in experiments. In Sect. 3, experimental analysis results and discussion are presented. Finally, conclusions of this paper and suggestions for future work are outlined in Sect. 4.

2 Materials and Methods

To make it more convenient to express opinions, some names are given as follows:

- **Original force**: the original force data directly obtained from force sensor;
- **Mechanical force**: the axial force resulting from mechanical vibration;
- **Insertion force**: the axial force obtained from original force data by filtering out its axial mechanical force part (noise);
- **Needle displacement**: the distance of needle motion from its original position to targets;
- **Insertion depth**: the distance of needle motion from tissue surface, into which the needle tip inserts, to targets.

2.1 Experimental Equipment

As shown in Fig. 1, the experimental equipment is utilized to perform needle insertion experiments. Two high-precision linear translation stages (PI Shanghai, M-L01 Series) are used for advancing the needle and setting the position of needle. The linear translation stages are all controlled by micro-displacement DC motor controllers (PI, Mercury C-863.10). In order to achieve real-time acquisition of insertion force acted on needle by tissue, a 6-DOF force/torque (F/T) sensor (ATI, Nano17 SI-50-0.5) is mounted at the end of needle base. Sensor's force resolution is 0.0125 N and torque resolution is 0.0625 Nmm. The F/T data from sensor are uploaded to computer through DAQ (Data Acquisition) Card (NI, PCI-6220). The beveled-tip needles used in experiments are all PTC needle (model: 18 G × 350 mm, diameter: 1.26 mm) made in Japan. To reduce the effects of needle bending on axial insertion force, large diameter needles are utilized in experiments. The shape of needle tip is shown in Fig. 1. All phantoms were placed within a

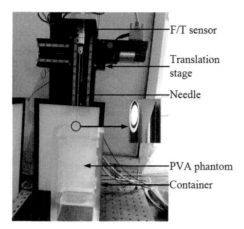

Fig. 1 Needle insertion experimental equipment

container made of acrylic sheets. Two small perforated acrylic sheets are used to limit the movement of tissues.

2.2 Experimental Material

As shown in Fig. 2, the experimental tissues include single-layer PVA phantom (not shown), double-layer PVA phantom, three-layer PVA phantom and porcine liver. Table 1 shows the components of PVA phantoms' each layer. The ratio of deionized water to dimethyl sulfoxide is constant. The PVA hydrogels' properties change with the different content of PVA. The content of first layer's PVA is more than second layer's but less than third layer's. The needle insertion force profile presented in Fig. 3 is needle insertion into three-layer PVA phantom at 5 mm/s.

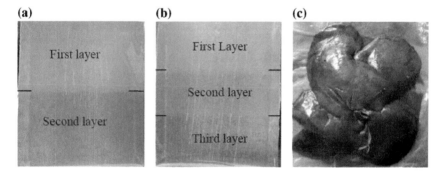

Fig. 2 The experimental tissues of double-layer PVA phantom (**a**), three-layer PVA phantom (**b**) and porcine liver (**c**)

Table 1 The components of different PVA phantoms

PVA phantom	First layer			Second layer			Third layer		
	PVA (g)	DW (ml)	DMSO (ml)	PVA (g)	DW (ml)	DMSO (ml)	PVA (g)	DW (ml)	DMSO (ml)
Single-layer	64	320	480	0	0	0	0	0	0
Double-layer	32	160	240	24	160	240	0	0	0
Three-layer	20	100	150	15	100	150	25	100	150
Ratio	4:20:30			3:20:30			5:20:30		

DW Deionized water; *DMSO* Dimethyl sulfoxide

Fig. 3 The force of insertion into the three-layer PVA phantom

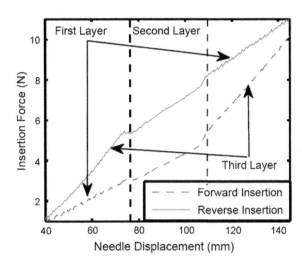

Forward insertion means needle insertion starting at the first layer of the multi-layer phantoms. Reverse insertion is performed in the opposite direction of forward insertion. For forward insertion, insertion force is largely decreased when the needle tip cuts the boundary between the first layer and second layer, which indicates that the stiffness of first layer is larger than that of second layer. The conclusion of the third layer's stiffness larger than second layer's can be obtained similarly. For the first layer and third layer are not linked together, this method cannot be used to compare their stiffness. However, the force profile's slope at third layer is obviously larger than that at first layer, showing that the third layer's stiffness is larger than first layer's. Consequently, the stiffness of PVA phantom increases with the increase of the content of PVA.

The PVA phantom preparation is shown in Fig. 4. To make PVA to dissolve in solvent, the mixtures of PVA, DMSO and deionized water are first stirred about 3 h at 90 °C. Natural cooling mixtures before freezing it are necessary to greatly decrease the surface tension of PVA phantoms and reduce bubbles inside phantoms. For the preparation of the multi-layer PVA phantoms, compared to the single-layer phantom preparation, the multi-layer phantoms preparation are more complex and

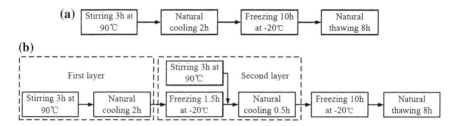

Fig. 4 The preparations of the single-layer PVA phantom (**a**) and double-layer PVA phantom (**b**)

time-consuming. One layer's material should be poured into container and then freezed into solid state before another layer's material poured into container. The preparation of three-layer PVA phantom (not shown in Fig. 4) is similar to that of double-layer PVA phantom. The boundaries between two layers are marked with short horizontal lines (Fig. 2).

The tissue properties and structures of different porcine liver are different. The experiment to avoid this difference affecting on insertion force, experimental liver was obtained from a pig. The porcine liver shown in Fig. 2c was bought from market in the early morning. To keep the properties of liver from not changing too much with time, all experiments on liver were finished as quickly as possible.

2.3 Methods

The insertions into the single-layer PVA phantom at different velocities mainly investigate the relationships between insertion stability and velocities. The insertion velocities were increased from 2 to 50 mm/s with an increment of 2 mm/s. The needle was inserted into phantom three times at each speed. Force data sampling frequency is 1 kHz for 2–20 mm/s, 2 kHz for 22–40 mm/s and 3 kHz for 42–50 mm/s. The change in data sampling frequency is necessary to increase date points for high speed. The multi-layer PVA phantoms experiments were performed at the same speed (5 mm/s) with the purpose of studying the insertion stability associated to tissue properties. Force data sampling frequency is 1 kHz. Forward and reverse insertion results are compared to study friction force's affect on insertion stability. Both forward and reverse insertions were repeated five times. The purpose of liver insertion experiments is similar to the single-layer PVA phantom but closer to clinical applications.

Different insertion equipments will generate different mechanical noise on insertion force, and mechanical noise extremely affects the force of insertion at high speeds. To only investigate the needle–tissue interaction force, mechanical force was first filtered out. Before performing insertion into tissues, insertions without tissues were implemented to capture mechanical force at each speed. Mechanical force can be used for spectrum analysis. The insertion force profile can be divided into several segments and some segments needed for analysis should be selected first. PVA phantoms are approximately homogeneous. There many vessels are distributed in porcine livers (Jiang et al. 2014), which make its structures more complex than PVA phantoms'. The characteristics of their insertion force profiles are obviously different. The insertion force vibrations of PVA phantoms are small and distributed evenly. In contrast, the insertion force vibrations of porcine livers are large and distributed randomly. The large vibrations are caused by inserting into vessels. Consequently, their analytical methods are different.

2.3.1 Separating Insertion Forces

FIR filters have exact linear phase and can meet the requirement of causing no signal distortion. FIR filters were used to filter out mechanical noise from original force. Kaiser window function (kaiserord()) and filter design function (fir1()) provided by MATLAB software were utilized to design filters. The principal parameters include sampling frequency, passband ripple, stopband attenuation, passband cutoff frequency and stopband cutoff frequency.

The sampling frequencies of filters are as same as force sampling frequencies. Passband ripple is 0.05 and stopband attenuation is 0.00001. Boundaries between very soft and very stiff structures can result in large force discontinuities and insertion instabilities (DiMaio and Salcudean 2005a). These boundaries are distributed on a macroscopic scale, so vibrational frequency of insertion force caused by boundaries is small. Needle insertion stability is caused by periodic elastic energy reserve and release (Alja'afreh 2010). The process of elastic energy reserve and release is obviously slow, and vibrational frequency of insertion force is small too. As shown in Fig. 5a, the frequencies of main insertion force amplitudes are smaller than 10 Hz. The main mechanical force amplitudes are located after 200 Hz in Fig. 5b, while mechanical force amplitudes between 30 and 40 Hz may not be ignored. The filtering effects of original force data are shown in Fig. 6. In order to distinguish original force from filtered force, original force–displacement curve has been moved 0.1 N upwards. The filtering effect shown in Fig. 6a is not very perfect, which indicates that mechanical noises between 30 and 40 Hz should be filtered out; in other words, cutoff frequencies should be between 10 and 30 Hz. After many tries, passband cutoff frequency of 20 Hz and stopband cutoff frequency of 22 Hz are ideal for all the experimental data (Fig. 6b).

2.3.2 Separating Force Data into Components

Needle insertion process includes insertion and removal. The process of insertion into tissue is generally studied. Insertion can be subdivided into two parts of pre-puncture and post-puncture (Abolhassani et al. 2007c). Pre-puncture starts from needle tip initially contacting with tissue surface and ends with penetration commencing, in which needle is acted on by stiffness force. In pre-puncture, insertion

Fig. 5 Amplitude spectrum.
a All experiments' original
force spectrum. **b** Mechanical
forces spectrum at all speeds

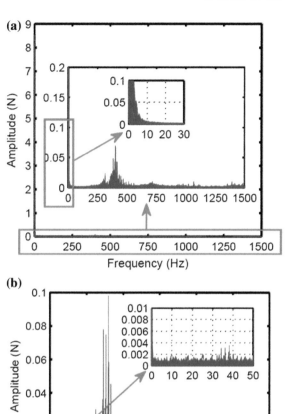

force steady rises and decreases sharply when needle tip initially penetrates tissue surface, which can be used to separate two processes of pre-puncture and post-puncture. In post-puncture, needle is acted on by friction and cutting forces (Okamura et al. 2004). For PVA phantoms are nearly homogeneous, insertion forces regularly vibrate. The forces caused by tip cutting homogeneous tissue are named as tissue forces including friction and cutting forces. Boundaries, including tissue surfaces, vascular walls and any other rigid surfaces, will appear among two structures with different tissue properties. When needle tips cut these boundaries, large force peaks are generated. Boundaries are randomly distributed in biological tissues and force peaks are randomly distributed too on insertion force curves. To reflect the structural characteristics of insertion forces generated on boundaries, they are named as structural forces.

Fig. 6 The comparison of filtering effect on the force of insertion into the single-layer PVA phantom at 2 mm/s.
a Passband cutoff frequency of 40 Hz and stopband cutoff frequency of 42 Hz.
b Passband cutoff frequency of 20 Hz and stopband cutoff frequency of 22 Hz

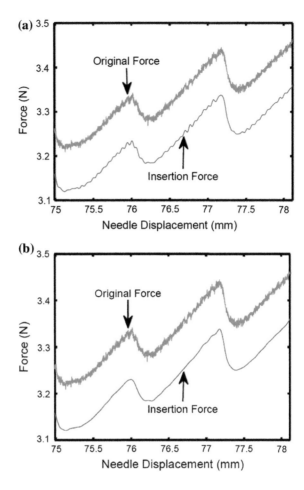

For the complexity of insertion force profile, it was separated manually. As shown in Fig. 7a, the segment for analysis is AB. Point A represents that penetration commences. Point B represents that needle tip commences getting out from the other side of the tissue. In Fig. 7b, A_1B_1, C_1D_1, A_2B_2 and C_2D_2 are analytical segments for both forward and reverse insertion into the double-layer PVA phantom. Transitional segments of B_1C_1 and B_2C_2 are similar to pre-puncture phase, but both forward and reverse insertions have their own features. The peak of B_1C_1 is smaller than previous one while the peak of B_2C_2 is larger than previous one. The points of A_1, D_1, A_2 and D_2 were obtained in a similar way to the single-layer PVA phantom's. The analytical segments of three-layer PVA phantom are shown in Fig. 7c, and their separation methods are similar to double-layer PVA phantom's. The separation results of porcine liver are shown in Fig. 7d. To simultaneously analyze the stabilities of tissue forces and structural forces of porcine liver, the phases of pre-puncture and needle tip getting out from the other side of tissue are all

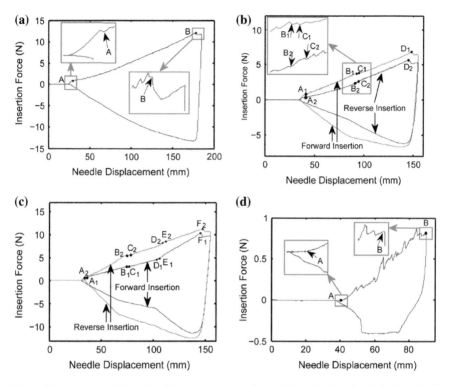

Fig. 7 The separation of insertion force curves. **a** The force of insertion into the single-layer PVA phantom at 2 mm/s. **b** The force of insertion into the double-layer PVA phantom at 5 mm/s. **c** The force of insertion into three-layer PVA phantom at 5 mm/s. **d** The force of insertion into the porcine liver phantom at 5 mm/s

included in analytical segments. Consequently, point A represents that needle tip initially touches tissue surface, and point B represents that the process of advancing needle is stopped.

2.3.3 Obtaining Unit Amplitudes

The stabilities of needle insertion forces of phantoms are reflected by unit amplitudes. Unit amplitude is the mean sum of amplitudes in per millimeter.

As shown in Fig. 8, the insertion force vibrations of PVA phantoms are very regular. In this case, insertion force profile approximately vibrates along a curve, and this curve can be obtained by interpolation. All maximum and minimum points were searched out firstly. Interpolation method is piecewise linear interpolation. The endpoints of every line are middle points of adjacent maximum points and minimum points. The interpolation curve is shown in Fig. 8. Insertion force vibration components were obtained by subtraction of interpolation curve from

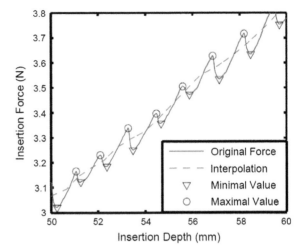

Fig. 8 The interpolation results of the single-layer PVA phantom's insertion force at speed of 2 mm/s

insertion force curve. The maximum and minimum points of Insertion force vibration components were all found out. The absolute values of these maximum and minimum points were summed to be divided by the length of analytical segment, and result is the unit amplitude. This method was used to obtain unit amplitudes for all kinds of PVA phantoms.

After finding out all maximum and minimum points shown in Fig. 9, all amplitudes were obtained through what each maximum value subtracts its followed minimum value. Amplitudes include both tissue force and structural force amplitudes. These two kinds of amplitudes should be separated. The number of tissue force amplitudes is far more than the number of structural force amplitudes at

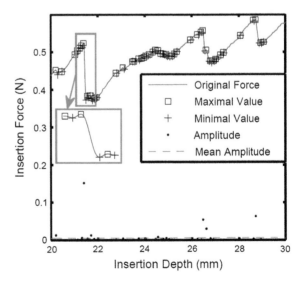

Fig. 9 The force amplitudes of insertion into the porcine liver at 2 mm/s

2 mm/s. The mean of all amplitudes as shown in Fig. 9 can be used for separating tissue force amplitudes from structural force amplitudes. With the increase of speed, the number of tissue force amplitudes rapidly decrease. The mean of all amplitudes is near to structural force amplitudes and cannot be used for separating tissue force amplitudes from structural force amplitudes anymore. In fact, tissue force amplitudes are generally near to their mean value. The two times of previous velocity's tissue force amplitudes' mean was used as separation tissue force amplitudes from structural force amplitudes when insertion speed is larger than 2 mm/s. Finally, their unit amplitudes were figured out.

3   Results and Discussion

3.1   Single-Layer PVA Phantom

When needle insertion velocities are larger than 20 mm/s, the amplitudes of insertion forces are so small that they are hardly obtained. As shown in Fig. 10, the unit amplitude versus velocity curve is presented at speeds of 2–20 mm/s. The curve can be divided into three phases including Phase I, Phase II and Phase III. When insertion velocities are less than or equal to v_{cs1} (8 mm/s), unit amplitudes are almost not affected by insertion velocity. In other words, tissue vibration is almost not changed with the increase of velocity. When insertion velocities are between v_{cs1} and v_{cs2} (16 mm/s), unit amplitudes rapidly decrease with the increase of velocity. When insertion velocities are equal or greater than v_{cs2}, unit amplitudes are very small and almost decease slowly with the increase of velocity. According to the need of insertion stability, v_{cs2} can be regarded as separation point after which insertion is stably performed. The critical speed has been ever obtained in literature too (Alja'afreh 2010). The stability of needle insertion can be improved by increasing velocity. In Phase II, the stability of insertion can be improved rapidly by

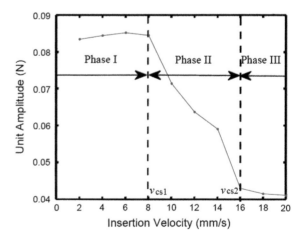

Fig. 10 The force unit amplitudes of insertion into the single-layer PVA phantom

increasing velocity. When insertion velocities are equal or greater than v_{cs2}, the stability of insertion is increased slowly. However, high insertion velocity is bad for real-time feedback control, so the velocities near v_{cs2} are proposed.

3.2 Multi-layer PVA Phantom

When the needle tip cuts the same layer in both forward and inverse insertions, cutting forces are the same and friction forces are different. As shown in Fig. 11a, unit amplitudes of forward and reverse cutting the same layer are unequal, which tell us that both cutting and friction force lead to insertion instability. The cutting force is generated by interactions between needle tip and its surrounding tissue. Consequently, in the interventional operations, the vibration of targets generated by cutting force is not affected by insertion path. The friction force is created from relative motion between needle and its surrounding tissue. The friction force is affected by insertion path, so the stability is affected by insertion path. To improve the stability of needle insertion, it is necessary to choose an optimal path (sometimes the shortest one).

For three-layer PVA phantom, the PVA content of first layer is more than that of second layer but less than that of third layer. As shown in Fig. 11b, unit amplitudes

Fig. 11 The unit amplitude curves of the multi-layer PVA phantoms. **a** Double-layer PVA phantom. **b** Three-layer PVA phantom

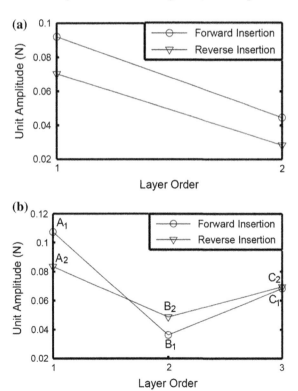

increase first then decrease with the increment of PVA content. In other words, there is a critical content that makes insertion most unstable. In an actual interventional operation, tissues can not be chose to improve the insertion stability. However, the PVA phantoms used in needle insertion experiments should be prepared far away from critical PVA content to improve insertion stability. In this case, we can focus on investigating on tissue deformation's effect on insertion accuracy.

3.3 Porcine Liver

In order to investigate the velocities' effects on the vibration of target when the needle is inserted into human livers, the needle is inserted into porcine liver at different velocities. As shown in Fig. 12, both tissue and structural force unit amplitude versus velocity curves are represented and have same change tendency.

Fig. 12 The unit amplitude curves of the porcine liver. **a** Tissue force unit amplitudes. **b** Structural force unit amplitudes

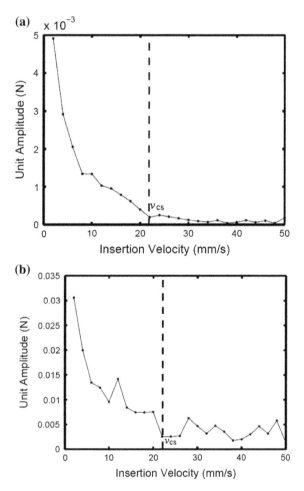

When insertion velocities are less than or equal to v_{cs} (20 mm/s), Their unit amplitudes rapidly decrease with the increase of velocity. When insertion velocities are equal or greater than v_{cs}, their unit amplitudes are very small and almost not affected by velocity. These results are consistent with experimental results of the single-layer PVA phantom. In both an actual interventional operation and the research on needle steering technologies, velocities that are equal or greater than v_{cs} can be used for largely improving insertion stability.

4 Conclusions and Future Work

This paper presented a method for the measurement of the stability of needle insertion into both artificial and biological phantoms. The stabilities of varied phantoms were analyzed quantitatively. The mechanical noise caused by the insertion devices was filtered for investigating the leading effect of insertion forces. The experiments with different velocities indicate that stable insertion can be achieved by raising the needle insertion velocity. The insertions into the multi-layer phantoms indicate that both cutting and friction forces cause unstable insertion. The tissue's properties largely affect insertion stability and insertion paths affect the insertion stability too. To improve the stability of needle insertion, velocity, tissue's property and insertion path should be all considered appropriately.

In the future, we will focus on modeling the tissue force and structure force in order to improve the accuracy of insertion into biological tissues. We will employ the force models to steer the needle and plan the trajectories.

Acknowledgments Project supported by the National Natural Science Foundation of China (No. 51165040) and the Natural Science Foundation of Qinghai Province (No. 2015-ZJ-906).

References

Abolhassani N, Patel R (2006) Deflection of a flexible needle during insertion into soft tissue. In: Proceedings of the 28th IEEE EMBS annual international conference, pp 3858–3861. doi:10.1109/IEMBS.2006.259519

Abolhassani N, Patel R, Ayazi F (2007a) Effects of different insertion methods on reducing needle deflection. In: Proceedings of the 29th annual international conference of the IEEE EMBS, pp 491–494. doi:10.1109/IEMBS.2007.4352330

Abolhassani N, Patel R, Ayazi F (2007b) Needle control along desired tracks in robotic prostate brachytherapy. In: 2007 IEEE international conference on systems, man and cybernetics, pp 3361–3366. doi:10.1109/ICSMC.2007.4413819

Abolhassani N, Patel R, Moallem M (2007c) Needle insertion into soft tissue: a survey. Med Eng Phys 29(4):413–431. doi:10.1016/j.medengphy.2006.07.003

Abolhassani N, Patel RV, Ayazi F (2007d) Minimization of needle deflection in robot-assisted percutaneous therapy. Int J Med Robot Comput Assist Surg 3(2):140–148. doi:10.1002/rcs.136

Alja'afreh T (2010) Investigating the needle dynamic response during insertion into soft tissue. J Eng Med 224(4):531–540. doi:10.1243/09544119JEIM698

Alterovitz R, Goldberg K, Okamura A (2005) Planning for steerable bevel-tip needle insertion through 2d soft tissue with obstacles. In: Proceedings of the 2005 IEEE international conference on robotics and automation, pp 1640–1645. doi:10.1109/ROBOT.2005.1570348

Alterovitz R, Branicky M, Goldberg K (2008) Motion planning under uncertainty for image-guided medical needle steering. Int J Robot Res 27(11–12):1361–1374. doi:10.1177/0278364908097661

Asadian A, Patel RV, Kermani MR (2014) Dynamics of translational friction in needle-tissue interaction during needle insertion. Ann Biomed Eng 42(1):73–85. doi:10.1007/s10439-013-0892-5

Chena JG, Zhang SW (2011) Liver cancer epidemic in china: past, present and future. Semin Cancer Biol 21(1):59–69. doi:10.1016/j.semcancer.2010.11.002

DiMaio SP, Salcudean SE (2005a) Interactive simulation of needle insertion models. IEEE Trans Biomed Eng 52(7):1167–1179. doi:10.1109/TBME.2005.847548

DiMaio SP, Salcudean SE (2005b) Needle steering and motion planning in soft tissues. IEEE Trans Biomed Eng 52(6):965–974. doi:10.1109/TBME.2005.846734

Duindam V, Xu JJ, Alterovitz R, Sastry S, Goldberg K (2010) Three-dimensional motion planning algorithms for steerable needles using inverse kinematics. Int J Robot Res 29(7):789–800. doi:10.1177/0278364909352202

Fukushima Y, Naemura K (2014) Estimation of the friction force during the needle insertion using the disturbance observer and the recursive least square. ROBOMECH J 1(1):1–8. doi:10.1186/s40648-014-0014-7

Fukushimaa Y, Saitoa K, Naemuraa K (2013) Estimation of the cutting force using the dynamic friction coefficient obtained by reaction force during the needle insertion. Procedia CIRP 5:265–269. doi:10.1016/j.procir.2013.01.052

Glozman D, Shoham M (2007) Image-guided robotic flexible needle steering. IEEE Trans Rob 23 (3):459–467. doi:10.1109/TRO.2007.898972

Jiang S, Li P, Yu Y, Liu J, Yang ZY (2014) Experimental study of needle-tissue interaction forces: effect of needle geometries, insertion methods and tissue characteristics. J Biomech 47 (13):3344–3353. doi:10.1016/j.jbiomech.2014.08.007

Lee H, Kim J (2014) Estimation of flexible needle deflection in layered soft tissues with different elastic moduli. Med Biol Eng Compu 52(9):729–740. doi:10.1007/s11517-014-1173-7

Lorenzo DD, Koseki Y, Momi ED, Chinzei K, Okamura AM (2011) Experimental evaluation of a coaxial needle insertion assistant with enhanced force feedback. In: 33rd annual international conference of the IEEE EMBS, pp 3447–3450. doi:10.1109/IEMBS.2011.6090932

Mallapragada VG, Sarkar N, Podder TK (2009) Robot-assisted real-time tumor manipulation for breast biopsy. IEEE Trans Rob 25(2):316–324. doi:10.1109/TRO.2008.2011418

Misra S, Reed KB, Douglas AS, Ramesh KT, Okamura AM (2008) Needle-tissue interaction forces for bevel-tip steerable needles. In: Proceedings of the 2nd Biennial IEEE/RAS-EMBS international conference on biomedical robotics and biomechatronics, pp 224–231. doi:10.1109/BIOROB.2008.4762872

Misra S, Reed KB, Schafer BW, Ramesh KT, Okamura AM (2010a) Mechanics of flexible needles robotically steered through soft tissue. Int J Robot Res 29(13):1640–1660. doi:10.1177/0278364910369714

Misra S, Reed KB, Schafer BW, Ramesh KT, Okamura AM (2010b) Mechanics of flexible needles robotically steered through soft tissue. Int J Robot Res 29(13):1640–1660. doi:10.1177/0278364910369714

O'Leary MD, Sirnone C, Washio T, Yoshinaka K, Okamura AM (2003) Robotic needle insertion: effects of friction and needle geometry. In: Proceedings of the 2003 IEEE international conference on robotics & automation, pp 1774–1780. doi:10.1109/ROBOT.2003.1241851

Okamura AM, Simone C, O'Leary MO (2004) Force modeling for needle insertion into soft tissue. IEEE Trans Biomed Eng 51(10):1707–1716. doi:10.1109/TBME.2004.831542

Park W, Wang YF, Chirikjian GS (2010) The path-of-probability algorithm for steering and feedback control of flexible needles. Int J Robot Res 29(7):813–830. doi:10.1177/0278364909357228

Webster RJIII, Kim JS, Cowan NJ, Chirikjian GS, Okamura AM (2006) Nonholonomic modeling of needle steering. Int J Robot Res 25(5–6):509–525. doi:10.1177/0278364906065388

The Metabolic Cost of Walking with a Passive Lower Limb Assistive Device

Jean-Paul Martin and Qingguo Li

Abstract Lower limb assistive devices capable of augmenting metabolic performance have typically targeted the ankle joint during push off. Here, a lower limb passive assistive device was evaluated that instead helps the user by performing negative work about the knee joint at the end of swing. A pilot study (n = 8), where subjects walked overground at a self-selected speed, revealed that the device is capable of reducing metabolic energy expenditure equal to the amount of additional energy required to carry the weight of the device. The cost of transport (COT) walking with the assistive device (COT = 2.55 ± 0.36 J/kg) showed no significant difference to normal walking (COT = 2.56 ± 0.33 J/kg) without the device.

Keywords Assistive device · Metabolic energy · Overground walking · Negative work assistance

1 Introduction

Lower limb assistive devices have typically targeted the ankle joint during push off (Collins et al. 2015; Asbeck et al. 2015; Mooney et al. 2014) because of the substantial amounts of positive mechanical work performed by the ankle during walking. Additionally, those assistive devices that have demonstrated a net beneficial metabolic effect (Collins et al. 2015; Mooney et al. 2014), all targeted the ankle. At the end of swing, however, knee flexors perform negative work about the knee to decelerate the shank in preparation for heel strike. Donelan et al. was the first to target this motion using energy harvesting knee braces (Donelan et al. 2008).

J.-P. Martin · Q. Li (✉)
Bio-Mechatronics and Robotics Laboratory, Queen's University, Kingston,
ON K7K 4A4, Canada
e-mail: ql3@queensu.ca

J.-P. Martin
e-mail: jeanpaul.martin@queensu.ca

© Zhejiang University Press and Springer Science+Business Media Singapore 2017   301
C. Yang et al. (eds.), *Wearable Sensors and Robots*, Lecture Notes in Electrical
Engineering 399, DOI 10.1007/978-981-10-2404-7_24

However, the assistive nature of the device was unable to overcome the metabolic burden of carrying the weight of the device. Shepertycky et al. developed a lower limb-driven energy harvester that similarly targeted negative work performed by the knee (Shepertycky and Li 2015). They instead located the device on a backpack, bringing the weight of the device closer to the centre of mass of the user. However, much like the knee brace of Donelan (Donelan et al. 2008), the device was not able to overcome the metabolic burden of carrying the weight of the device walking at 1.5 m/s. Instead, in this study we used the same lower limb-driven energy harvester (Shepertycky and Li 2015), but in a passive mode where no electricity is being generated. In a previous treadmill study using the lower limb-driven harvester, it has been shown that with increased resistance on the user, metabolic benefit is reduced (Shepertycky and Li 2015). This study was therefore a pilot to understand if a device applied only passive mechanical resistance to the user during swing, in a walking environment similar to its intended use of overground walking at a self-selected speed, whether there would be a metabolic benefit to wearing the device.

2 Methods

The lower limb-driven energy harvester (Shepertycky and Li 2015), shown in Fig. 1, consists of the harvesting unit mounted to the bottom of a backpack frame. Two cables extend from the device and attach to foot harnesses worn at the ankle. During the swing phase of gait, the cables are extended and a resistance is applied to the user proportional to the inertial force and constant spring force. For this study, an open electronic circuit condition was used and therefore the generator was not applying a back EMF force or generating electricity.

Eight (n = 8) young, healthy, adult male subjects of age 20–24 years (age = 22.7 ± 1.6 years, weight = 78.3 ± 11.4 kg, height = 185.5 ± 10.3) participated in the study. Ethical approval from Queen's University General Research Ethics

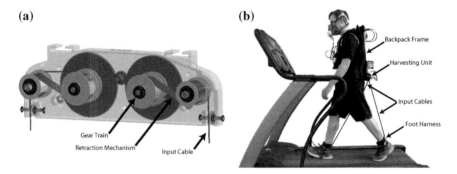

Fig. 1 **a** The harvesting unit with front shell removed. **b** The subject walking with the device on a treadmill during the familiarization period

Board (GMECH- 015-12 Biomechanical Energy Harvesting Backpack, ROMEO# 6006569) was obtained. Each subject walked a 10-minute familiarization period with the device on a treadmill. Subject's were shod, wore athletic clothing, and were asked to refrain from consuming food or beverages other than water 2 h prior to the test.

Subjects walked a total of two 10-minute trials: normal walking and walking with the device. Subjects walked in an indoor track of 60 m in length. A 3-minute rest preceded each trial. Volume flow rate of oxygen consumption and carbon dioxide production was measured from 5 to 7.5 min mark of each trial using a portable open respirometry unit (K4b2, COSMED, Italy). The metabolic cost of walking was determined using (Brockway 1987) from which the resting metabolic rate was then subtracted. The metabolic cost of transport (COT) was then found by normalizing the metabolic power to both subject body weight and average walking speed during the 5–7.5 min mark. Comparison of the COT and walking speed between conditions was performed using paired t-tests ($P < 0.05$).

3   Results

There was no significant difference between the cost of transport during normal walking (COT = 2.56 ± 0.33 J/kgm) and that while walking with the device (COT = 2.55 ± 0.36) as seen in Fig. 2. There was also no significant difference between the net metabolic cost of walking (normal = 250.4 ± 54.6 W, device = 254.7 ± 38.3 W) between the two conditions. There was no significant difference between the walking speed of both trials (normal = 1.28 ± 0.11 m/s, device = 1.29 ± 0.10 m/s) as seen in Fig. 3.

Fig. 2 The metabolic cost of transport (COT) of the two testing conditions: normal walking and passive assistive walking with the device. Error bars indicate 1 standard deviation from the mean

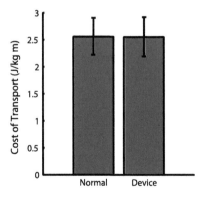

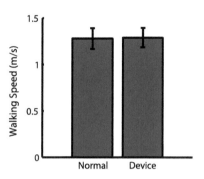

Fig. 3 The average walking speed of the two testing conditions: normal walking and passive assistive walking with the device

4 Discussion

The device was found to provide a metabolic benefit equal to the amount of metabolic energy required to carry the weight of the device. The estimated metabolic cost of carrying the weight of the device (2.53 kg at the waist and 0.13 kg at the ankle) was 10.3 W based on the regression equations of previous load carriage studies (Browning et al. 2007; Tzu-wei and Kuo 2014). These findings coincide with what both Donelan (Donelan et al. 2008) and Shepertycky (Shepertycky and Li 2015) found, that helping a user perform negative work about the knee during swing can have positive metabolic benefits. As the lower limb-driven energy harvesting device has also shown to apply an optimal amount of resistance on the user in a passive mode compared to other power generation conditions (Shepertycky and Li 2015), to achieve metabolic benefit while wearing the device the cost of carrying the device's weight must be reduced. This could be achieved by designing a device with reduced weight from eliminating the device's ability to generate electricity. An optimization of moment of inertia and spring constant may also produce even more metabolic reductions. Experiments in this study were conducted overground at a self-selected pace to recreate how the assistive device might be used in its intended setting: outdoors and at a self-selected pace.

5 Conclusion

A pilot study having subjects walk an overground track at a self selected speed has shown that a lower limb-driven energy harvesting device in a passive state can provide metabolic assistance equivalent to the cost of carrying the device's weight. Further work should reduce the device's weight but provide the ability to alter the moment of inertia and spring force. These results indicate that by doing so, perhaps a metabolic reduction could be achieved through helping users perform negative work about the knee during swing.

Acknowledgments This work was supported by a NSERC Discover Grant to Q. L.

References

Asbeck AT, De Rossi SMM, Holt KG, Walsh CJ (2015) A biologically inspired soft exosuit for walking assistance. Int J Robot Res

Brockway JM (1987) Derivation of formulae used to calculate energy expenditure in man. Hum Nutr Clin Nutr 41(6):463–471

Browning RC, Modica JR, Kram R, Goswami A et al (2007) The effects of adding mass to the legs on the energetics and biomechanics of walking. Med Sci Sports Exerc 39(3):515

Collins SH, Wiggin MB, Sawicki GS (2015) Reducing the energy cost of human walking using an unpowered exoskeleton. Nature

Donelan JM, Li Q, Naing V, Hoffer JA, Weber DJ, Kuo AD (2008) Biomechanical energy harvesting: generating electricity during walking with minimal user effort. Science 319(5864):807–810

Mooney LM, Rouse EJ, Herr HM (2014) Au- tonomous exoskeleton reduces metabolic cost ofhuman walking during load carriage. J Neuroeng Rehabil 11(1):80

Shepertycky M, Li Q (2015) Generating electricity during walking with a lower limb-driven energy harvester: targeting a minimum user effort. PLoS ONE 10(6):e0127635

Tzu-wei PH, Kuo AD (2014) Mechanics and energetics of load carriage during human walking. J Exp Biol 217(4):605–613

Part III
Advanced Control System

A Novel Method for Bending Stiffness of Umbilical Based on Nonlinear Large Deformation Theory

Zuan Lin, Lei Zhang and Can-jun Yang

Abstract In this paper a novel method is introduced to test the bending stiffness of umbilical. The four point bending test is studied both theoretically and experimentally. The nonlinear deflection model has been established because of large deformation. Half of the umbilical can be considered as a cantilever beam. By applying MATLAB optimization toolbox, we deal with differential equations with a nonlinear term pretty fast. Finally, an automatic testing system was designed to test the bending stiffness, which allows us to experimentally study the deflections of umbilical by means of a series of measurements.

Keywords Umbilical · Bending stiffness · Large deformation · Nonlinear

1 Introduction

Subsea umbilical systems provide the vital supply and control link from platforms or topside vessels to subsea oil and gas equipment. A steel tube umbilical consists of steel tube fluid conduits, electrical cables, and fiber optic cables arranged in contra-helically wound layers as shown in Fig. 1. The bundle is contained by binding tape providing radial reinforcement and protected by a polymer outer sheath. The primary function is to provide chemical injection for flow assurance; electrical signals for valve control and monitoring; hydraulic pressure for valve actuation; electrical power for subsea pumping; fiber optics for data acquisition and monitoring.

Z. Lin · C. Yang (✉)
School of Mechanical Engineering, Zhejiang University, Hangzhou 310027, China
e-mail: ycj@zju.edu.cn

Z. Lin
e-mail: lzuan@nit.zju.edu.cn

Z. Lin · L. Zhang
Ningbo Institute of Technology, Zhejiang University, Ningbo 315100, China
e-mail: nitzhanglei@126.com

© Zhejiang University Press and Springer Science+Business Media Singapore 2017
C. Yang et al. (eds.), *Wearable Sensors and Robots*, Lecture Notes in Electrical Engineering 399, DOI 10.1007/978-981-10-2404-7_25

309

Fig. 1 Typical steel tube
umbilical

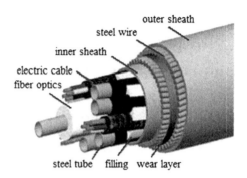

Tensile and bending load acts on steel tube umbilical under environment load; such as wave, ocean current, and float bodies. As the unbounded winding structure components can slip each other, and steel tube umbilical has enough tensile strength and good flexibility. So bending stiffness is an important index to evaluate the bending performance of steel tube umbilical cable, (Parsinejad 2013; Quan 2013; Dong et al. 2015).

Many researchers studied the respond of helical structure under bending load. (Costello 1975) determined analytically the axial stiffness of single-layer cables composed of an arbitrary number of smooth wires from the general nonlinear theory for the bending and twisting of thin helical rods. (Witz and Tan 1992) established an analytical model to predict the flexural structural behavior of multiple layer flexible structures. (Ramos 2004) assumes that umbilical has the "no-slip" model and the "full-slip" model under bending load. (Feret 1987) established a theoretical approach to evaluating the stresses due to bending and the relative slip between layers due to bending. (Kraincanic and Kebadze 2001) took into account the nonlinearity of the layer caused by sliding of individual helical elements between the surrounding layers. Bending stiffness of a helical layer is a function of bending curvature, interlayer friction coefficients, and interlayer contact pressures (Caire 2007).

The bending stiffness theories mentioned above are based on the hypothesis of small deflection and linearly elastic mechanics, wherefore accuracy is weakness, although computing speed is pretty quick. With the development of computer hardware and computational mechanics, numerical simulation is more and more popular such as FEA (finite element analysis) (Fang 2013). The large problem size and the large number of time increments lead to time-consuming simulations in FEA. So a method is urgent, which can solve the problem of large bending deformation of umbilical with fairly fast speed under the condition of test sites.

2   Theoretical Analysis

In this paper the bending deformation of umbilical is simplified as a large deflection of plane beam, and analyzed by the theory of nonlinear large deformation member.

Fig. 2 Top view of small bending deformation of umbilical

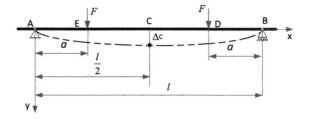

The bending stiffness test of umbilical uses four-point bending test method. Figure 2 is the top view of the bending deformation of umbilical, which is supported by basement in experiment. The umbilical (the bold beam) was held at both of its extremities by two support blocks (A and B point). The length of the specimen is l, two tension straps pulled the umbilical at D and E point. The distance of AE and BD is a. C point is the center of umbilical, so Δc is the maximum deflection of umbilical. The pull load and the umbilical deflection near E, C, and D points were monitored. The curvature was given by the relative displacement of three points marked on the umbilical.

Initially small bending deformation of umbilical occurred by pull load, and bending stiffness can be calculated using simply supported beam model. From Eq. (1) it is easy to obtain the bending stiffness:

$$EI = \frac{Fa(3l^2 - 4a^2)}{24\Delta c} \tag{1}$$

where F is pull load, E is modulus of elasticity, I is moment of inertia of the beam cross-section about the neutral axis.

The pull load increases gradually, and large bending deformation occurred (Fig. 3). A and B point on the umbilical will move from A(B) to A'(B'), so Eq. (1) is not suitable. Nonlinear large deflection model need to be established.

In order to simply the model, half of the umbilical can be considered as a long, thin, cantilever beam of uniform circle cross-section made of a linear elastic material that is homogeneous and isotropic (Fig. 4), in which the law of behavior of the material is represented by the linear relation:

$$\sigma = \varepsilon E \tag{2}$$

That it is known as Hooke's law (Feynman et al. 1989), and where σ is the normal stress, ε is the strain and E is the modulus of elasticity or Young's modulus (McGill 1995). The deflection of a cantilever beam is essentially a three-dimensional problem. An elastic stretching in one direction is accompanied by a compression in the perpendicular directions: the ration is known as Poisson's ratio. However, we can ignore this effect when the length of the beam is larger than the thickness of the perpendicular cross-section and this is shorter than the curvature radius of the beam (Feynman et al. 1989). In this study, we assume that the beam is non-extensible and

Fig. 3 Top view of large
bending deformation of
umbilical

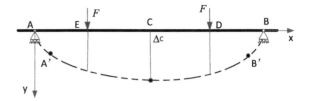

Fig. 4 Top view of
cantilever beam model of half
of the umbilical

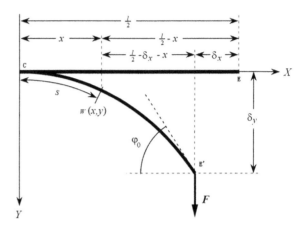

strains remain small, and that Bernoulli–Euler's hypothesis is valid, i.e., plane
cross-sections which are perpendicular to the neutral axis before deformation remain
plane and perpendicular to the neutral axis after deformation. Next, we also assume
that the plane-sections do not change their shape or area. Following, for instance, the
analysis proposed by Feynman regarding the study of the deflection of a cantilever
beam, it is possible to write the Bernoulli–Euler bending moment–curvature relationship for a uniform-section rectangular beam of linear elastic material as follows
(Feynman et al. 1989):

$$EI\frac{\mathrm{d}\varphi}{\mathrm{d}s} = M \tag{3}$$

where M and $\kappa = \mathrm{d}\varphi/\mathrm{d}s$ are the bending moment and the curvature at any point of
the beam respectively, and I is the moment of inertia of the beam cross-section
about the neutral axis. We will consider the deflections of a cantilever beam subjected to one vertical concentrated load at the free end, by supposing that the
deflection due to its self-weight is null (the umbilical is supported by basement
above mentioned). This implies considering a mass less beam.

Figure 4 shows a cantilever beam of length $l/2$ with a concentrated load
F applied at the free end of the beam. In this figure δ_x and δ_y are the horizontal and
vertical displacements at the free end, respectively, and φ_0 takes into account the
maximum slope of the beam. We take the origin of the Cartesian coordinate system
at the fixed end of the beam and let (x, y) be the coordinates of point W, and s the

arc length of the beam between the fixed end and point W. If we differentiate Eq. (3) once with respect to s, we obtain

$$EI\frac{d^2\varphi}{ds^2} = \frac{dM}{ds} \tag{4}$$

where the bending moment M at a point W with Cartesian coordinates (x, y) is given by the equation (Fig. 4):

$$M(s) = F(l/2 - \delta_x - x) \tag{5}$$

By differentiating Eq. (5) once with respect to s, taking into account the relation $\cos\varphi = dx/ds$ and substituting in Eq. (4), we obtain the nonlinear differential equation that governs the deflections of a cantilever beam made of a linear material under the action of a vertical concentrated load at the free end:

$$EI\frac{d^2\varphi}{ds^2} + F\cos\varphi = 0 \tag{6}$$

Equation (6), although straightforward in appearance, is in fact rather difficult to solve because of the nonlinearity inherent in the term $\cos\varphi$. In order to obtain the solution of Eq. (6), this equation is multiplied by $d\varphi/ds$, so that it becomes

$$EI\frac{d^2\varphi}{ds^2}\frac{d\varphi}{ds} + F\cos\varphi\frac{d\varphi}{ds} = 0 \tag{7}$$

which can be written as

$$\frac{d}{ds}\left[\frac{1}{2}EI\left(\frac{d\varphi}{ds}\right)^2 + F\sin\varphi\right] = 0 \tag{8}$$

Equation (8) is immediately integrable taking into account that at the free end $\varphi(l/2) = \varphi_0$, where φ_0 is the unknown slope at the free end of the beam (see Fig. 4), and from Eqs. (3) and (5), it follows that $(d\varphi/ds)_{s=l/2} = 0$. From Eq. (8) we can obtain

$$\left(\frac{d\varphi}{ds}\right)^2 = \frac{2}{EI}(F\sin\varphi_0 - F\sin\varphi) \tag{9}$$

and by integrating it we can obtain the following equation for the arc length s as a function of the slope φ:

$$s = \sqrt{\frac{EI}{2F}}\int_0^\varphi \frac{d\varphi}{\sqrt{\sin\varphi_0 - \sin\varphi}} \tag{10}$$

The total length $l/2$ corresponds to the unknown angle φ_0 at the free end of the beam:

$$\frac{l}{2} = \sqrt{\frac{EI}{2F}} \int_0^{\varphi_0} \frac{d\varphi}{\sqrt{\sin \varphi_0 - \sin \varphi}} \tag{11}$$

Equation (11) allows us to obtain the angle φ_0 at the free end of the beam as a function of the length, $l/2$, the modulus of elasticity, E, the moment of inertia of the cross-section of the beam, I, and the external load, F. After obtaining the arc length s as a function of φ [Eq. (10)], and taking into account that $\cos \varphi = dx/ds$ and $\sin \varphi = dy/ds$, the x and y coordinates of the horizontal and vertical deflection at any point along the neutral axis of the cantilever beam are found as follows:

$$x = \sqrt{\frac{2EI}{F}} (\sqrt{\sin \varphi_0} - \sqrt{\sin \varphi_0 - \sin \varphi}) \tag{12}$$

$$y = \sqrt{\frac{EI}{2F}} \int_0^{\varphi} \frac{\sin \varphi \, d\varphi}{\sqrt{\sin \varphi_0 - \sin \varphi}} \tag{13}$$

From Fig. 4, it is easy to see that the horizontal and vertical displacements at the free end can be obtained from Eqs. (12) and (13) taking $\varphi = \varphi_0$:

$$\delta_x = l/2 - x(\varphi_0) \tag{14}$$

$$\delta_y = y(\varphi_0) \tag{15}$$

We introduce the nondimensional load parameter α, and the nondimensional coordinates ξ and η defined as follows:

$$\alpha = \frac{F(l/2)^2}{2EI} \quad \xi = \frac{x}{l/2} \quad \eta = \frac{y}{l/2} \tag{16}$$

As well as the nondimensional tip deflection ratios

$$\begin{aligned} \beta_x &= \delta_x/(l/2) = 1 - \xi(\varphi_0) \\ \beta_y &= \delta_y/(l/2) = \eta(\varphi_0) \end{aligned} \tag{17}$$

Using the parameter α, it is possible to obtain a more general view of the results, because cantilever beams with different combinations of E, I, F, and l may give the same value of α and, consequently, they would have the same behavior. Taking into account the definitions of α, ξ, and η, Eqs. (11), (12), and (13) can be written as

$$2\sqrt{\alpha} = \int_0^{\varphi_0} \frac{d\varphi}{\sqrt{\sin\varphi_0 - \sin\varphi}} \tag{18}$$

$$\xi = \frac{1}{\sqrt{\alpha}}\left(\sqrt{\sin\varphi_0} - \sqrt{\sin\varphi_0 - \sin\varphi}\right) \tag{19}$$

$$\eta = \frac{1}{2\sqrt{\alpha}}\int_0^{\varphi} \frac{\sin\varphi\, d\varphi}{\sqrt{\sin\varphi_0 - \sin\varphi}} \tag{20}$$

Equation (18) allows us to obtain φ_0 as a function of the nondimensional load parameter α. However, Eqs. (19) and (20) are elliptic integrals that may be evaluated numerically.

3 Numerical Analysis

As mentioned above, in order to study large deflections of a cantilever beam subjected to a vertical concentrated load at the free end, it is necessary to know the angle φ_0. To do this, it is necessary to solve Eq. (18) in order to obtain φ_0 as a function of α. Instead of writing Eqs. (18) and (20) in terms of elliptic functions, by means of complex changes of variable which give little insight as to the nature of the solutions, it is easier to solve them numerically using one of the packages of commercial software available such as MATLAB. Taking this into account, we used the MATLAB program, with the aid of the "Optimization Toolbox," to solve the improper integrals in Eqs. (18) and (20).

From Fig. 4, it is easy to measure the horizontal and vertical displacements at the free end δ_x and δ_y, so some program variables are defined as follows from Eq. (17) and (20):

$$\text{betax} = \beta_x = 1 - \xi(\varphi_0) \quad \text{betay} = \beta_y = \eta(\varphi_0)$$

$$F1 = \frac{1}{\sqrt{\sin\varphi_0 - \sin\varphi}} \quad i1 = \int_0^{\varphi_0} \frac{d\varphi}{\sqrt{\sin\varphi_0 - \sin\varphi}}$$

$$F2 = \frac{\sin\varphi}{\sqrt{\sin\varphi_0 - \sin\varphi}} \quad i2 = \int_0^{\varphi_0} \frac{\sin\varphi\, d\varphi}{\sqrt{\sin\varphi_0 - \sin\varphi}}$$

As δ_x and δ_y is known, then we establish a optimization function Eq. (21) and use "fsolve" command in "Optimization Toolbox" to find a value of φ_0 between 0 and 90°. The value of Eq. (21) is zero when the value of φ_0 is applied in Eq. (21). Equations (21) can be written as

$$F = \frac{\beta_y}{\beta_x} = \frac{\eta(\varphi_0)}{1 - \xi(\varphi_0)}$$

$$= \frac{\int_0^{\varphi_0} \frac{\sin\varphi \, d\varphi}{\sqrt{\sin\varphi_0 - \sin\varphi}}}{\int_0^{\varphi_0} \frac{d\varphi}{\sqrt{\sin\varphi_0 - \sin\varphi}} - 2\sqrt{\sin\varphi_0}} \qquad (21)$$

$$= \frac{i2}{i1 - 2\sqrt{\sin\varphi_0}}$$

The MATLAB algorithm script file is name as cantilevr_beam.m, can be written as

```
function Fun=cantilever_beam(phi0)
F1=@(phi)1./sqrt(sin(phi0)-sin(phi));
I1=integral(F1,0,phi0);
F2=@(phi)sin(phi)./sqrt(sin(phi0)-sin(phi));
I2=integral(F2,0,phi0);
Fun=(betay/betax)-I2./(I1-2*sqrt(sin(phi0)));
```

To prove the algorithm, a calculating example can be given out. Many examples were illustrated in Table 1, Page 68 (Cheng 1994). We chose a calculating example randomly $\beta_x = 0.0764$, $\beta_y = 0.349$, and $\varphi_0 = 0.5373$. The following command is used to calculate the value of φ_0:

```
betax = 0.0764, betay = 0.349;
tic
phi0=fsolve(@myfun,pi/4)
toc
```

The result is phi0=0.5373, and elapsed time is 0.033838 s(Intel Core i5 processor).

Knowing φ_0 as a function of α, it is easy to calculate the horizontal nondimensional tip deflection ratio using Eq. (17), and integrating Eq. (18) with the aid of the MATLAB program, the vertical nondimensional tip deflection ratio can also be calculated. Figure 5 shows the results obtained.

4   Experimental Results

In the laboratory, we applied the above mentioned method to design an automatic bending stiffness testing system (Fig. 6). Automatic pull load, data acquisition, and calculations are available using sensors and computer control system. The length of the specimen is 8 m, the distance of AB is 4.8 m ($l = 4.8$), and the distance of AE is 1 m ($a = 1$).

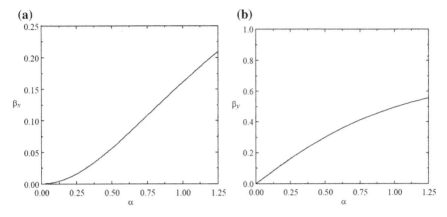

Fig. 5 The horizontal nondimensional tip deflection ratio, β_x, as a function of the nondimensional load parameter α (**a**), and the vertical nondimensional tip deflection ratio, β_y, as a function of the nondimensional load parameter α (**b**)

Fig. 6 Automatic bending stiffness testing system

Fig. 7 Experimental and theoretical elastic curves for the umbilical

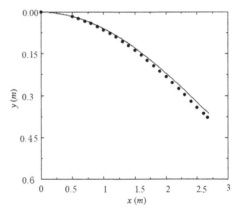

Fig. 8 The relation between bending stiffness and curvature

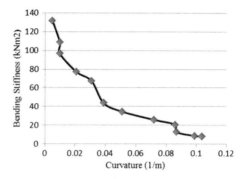

The bending stiffness and curvature of the specimen can be calculated according to δ_x, δ_y and pull load F acquired by testing system. Figure 7 shows the experimental elastic curve as well as the one calculated with the aid of Eqs. (18)–(20). The difference between both is due to the friction of the umbilical slip, which has not been considered in the theoretical treatment. Figure 8 shows the relation between bending stiffness and curvature.

5 Conclusions

We have studied the four-point bending test method of umbilical both theoretically and experimentally. Because of large deformation, nonlinear deflection model need to be established. We have shown that half of the umbilical can be considered as a long and thin cantilever beam. We deal with the physical system, which is described by differential equations with a nonlinear term. Although the solutions to the elastica equations could be expressed in terms of elliptic functions, it is much more convenient to use numerical integration as outlined in this paper. Numerical integration was performed using the MATLAB optimization toolbox obtain the exact solution. Finally, an automatic testing system was designed to test the bending stiffness of the specimen, which allows us to experimentally study the deflections of umbilical by means of a series of measurements.

Acknowledgments Project supported by Ningbo People's Livelihood, Science and Technology (No.2013C11037).

References

Caire M (2007) The effect of flexible pipe non-linear bending stiffness behavior on bend stiffener analysis, pp 103–109

Cheng Z (1994) Beam, plates, and shells large deformation theory. Science Press, China Science Publishing & Media Ltd (in Chinese)

Costello GA (1975) Effective Modulus of Twisted Wire Cables. T.&A.M. Report ,University of Illinois at Urbana—Champaign, Department of Theoretical and Applied Mechanics(398)

Dong L, Huang Y et al (2015) Bending behavior modeling of unbonded flexible pipes considering tangential compliance of interlayer contact interfaces and shear deformations. Mar Struct 42:154–174. doi:10.1016/j.marstruc.2015.03.007

Fang Z-F (2013) A finite element cable model and its applications based on the cubic spline curve. China Ocean Eng 27(5):683–692. doi:10.1007/s13344-013-0057-1

Feret JJ (1987) Calculation of stresses and slip in structural layers of unbonded flexible pipes. J Offshore Mech Arct Eng 109(3):263–269

Feynman R, Leighton RB, Sands M (1989) Volume II: Mainly electromagnetism and matter

Kraincanic I, Kebadze E (2001) Slip initiation and progression in helical armouring layers of unbonded flexible pipes and its effect on pipe bending behaviour

McGill DJ, King WW (1995) Engineering mechanics: statics Chap. 5. PWS Publishing Company, Boston

Parsinejad F (2013) Friction, contact pressure and non-linear behavior of steel tubes in subsea umbilicals: V04AT04A004. doi:10.1115/omae2013-10056

Quan WC (2013) Three-dimensional dynamic behavior of the flexible umbilical cable system. Appl Mech Mater 390:225–229. doi:10.4028/www.scientific.net/AMM.390.225

Ramos R (2004) A consistent analytical model to predict the structural behavior of flexible risers subjected to combined loads. J Offshore Mech Arct Eng 126(2):141–146. doi:10.1115/1.1710869

Witz JA, Tan Z (1992) On the flexural structural behaviour of flexible pipes, umbilicals and marine cables. Marine Struct 5(2–3):229–249. doi:10.1016/0951-8339(92)90030-S

Silicon Micro-gyroscope Closed-Loop Correction and Frequency Tuning Control

Xingjun Wang, Bo Yang, Bo Dai, Yunpeng Deng and Di Hu

Abstract In this paper, we present the improved performance of silicon micro-gyroscope under closed-loop correction and frequency tuning. First, this paper analyzes the source of the quadrature error in the micro-gyroscope. Second, the mode-matching state of the silicon vibratory gyroscope is studied. The sense mode resonant frequency is tuned to be closed to the drive mode resonant frequency under the mode-matched condition. Then, the closed-loop system correction is realized under the mode-matched condition and the proportion-integral differential correction with a passive impedance network. Finally, the experiment results demonstrate the feasibility of the closed-loop system correction scheme under the mode-matched condition.

Keywords Quadrature error · Frequency tuning · Mode-matched · PID · Closed-loop

1 Introduction

Micro-gyroscopes are one of the fastest growing segments in the field of the micro sensor market. However, their performance is still lower than the conventional mechanical, such as the fiber optic gyroscope, the laser gyroscopes and so on (Yazdi et al. 1998). For the improvement of performance, various methods have been introduced in the literature (Xu et al. 2014).

X. Wang · B. Yang (✉) · B. Dai · Y. Deng · D. Hu
School of Instrument Science & Engineering, Southeast University, Nanjing 210096, People's Republic of China
e-mail: yangbo20022002@163.com

X. Wang
e-mail: wangxingjun2000@126.com

X. Wang · B. Yang · B. Dai · Y. Deng · D. Hu
Key Laboratory of Micro-inertial Instrument and Advanced Navigation Technology, Ministry of Education, Nanjing 210096, People's Republic of China

© Zhejiang University Press and Springer Science+Business Media Singapore 2017
C. Yang et al. (eds.), *Wearable Sensors and Robots*, Lecture Notes in Electrical Engineering 399, DOI 10.1007/978-981-10-2404-7_26

Quadrature error is one of the major error sources in MEMS gyroscopes. Owing to the micro fabrication technologies imperfection, the quadrature error is inevitable. The quadrature motion can be reduced with the laser trimming of proof mass or can be mechanically suppressed by carefully designed levers (Yazdi et al. 1998; Geen et al. 2002). But this method cannot remove the quadrature error completely. A better approach that can eliminate the overall quadrature error by applying differential DC potentials to the mechanical electrodes on the sensor is the electrostatic quadrature cancellation. Electrostatic quadrature suppression method is the most effective technique (William and Roger 1996). Electrostatic technique completely removes the quadrature error. The high performance gyroscopes reported in this paper makes use of electrostatic quadrature cancellation.

Mode-matching is the resonant frequency of the sense mode and is tuned to equal with that of the drive mode, which can greatly improve the overall performance of the gyroscope. In order to reduce the frequency mismatch, some methods have been introduced in the literature, such as the selective deposition of poly silicon, the effect of localized thermal stressing, and the effect of negative electrostatic stiffness (Joachim and Lin 2003; Remtema and Lin 2001). In these methods, the frequency-tuning based on the effect of negative electrostatic spring constant is an effective method.

2 Gyroscope System

Figure 1 shows the simplified structure of the gyroscope which is comprised of two bilaterally symmetric single mass gyroscopes. The drive frame, sense frame and a proof mass constitute each single mass gyroscope. In the proposed gyroscope design, the quadrature error is cancelled by DC potentials, whose amplitudes are automatically adjusted with the help of dedicated closed-loop quadrature cancellation, applied to the quadrature electrodes. The mode matching system is using the electrostatic tuning capability of the sense mode. The sense mode resonance frequency is adjusted by changing the tuning voltage.

2.1 Quadrature Error Analysis

Quadrature error is defined as the direct coupling of drive motion into sense mode of the gyroscope, is one of the major error sources in MEMS gyroscopes (Shkel et al. 1999; Tatar et al. 2012). Quadrature error occurs due to poor micro fabrication tolerances. Different from Coriolis signal which depends on drive mode velocity, quadrature signal directly depends on the drive mode displacement itself. Owing to the amplitude of quadrature signals is much larger than the amplitude of Coriolis signal, even small phase errors result in large errors at the output after demodulation. Thus, the sources of the quadrature error should be well identified.

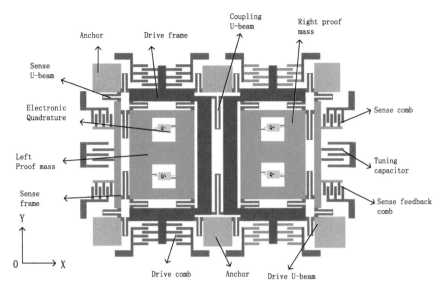

Fig. 1 Schematic structure of the gyroscope

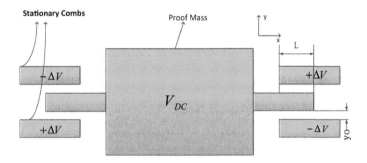

Fig. 2 Configuration for electrostatic quadrature suppression

Figure 2 shows the configuration for electrostatic quadrature suppression. Assume proof mass displaces in the drive direction (position × direction) for an amount of X, and a small amount of y in positive y direction.

Then total force acting on the proof mass in y direction for a device thickness of h is found as

$$F = -\frac{4V_{\mathrm{DC}}\Delta V \varepsilon n X h}{y_0^2} \tag{1}$$

where n is the number of quadrature electrode combs, h is the thickness of combs, ε is permittivity The configuration given in Fig. 2 generates a force to stop this movement.

Fig. 3 Part of the tuning combs

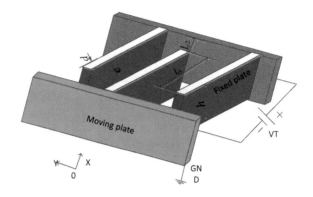

2.2 *Frequency Tuning Analysis*

Mode-matching tunes the resonant frequency of the sense mode to equal of the drive mode by using the electrostatic tuning capability. Figure 3 shows a part of the tuning combs. When the tuning voltage VT is loaded on the fixed plate, the moving plate will be only converted in positive y direction.

When the tuning voltage VT loaded on the tuning combs, the frequency of the sense direction can be derived simple as

$$f_{y\,\mathrm{eff}} = \frac{\sqrt{(2\pi \times f_{y0})^2 - b \cdot \mathrm{VT}^2}}{2\pi} \tag{2}$$

where b is a mechanical spring constant of the sense mode, $b = 3.0814 \times 10^5$, and f_{y0} is the resonant frequency of sense mode.

3 Control System

In order to extract the rotation-rate information from the sense electrodes, a considerable amount of signal processing is required. The signal processing circuit has been implemented and consists of two main blocks: (1) drive loop to maintain the resonance frequency of drive mode and (2) sense circuit eliminate the quadrature error and extracts the Coriolis signal from the output of the sense electrodes. The drive loop uses a phase-locked loop (PLL) to lock the drive resonant frequency and supplies the actuation voltage for the drive-comb electrodes.

The quadrature electrodes are set on the structure of silicon micro-gyroscope to eliminate the quadrature error. The voltage applied on the quadrature is regulated from the detected quadrature signal, which will generate the electrostatic force to cancel the quadrature error. Figure 4 shows the quadrature error correction system

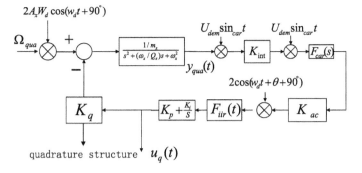

Fig. 4 The quadrature error correction system of silicon micro-gyroscope

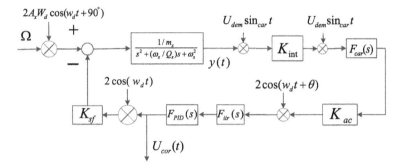

Fig. 5 The closed-loop detection principle diagram of Coriolis signal

of sense mode. $F_{car}(s)$ and $F_{iir}(t)$ are the filter. Thus, the quadrature error is eliminated at its source.

The Coriolis signal is extracted from the output signal when the quadrature error is suppressed through the quadrature error correction system. Figure 5 is the closed-loop detection for the Coriolis signal. In Fig. 5, $F_{PID}(s)$ is the transfer function of proportion-integral differential (PID) control; $F_{iir}(s)$ is the transfer function of low-pass filter after Coriolis signal demodulation. K_{sf}, K_{int} and K_{ac} are the circuit gain.

Then the transfer function of the Coriolis signal is

$$H_{cor}(s) = \frac{G(s, w_d)K_{cor}F_{cor}(s)F_{crt}(s)}{1 + G(s, w_d)K_{cor}F_{cor}(s)F_{crt}(s)K_{for}K_{sf}/A_x w_d} \tag{3}$$

where

$$K_{cor} = \frac{U_{car}U_{dem}K_{int}K_{ac}A_x w_d}{2}$$

The resonant frequency of sense mode is changed in the presence of tuning voltage. The mode-matched is performed by increasing the DC tuning voltage

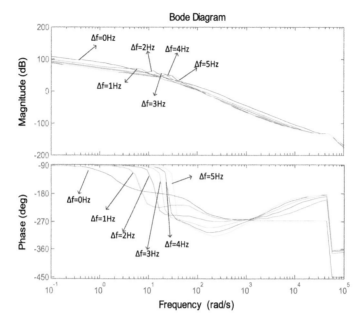

Fig. 6 The curve of the open-loop under different frequency offset

(VT) on the MEMS structure until electrostatic spring softening decreases the sense-mode frequency to become equal to the drive-mode frequency. The MATLAB simulation of the closed-loop correction system is shown in Fig. 6, where Δf is the frequency difference between the sense mode and drive mode.

The simulation results in Fig. 6 show that when the frequency difference of two modes is larger than 2 Hz, there will exist a peak value which also increases along with the growing of frequency difference. Hence, the frequency difference must be controlled to less than 2 Hz by the tuning voltage.

The compensation $F_{PID}(s)$ in Fig. 5 is a decisive factor of dynamic characteristic in closed-loop system, such as bandwidth, phase and amplitude margin, response speed, and so on.

When two mode frequency offset is zero, in theory, even Q_y increases, there is no resonance peak in the open system. Therefore, it can fully reflect the superiority of the high quality factor. The width of the open-loop system is very narrow, and the correction is called narrow-band correction.

When the proportional integral control is used, the bandwidth of the closed-loop system will be too narrow or cannot meet the appropriate phase angle margin and gain margin. Thus, the differentiation control is added to achieve the PID (proportion-integration-differentiation) control. Figure 7 shows that the bandwidth of PID correction is expanded after correction.

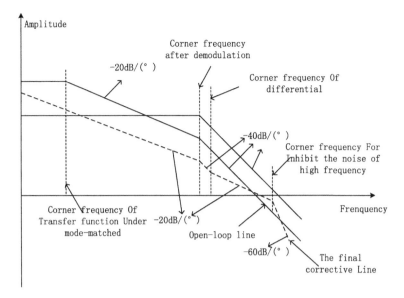

Fig. 7 The bode diagrams of narrow-band correction

4 Experiment Measurement

Based on the above theoretical analysis and simulation results, the closed-loop correction system experiments of the silicon micro-gyroscope are implemented, as shown in Fig. 8. It mainly includes the mode-matching loop, the closed-loop drive, and the closed-loop sense together.

The effect of quadrature cancellation can be seen directly from Fig. 9. Figure 9a shows that open-loop without quadrature cancellation, the output signal contains a

Fig. 8 Vacuum packaged gyroscope with PCB test circuits

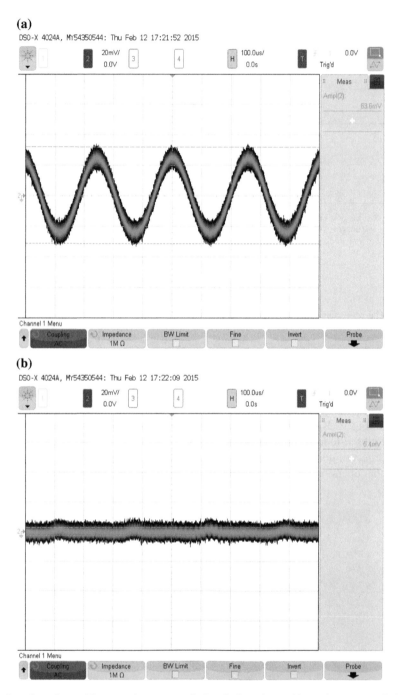

Fig. 9 **a** Open-loop without quadrature cancellation. **b** Open-loop with quadrature cancellation

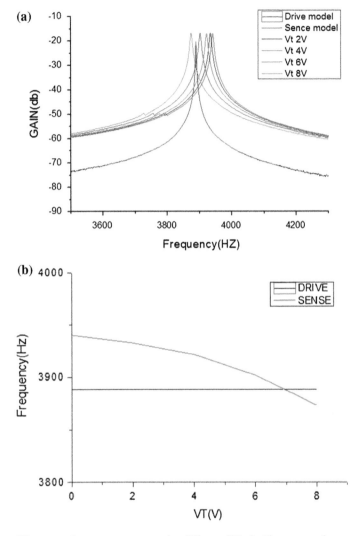

Fig. 10 a The sense frequency sweep under different VT. **b** The curve of sense resonant frequency with the *tuning voltage*

larger quadrature component. The quadrature error signal will be suppressed completely through the quadrature cancellation, the remaining is only a small in-phase error which can be seen from Fig. 9b. The design of quadrature error correction scheme can effectively eliminate the quadrature signal and improve the performance of silicon micro-gyroscope.

Figure 10a shows the state of sense frequency under the different tuning voltage. It can be seen that mode-matched is performed by change the DC polarization voltage (VT) on the MEMS structure. Figure 10b shows that the natural frequencies

VT (V)	0	2	4	6.6
Bias instability (°/h)	3.26	1.93	1.44	0.79

Table 1 The bias instability under different VT

of the sense mode decrease when the tuning voltage VT increases. When the tuning voltage is 6.6 V, the gyroscope is working under the mode-matched.

Table 1 shows the state of the gyroscope in different tuning voltage. When the tuning voltage is 6.6 V, the micro-gyroscope will work under the mode-matched condition. The bias instability in mode-matched is 0.79°/h, which is improved four times than the bias instability without tuning voltage.

5   Conclusion

In this paper, the correction method based on negative stiffness effect for the micro-gyroscopes is analyzed. By detecting the quadrature signal and regulating the input voltage applied on the quadrature electrodes, the quadrature error is eliminated at its source. By taking advantage of the tuning combs and the negative stiffness effect, the resonance frequency of the sense mode is adjusted by the tuning voltage. When the gyroscope works on the mode-matched model, it greatly improves the gyroscope sensitivity. In order to realize the closed-loop control, the PID correction system is designed to broaden the bandwidth of the open loop. The simulation and experiment show the feasibility of correction circuit. In the presence of quadrature correction and frequency tuning, the bias stability is dramatically improved.

Acknowledgments   This work is supported by NSAF (Grant No:U1230114) and National Natural Science Foundation of China (Grant No: 61571126 and 61104217).

References

Geen JA, Sherman SJ, Chang JF, Lewis SR (2002) Single-chip surface micro machined integrated gyroscope with 50°/h Allan deviation. IEEE J Solid-State Circuits 37(12):1860–1866
Joachim D, Lin L (2003) Characterization of selective polysilicon deposition for MEMS resonator tuning. J Microelectromech Syst 12(2):193–200
Remtema T, Lin L (2001) Active frequency tuning for micro resonators by localized thermal stressing effects. Sens Actuators A 91(2001):326–332
Shkel AM, Seshia AA, Horowitz R (1999) Dynamics and control of micromachined gyroscopes. In: Proceedings of the American control conference, San Diego, USA, pp 2119–2124

Tatar E, Alper SE, Member IEEE, Akin T (2012) Quadrature-error compensation and corresponding effects on the performance of fully decoupled MEMS gyroscopes. J Microelectromech Syst 21(3):656–667

William AC, Roger TH (1996) Surface micromachined z-axis vibratory rate gyroscope. In: Solid state sensor and actuator workshop, Hilton Head, USA

Xu L, Li HS, Ni YF, Liu J, Huang LB (2014) Frequency tuning of work modes in z-axis dual-mass silicon microgyroscope. J Sens 2014(3):1–13

Yazdi N, Ayazi F, Najafi K (1998) Micromachined inertial sensors. Proc IEEE 86(8):1640–1658

Autofocus for Enhanced Measurement Accuracy of a Machine Vision System for Robotic Drilling

Biao Mei, Wei-dong Zhu and Ying-lin Ke

Abstract Erroneous object distance often causes significant errors in vision-based measurement. In this paper, we propose to apply autofocus to control object distance in order to enhance the measurement accuracy of a machine vision system for robotic drilling. First, the influence of the variation of object distance on the measurement accuracy of the vision system is theoretically analyzed. Then, a Two Dimensional Entropy Sharpness (TDES) function is proposed for autofocus after a brief introduction to various traditional sharpness functions. Performance indices of sharpness functions including reproducibility and computation efficiency are also presented. A coarse-to-fine autofocus algorithm is developed to shorten the time cost of autofocus without sacrificing its reproducibility. Finally, six major sharpness functions (including the TDES) are compared with experiments, which indicate that the proposed TDES function surpasses other sharpness functions in terms of reproducibility and computational efficiency. Experiments performed on the machine vision system for robotic drilling verify that object distance control is accurate and efficient using the proposed TDES function and coarse-to-fine auto-focus algorithm. With the object distance control, the measurement accuracy related to object distance is improved by about 87 %.

Keywords Object distance · Measurement accuracy · Sharpness function · Autofocus · Vision-based measurement · Robotic drilling

1 Introduction

Robotic drilling system has become a feasible option for fastener hole drilling in aircraft manufacturing because of its low investment, high flexibility, and satisfactory drilling quality. Research and development of flexible robotic drilling

B. Mei · W. Zhu (✉) · Y. Ke
The State Key Lab of Fluid Power Transmission and Control, School of Mechanical
Engineering, Zhejiang University, Hangzhou 310027, China
e-mail: wdzhu@zju.edu.cn

© Zhejiang University Press and Springer Science+Business Media Singapore 2017   333
C. Yang et al. (eds.), *Wearable Sensors and Robots*, Lecture Notes in Electrical
Engineering 399, DOI 10.1007/978-981-10-2404-7_27

systems have been conducted by universities and aircraft manufacturers since the year 2000. Lund University developed an effective robotic drilling prototype system, which demonstrated the great potential of applying robotic drilling in the field of aircraft assembly (Olsson et al. 2010). Electroimpact (EI) Corporation developed the One-sided Cell End effector (ONCE) robotic drilling system, and successfully applied it to the assembly of Boeing's F/A-18E/F Super Hornet (DeVlieg et al. 2002). Through scanning a workpiece with a vision system, the ONCE system can measure the workpiece's position, and correct drilling positions by comparing the nominal and actual positions of the workpiece. Beijing University of Aeronautics and Astronautics (BUAA) developed an end-effector MDE60 for robotic drilling (Bi and Liang 2011), which integrated a vision unit for measuring the locations of workpieces and welding seams. A robotic drilling system developed at Zhejiang University (ZJU) also integrated a vision system, which was applied to measure reference holes for correcting the drilling positions of a workpiece, refer Fig. 1.

In robotic drilling, the mathematical model of the work cell (including the robot, workpiece, jigs, etc) is used as the basis for the generation of robot programs. However, the mathematical model is not accurately coincident with reality regarding the shape, position, and orientation of the workpiece, which leads to the position errors of drilled fastener holes. In order to enhance the position accuracy of drilled holes, the robot programs should be created in accordance with the actual assembly status of the workpiece. This is usually achieved through creating some reference holes on the workpiece, whose actual positions are typically measured with a vision system, and correcting the drilling positions of fastener holes according to the differences of the actual and nominal positions of the reference holes (Zhu et al. 2013).

High quality standards in aerospace industry require that fastener holes drilled by a robotic system have a position accuracy of ±0.2 mm (Summers 2005), which in turn places a stringent requirement on the measurement accuracy of the vision system integrated into the robotic system. Since the image of an object being photographed changes with the variation of object distance (distance between the

Fig. 1 ZJU's robotic drilling system equipped with a vision system

camera lens and object being shot), measurement errors occur when the object is not in-focus in vision-based measurements. Therefore, it is important to ensure that object distance is correct and consistent during vision-based measurements. Autofocus, which automatically adjusts the distance between the object being shot and the camera lens to maximize focus of a scene (Chen et al. 2013), can be used to locate the in-focus position to achieve the correct object distance. Focusing has been widely used in bio-engineering, medicine, and manufacturing (Firestone et al. 1991; Geusebroek et al. 2000; Mateos-Pérez et al. 2012; Santos et al. 1997; Liu et al. 2007; Osibote et al. 2010; Handa et al. 2000), where it is used to improve image sharpness. In this paper, we propose to use autofocus to improve the measurement accuracy of a machine vision system.

The rest of the paper is organized as follows: In Sect. 2, measurement errors of a machine vision system with respect to the variation of object distance are theoretically analyzed. Several traditional sharpness functions and a new sharpness function based on two-dimensional entropy are presented in Sect. 3. In Sect. 4, a coarse-to-fine autofocus algorithm for fast autofocus in the robotic drilling environment is discussed. Section 5 addresses experimental evaluations of various sharpness functions. Experiments of object distance control in a machine vision system with the proposed autofocus method are presented in Sect. 6. Finally, conclusions are drawn in Sect. 7.

2 Measurement Errors Induced by Erroneous Object Distance

In robotic drilling, positions of reference holes are measured with a 2D vision system. In order to achieve the required measurement accuracy, error sources of the vision system should be analyzed and controlled. Since the drill axis is perpendicular to the workpiece in robotic drilling, the deviation along the drill axis does not introduce positioning errors of fastener holes. Thus, major factors regarding measurement accuracy of the vision system mainly include: flatness of the workpiece surface being measured, perpendicularity of the optical axis of the camera with respect to the workpiece surface, and variation of object distance. In robotic drilling for aircraft assembly, aeronautical components dealt with are often panel structures such as wing and fuselage panels, which are small in surface curvature. In addition, a region shooted by a camera in a single shoot is small. Therefore, it can be reasonably assumed that the region captured by the camera is planar. The optical-axis-to-workpiece perpendicularity can be guaranteed by accurate installation and extra perpendicularity sensors. Due to improvement of lens production, use of a camera with small field of view of $18°$ (the visible area is 28 mm × 21 mm) and short object distance, skewness and lens distortion can be reasonably ignored (Zhan and Wang 2012). So, it is suitable to analyze influence of object distance on measurement accuracy with the pinhole imaging model.

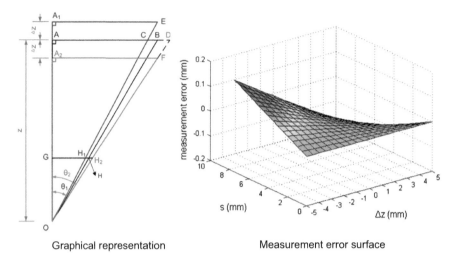

Graphical representation   Measurement error surface

Fig. 2 Measurement errors induced by erroneous object distance

Measurement errors caused by erroneous object distance in a machine vision system for robotic drilling are illustrated in Fig. 2a. Suppose O is the optical center of a camera, OA is the optical axis. GH is the image plane, AB is an object plane perpendicular to the optical axis. Ideally, the projection of an object point B onto the image plane is H. The corresponding image point moves to H_1, when object distance increases to OA_1; accordingly, the corresponding image point moves to H_2, when object distance decreases to OA_2. Thus, measurement errors of the vision system occur when the geometric dimension of the object being measured is calculated from an image acquired at some erroneous object distance.

Let in-focus object distance OA be z, and suppose object distance fluctuates by Δz, hence, $OA_1 = z + \Delta z$, $OA_2 = z - \Delta z$. For clarity, define $AB = A_1E = A_2F = s$. From the similarity $\triangle OAC \cong \triangle OA_1E$, it follows that

$$\tan \theta_1 = \frac{A_1E}{OA_1} = \frac{AC}{OA}, \quad AC = \frac{OA \cdot A_1E}{OA_1} = \frac{zs}{z + \Delta z} \tag{1}$$

and the measurement error is

$$\Delta s = AC - AB = \frac{zs}{z + \Delta z} - s < 0 \tag{2}$$

Similarly, from $\triangle OAD \cong \triangle OA_2F$, we can obtain

$$\tan \theta_2 = \frac{A_2F}{OA_2} = \frac{AD}{OA}, \quad AD = \frac{OA \cdot A_2F}{OA_2} = \frac{zs}{z - \Delta z} \tag{3}$$

thus the measurement error is

$$\Delta s = AD - AB = \frac{zs}{z - \Delta z} - s > 0 \qquad (4)$$

Assuming that Δz is positive when object distance increases (and vice versa), Eqs. (2) and (4) can be combined into a unified model of measurement errors as follows.

$$\Delta s = \frac{zs}{z + \Delta z} - s \qquad (5)$$

The surface model of Eq. (5) is shown in Fig. 2b. If the in-focus object distance of the vision system is $z = 225$ mm, object distance deviates by $\Delta z = \pm 2$ mm (accuracy of industrial robots and setup accuracy of aircraft panel structures are usually at the mm level), and the distance of a geometric feature from the optical axis is $s = 7.5$ mm, then the maximum measurement error is 0.07 mm. Such an error is significant for vision-based measurement in robotic drilling because the target accuracy of which is typically ± 0.1 mm. Moreover, object detection algorithms demand sharply focused images to extract useful information (Bilen et al. 2012). So, it is essential to use autofocus to achieve the correct object distance and make a workpiece being shot in-focus in vision-based measurements in robotic drilling.

3 Sharpness Functions and Their Performance Indices

3.1 Sharpness Functions

Autofocus is the process of adjusting object distance or focal length based on the sharpness of acquired images so that the object being photographed is at the focal position. In robotic drilling, camera internal parameters including the focal length are constants in order to take correct measurements of geometric dimensions of workpieces. Therefore, autofocus is achieved through adjusting the distance between the camera lens and workpiece surface being photographed on the basis of image sharpness measures. Sharpness of an image reflects the amount of high frequency components in the frequency field of the image (Tsai and Chou 2003), as well as the richness of edges and details in the image. Defocused images inherently have less information than sharply focused ones (Krotkov 1988). Based on the described phenomena, various sharpness functions have been proposed in literature. Considering the frequency of usage and variety, the following classical sharpness functions are selected for comparison with the proposed new sharpness function.

A sharpness function based on Discrete Cosine Transform (DCTS) (Zhang et al. 2011) uses the amount of high frequency components in an image as the sharpness index of the image, and the greater the amount of high frequency components, better degree of focus of the image. A sharpness function based on Prewitt Gradient Edge Detection (PS) (Shih 2007) evaluates image sharpness by detecting the edge gradient features of an image using an operator simpler than the Sobel operator. Laplacian sharpness (LS) (Mateos-Pérez et al. 2012) function instead uses the Laplacian operator to detect gradient features of edges in an image. Variance sharpness (VS) function (Mateos-Pérez et al. 2012) uses the square of the total standard deviation of an image as its sharpness criterion. One dimensional entropy sharpness (ODES) function is based on Monkey Model Entropy (MME) (Firestone et al. 1991). MME is the first simple and widely used method for the calculation of the entropy of an image (Razlighi and Kehtarnavaz 2009), which reflects the average amount of information or uncertainty in the image. According to the definition of information entropy (Shannon 2001), the entropy of an image characterizes its overall statistical property, or more specifically, the overall measure of uncertainty. In-focus image often has wider distribution of gray values than out-of-focus images, and the distribution of gray values reflect image sharpness to some extent. Therefore, the entropy of an image can be used as the sharpness measure of the image.

For an 8-bit gray-scale image, the range of gray values is $\chi_1 = \{0, 1, 2, \ldots, 255\}$. It is assumed that all gray levels have the same and independent distribution. The frequency of a gray level $x(x \in \chi_1)$ is N_x in the image with N pixels. A pixel in the image is denoted X_{ij}, where i and j are the row and column numbers, respectively. Thus, the one dimensional probability density function $P(x) = \Pr\{X_{ij} = x\}$ can be constructed from the histogram of the image using $P(x) = N_x/N$. Based on Shannon's definition of information entropy (Shannon 2001), image entropy can be similarly defined by

$$H(X) = -\sum_{x \in \chi_1} P(x) \log P(x) \tag{6}$$

and one-dimensional entropy sharpness function (Firestone et al. 1991) is:

$$F = -\sum_{x=0}^{255} P(x) \log P(x) \tag{7}$$

Equation (6) is computationally efficient, however, it is not accurate enough for estimating image entropy. In Eq. (6), each pixel is used without considering its adjacent pixels in the image, and the spatial relationships between the pixels are also neglected. Therefore, less information is available in this statistical property of gray values, and the resulting image entropy is lower than its actual magnitude. In order to describe image entropy more accurately, it is necessary to consider the spatial relationships between pixels in image entropy calculation. However, the dimension of the random distribution field of gray values increases when more relationships between pixels are included in image entropy estimation, leading to increased computational complexity. A good balance between accuracy and

Table 1 The range of joint gray values of an 8-bit image

(0,0)	(0,1)	(0,2)	...	(0,254)	(0,255)
(1,0)	(1,1)	(1,2)	...	(1,254)	(1,255)
(2,0)	(2,1)	(2,2)	...	(2,254)	(2,255)
...	...	...	...	...	...
(254,0)	(254,1)	(254,2)	...	(254,254)	(254,255)
(255,0)	(255,1)	(255,2)	...	(255,254)	(255,255)

complexity of entropy calculation is needed because image processing speed is important for the in-process vision system integrated in the robotic drilling system. Thus, we propose to use the two-dimensional joint entropy of vectors defined by a pixel and its right adjacent in an image, which was initially used for object extraction (Pal and Pal 1991), as the measure of image entropy.

The joint gray value of a pixel and its right adjacent in an image can be represented by (x, y), and its range is shown in Table 1. Denote the two-dimensional probability distribution of the joint gray values as $P(x, y) = \Pr\{X_{ij} = x, X_{i+1,j} = y\}$, where x and y are the gray values of a pixel X_{ij} and its right adjacent pixel $X_{i+1,j}$, respectively. This probability distribution can be obtained from the histogram of two-dimensional joint gray values by using $P(x, y) = N_{xy}/N$, where N_{xy} is the frequency of a joint gray value (x, y) in the image with N pixels. The two-dimensional spatial entropy of the image is defined as

$$H(X, Y) = - \sum_{(x,y) \in \chi_2} P(x, y) \log P(x, y) \tag{8}$$

Based on Eq. (8), a Two-Dimensional Entropy Sharpness (TDES) function for autofocus can be given as follows:

$$F = - \sum_{x=0}^{255} \sum_{y=0}^{255} P(x, y) \log P(x, y) \tag{9}$$

3.2 Performance Indices of Sharpness Functions

Besides performance indices of sharpness functions such as accuracy, focusing range, unimodality, half width, and sensitivity to environmental parameters summarized in (Firestone et al. 1991; Shih 2007), two additional performance indices: reproducibility and computation efficiency for the evaluation of various sharpness functions, are presented in this section.

Since variation of object distance causes significant errors in vision-based measurements, control of object distance is important to achieve the same in-focus

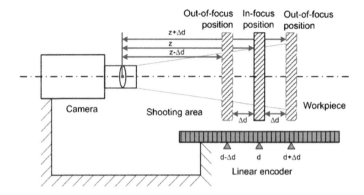

Fig. 3 The relationship of object distance z and controllable position variable d

position in multiple autofocus operations. Therefore, reproducibility is used as a performance index in autofocus in the vision system. Reproducibility is different from the index "accuracy" in literature (Firestone et al. 1991) which is defined as twice the standard deviation (the radius of a 95 % confidence interval) of the resulting in-focus positions in multiple autofocus operations. The lower reproducibility is easier and the same in-focus position can be achieved, and the better object distance can be controlled in multiple autofocus operations. The index "reproducibility" is treated as the most important performance index for autofocus due to its direct and significant impact on measurement accuracy of a vision system.

For verifying the effectiveness regarding control of object distance in the vision system integrated in the robotic drilling system, encoder reading of the end-effector can be used as the as the controllable variable during the autofocus processes of the vision system. The relationship of object distance and the encoder reading is illustrated in Fig. 3. Object distance z is the distance between an object plane and the optical center of camera lens, which is difficult to measure. However, variation of object distance in autofocus operations can be directly reflected by the encoder reading d, which can be controlled using the control system of robotic drilling.

Another important performance index for online vision-based measurements is computation efficiency which is treated as the time cost for images sharpness estimation.

4 Coarse-to-Fine Autofocus Algorithm

The time cost for image sharpness estimation contributes to the total time cost of an autofocus process with a certain proportion. A coarse-to-fine autofocus algorithm based on global search (Kehtarnavaz and Oh 2003) is proposed to reduce the time cost and eliminate negative influence of local extremums, which often occur in the traditional hill-climbing autofocus algorithm. The coarse-to-fine autofocus algorithm

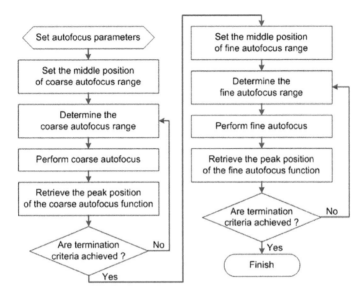

Fig. 4 The flow chart of the proposed autofocus algorithm

is shown in Fig. 4, in which autofocus is achieved by adjusting object distance z according to the spindle position d of the drilling end-effector.

The detailed procedures of the algorithm are described as follows:

Step 1: Set autofocus parameters:

- Range width of coarse autofocus, W_{coarse};
- Range width of fine autofocus, W_{fine};
- Increment of coarse autofocus, Δd_{coarse};
- Increment of fine autofocus, Δd_{fine};
- Threshold of the difference between the peak position of sharpness function and boundaries of coarse autofocus range in coarse autofocus, T_{coarse};
- Threshold of the difference between the peak position of sharpness function and boundaries of fine autofocus range in fine autofocus, T_{fine}.

Step 2: Set the middle position of coarse autofocus range d_{coarse}^{mid} as the home position of the spindle, where the object being photographed can be roughly seen.
Step 3: Determine coarse autofocus range $\left(d_{coarse}^{min}, d_{coarse}^{max}\right)$, where $d_{coarse}^{min} = d_{coarse}^{mid} - 0.5 \cdot W_{coarse}$, and $d_{coarse}^{max} = d_{coarse}^{mid} + 0.5 \cdot W_{coarse}$;
Step 4: Perform coarse autofocus with the starting position d_{coarse}^{min} and the increment Δd_{coarse} until the destination d_{coarse}^{max} is reached;
Step 5: Determine d_{coarse}^{peak}, which is the peak position of the sharpness function used in coarse autofocus;

Step 6: Set the middle position of fine autofocus range $d_{\text{fine}}^{\text{mid}}$ as $d_{\text{coarse}}^{\text{peak}}$, if termination criteria $d_{\text{coarse}}^{\text{peak}} - d_{\text{coarse}}^{\text{min}} > T_{\text{coarse}}$ and $d_{\text{coarse}}^{\text{max}} - d_{\text{coarse}}^{\text{peak}} > T_{\text{coarse}}$ are achieved. Otherwise, set $d_{\text{coarse}}^{\text{mid}}$ as $d_{\text{coarse}}^{\text{peak}}$, and go back to Step 3;

Step 7: Determine fine autofocus range $\left(d_{\text{fine}}^{\text{min}}, d_{\text{fine}}^{\text{max}}\right)$, where $d_{\text{fine}}^{\text{min}} = d_{\text{fine}}^{\text{mid}} - 0.5 \cdot W_{\text{fine}}$, and $d_{\text{fine}}^{\text{max}} = d_{\text{fine}}^{\text{mid}} + 0.5 \cdot W_{\text{fine}}$;

Step 8: Perform fine autofocus with the starting position $d_{\text{fine}}^{\text{min}}$ and the increment Δd_{fine} until the destination $d_{\text{fine}}^{\text{max}}$ is reached;

Step 9: Determine $d_{\text{fine}}^{\text{peak}}$, which is the peak position of the sharpness function used in fine autofocus;

Step 10: Accept $d_{\text{fine}}^{\text{peak}}$ as the result of coarse-to-fine autofocus, if termination criteria $d_{\text{fine}}^{\text{peak}} - d_{\text{fine}}^{\text{min}} > T_{\text{fine}}$ and $d_{\text{fine}}^{\text{max}} - d_{\text{fine}}^{\text{peak}} > T_{\text{fine}}$ are reached. Otherwise set $d_{\text{fine}}^{\text{mid}}$ for fine autofocus as $d_{\text{fine}}^{\text{peak}}$, and go to Step 7.

5 Experimental Evaluation of Sharpness Functions

5.1 Setup and Procedures of the Experiments

The experimental platform used for sharpness function evaluation is shown in Fig. 5, which consists of a Coord3 coordinate measuring machine (CMM), a Baumer TXG12 CCD camera, an annular light, a clamping unit, a test piece for reference hole detection, and a computer installed with the developed machine vision software.

The CCD camera, which has a lens of fixed focal length, is hold by the clamping unit. The CCD camera is connected to the computer using an Ethernet cable. The test piece is fixed on the Z-axis of the CMM. The position of the reference hole on the image plane can be adjusted by moving the CMM along its X- and Z-axis. Object distance can be adjusted by moving the CMM along its Y-axis. However,

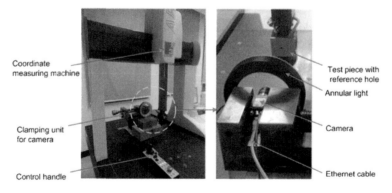

Fig. 5 The experimental platform for evaluation of sharpness functions

Table 2 Experimental conditions for image acquisition

Experimental conditions	Exposure time (μs)	Distribution of light intensity	Position of the reference hole	Type of the reference hole
1	13,793	Uniform	Centered	Countersunk
2	23,793	Uniform	Centered	Countersunk
3	38,793	Uniform	Centered	Countersunk
4	53,793	Uniform	Centered	Countersunk
5	68,793	Uniform	Centered	Countersunk
6	83,793	Uniform	Centered	Countersunk
7	38,793	Non-uniform	Centered	Countersunk
8	38,793	Uniform	Not centered	Countersunk
9	38,793	Uniform	Centered	No hole
10	38,793	Uniform	Centered	Not countersunk

the exact value of object distance is not known. Therefore, we use the position d, which is the reading of Y-axis coordinate of the CMM, to achieve autofocus.

During the image acquisition processes in focusing operations, various environmental conditions listed in Table 2 were designed to cover as many as possible real scenarios in robotic drilling. And the main environmental factor is light intensity which have a directly impact on brightness of acquired images. Low light intensity lower the intensity of each pixel so that all intensities in images are within a narrow range (Jin et al. 2010). However, light intensity is influenced by many factors, and is difficult to control and quantify. With light intensity being fixed, image brightness can be adjusted by exposure time, and short exposure time leads to darker images. Hence, exposure time was used as a major influential condition for evaluation of sharpness functions. Besides, four supplemental influence factors: distribution of light intensity, position of reference holes in images (centered or not), type of reference holes (countersunk or not, refer Fig. 6), were also included to test robustness of various sharpness functions. Steps for sharpness function evaluation are presented as follows:

(1) Adjust object distance by moving the CMM along its Y-axis to approximately make the vision system in-focus, and the reading of Y-axis coordinate of the CMM $d = 215$ mm. Set focusing range as [205, 225] mm based on this roughly in-focus position d.
(2) Conduct focusing with the global search strategy (Kehtarnavaz and Oh 2003). Firstly, move the test piece to the position $d = 205$ mm; then move the test piece forward with an increment 0.2 mm; capture an image after each moving, and compute the sharpness of the image used various sharpness functions; repeat this process until the position $d = 225$ mm.
(3) Plot curves of various sharpness functions in the XY coordinate system, where X-axis and Y-axis stands for stands for the position d of the images and the sharpness values of the images normalized to [0, 1], respectively.

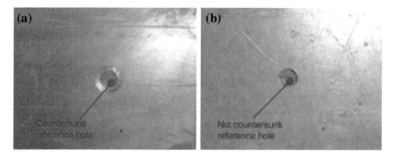

Fig. 6 Test pieces with **a** countersunk and **b** non-countersunk reference holes used in experiments

5.2 Performance Evaluation of Sharpness Functions

Curves of various sharpness functions obtained in experiments are shown in Fig. 7, and the characteristics of the sharpness functions are analyzed as follows:

All sharpness functions have their best characteristics when the imaged area of the test piece has no reference hole, refer to Fig. 7i. Sharpness functions fluctuate when distribution of light intensity is nonuniform, refer to Fig. 7g; however, almost all sharpness functions do not fluctuate in the range close to the in-focus position except the VS function, which lost the bell-shaped property in this case, and cannot be used in autofocus. Thus, the VS function is not further considered.

Form the curves of various sharpness functions obtained with different exposure times (Fig. 7a–f), it can be seen that ODES is sensitive to the change of exposure time, while the DCTS, PS, LS, and TDES functions are stable with respect to various exposure times used.

The resulting in-focus positions obtained using various sharpness functions are shown in Fig. 8, and reproducibility values of various functions DCTS, PS, LS, ODES, TDES, VS are 0.454, 0.430, 0.535, 0.386, 0.368, 8.909 mm, respectively, refer Fig. 8. Thus, it can be seen: the reproducibility value of VS function is 8.908 mm, which is coincident with the conclusion above that focusing is hard to achieved using VS function; the reproducibility value of TDES function is 0.368 mm as the smallest, which indicates that the same in-focus position is most likely to be obtained using this function in multiple autofocus operations under various experimental conditions.

The computer used in the experiments was a DELL OPTIPLEX 380 with a 2.93 GHz Intel Core CPU and 2G RAM. And the computer was installed with Microsoft Windows XP Professional and Microsoft Visual Studio 2005. The algorithms were implemented with C++ and OpenCV (Bradski and Kaehler 2008). Time costs for sharpness estimation with various sharpness functions are shown in Table 3, and a time cost is the time for calculating sharpness values for 101 images with 1296×966 pixels in size and acquired under experimental condition 3. From Table 3, the time costs of PS and DCTS are significantly longer than the other sharpness functions'.

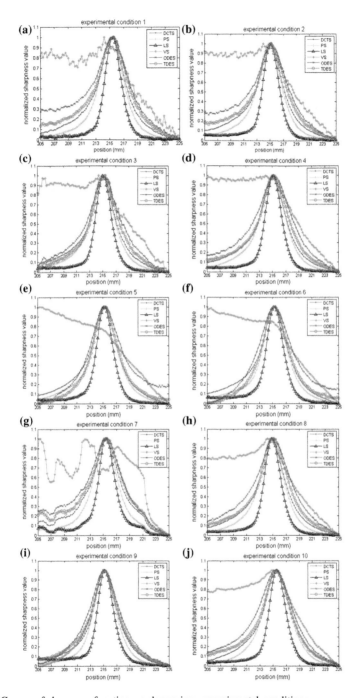

Fig. 7 Curves of sharpness functions under various experimental conditions

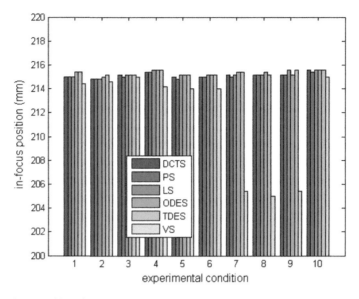

Fig. 8 In-focus positions for various sharpness functions

Table 3 Time costs for sharpness estimation with various sharpness functions

Sharpness functions	PS	DCTS	LS	VS	ODES	TDES
Time (s)	16.391	9.578	1.829	1.625	0.563	2.032

From the above discussions, it can be concluded that the TDES sharpness function has the smallest reproducibility value, reasonably small time cost, suitable focusing range, and stable focusing curve. Therefore, the TDES sharpness function is selected for the vision system used in robotic drilling, since both accuracy and efficiency of vision-based measurement are important in real assembly scenarios.

6 Experiments of Object Distance Control with Autofocus on a Vision System for Robotic Drilling

The proposed autofocus method has been adopted in the vision-based measurement system. Autofocus experiments have been performed on a robotic drilling end-effector to evaluate the effectiveness and reproducibility of the developed autofocus method. Figure 9 shows the robotic drilling platform for autofocus experiments, including a KUKA industrial robot, an drilling end-effector, a calibration board and its spindle mandril, a Baumer TXG12 CCD camera, an annular light, and a computer system (with vision-based measurement software refer to Fig. 10).

Fig. 9 The robotic drilling system used in experiments

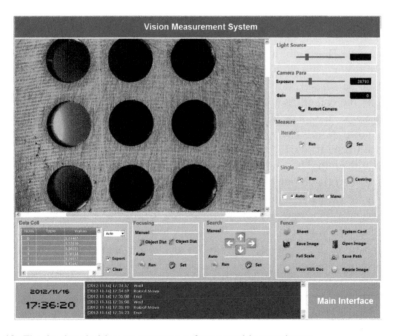

Fig. 10 The developed vision measurement software used in experiments

The CCD camera and the annular light are installed in a housing box of the drilling end-effector. The tool holder of the spindle clamps the spindle mandril, which is connected to the calibration board. The position of the calibration board is

Fig. 11 The calibration board and test pieces with reference holes

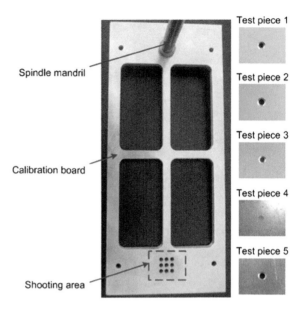

adjusted so that the shooting area (3 by 3 hole array) is visible to the camera, refer to Fig. 11. Once the adjustment is completed, the rotation of the spindle is restricted using locknut. Thus, only motion between the calibration board and the camera is along the drilling direction. The distance between the calibration board and the camera can be adjusted through moving the spindle along the drilling direction by controlling the end-effector. Since the position of the spindle directly reflects the object distance, autofocus of the calibration board is achieved through adjusting the spindle position d. The parameters of the autofocus algorithm are set as $W_{\text{coarse}} = 20\,\text{mm}$, $W_{\text{fine}} = 2\,\text{mm}$, $\Delta d_{\text{coarse}} = 1\,\text{mm}$, $\Delta d_{\text{fine}} = 0.2\,\text{mm}$, $T_{\text{coarse}} = 2\,\text{mm}$, and $T_{\text{fine}} = 0.2\,\text{mm}$. Then move the spindle such that the linear encoder reading (the spindle position) is 90 mm, and set $d_{\text{coarse}}^{\text{mid}}$ as this value.

The curve of a coarse-to-fine autofocus of the calibration board is shown in Fig. 12. Totally 32 images are captured and used for sharpness calculation with the TDES sharpness function in the autofocus process. According to Fig. 12, the spindle positions $d_{\text{coarse}}^{\text{peak}}$ and $d_{\text{fine}}^{\text{peak}}$ are 96.003 and 96.203 mm in the coarse and fine autofocus stage, respectively, during the autofocus of the calibration board. The time cost for computing the sharpness of the 32 images is 0.64 s.

In the experiments, five test pieces with reference holes were used to verify the reproducibility and robustness of the proposed sharpness function and coarse-to-fine autofocus algorithm. After camera calibration, each of the five test pieces was fixed on the calibration board so that it was visible to the camera, and autofocus experiments were performed. The in-focus spindle positions with five different test pieces are shown in Fig. 13, and the reproducibility value is around

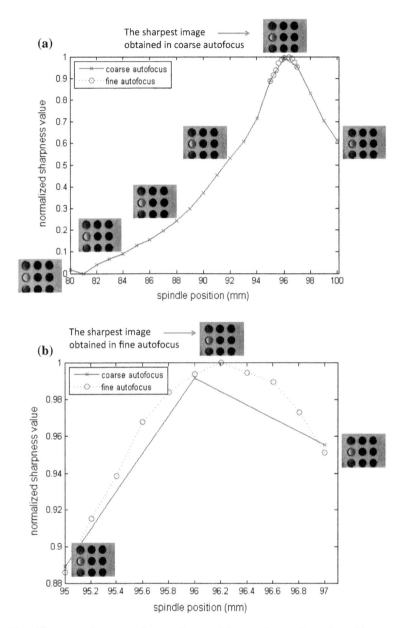

Fig. 12 **a** The curve of coarse-to-fine autofocus and **b** zoom in near the peak position

0.275 mm. Thus, the measurement error caused by the inconsistency of object distance is less than 0.009 mm when the in-focus object distance is 225 mm. And the measurement accuracy related to object distance is improved by about 87 %.

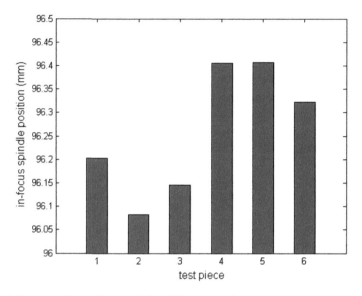

Fig. 13 In-focus spindle positions with five different test pieces

7 Conclusions

In this paper, the influence of erroneous object distance on the accuracy of vision-based measurement system is analyzed, and autofocus is applied to control the object distance to improve the measurement accuracy. A TDES function has been proposed for autofocus of the vision-based measurement system for robotic drilling applications. Experimental results show that overall performance of the proposed TDES function surpasses the traditional sharpness functions, and is most appropriate for vision-based measurement in robotic drilling. A coarse-to-fine autofocus algorithm has also been proposed to shorten the time cost of autofocus. Experiments performed on a robotic drilling end-effector show that the variation of object distance can be controlled to be within 0.275 mm using the proposed TDES function and autofocus algorithm, which ensures measurement error induced by erroneous object distance to be below 0.009 mm. The method of object distance control based on autofocus for improved vision-based measurement accuracy has been applied in robotic drilling for flight control surface assembly at Xi'an aircraft industry (group) company Ltd., Aviation Industry Corporation of China.

Acknowledgments Project supported by the National Natural Science Foundation of China (No. 51205352).

References

Bi S, Liang J (2011) Robotic drilling system for titanium structures. Int J Adv Manuf Technol 54 (5–8):767–774. doi:10.1007/s00170-010-2962-2

Bilen H, Hocaoglu M, Unel M et al (2012) Developing robust vision modules for microsystems applications. Mach Vis Appl 23(1):25–42. doi:10.1007/s00138-010-0267-y

Bradski G, Kaehler A (2008) Learning OpenCV: computer vision with the OpenCV library. O'Reilly Media, Incorporated, Sebastopol

Chen C-C, Chiu S-H, Lee J-H et al (2013) A framework of barcode localization for mobile robots. Int J Robot Autom 28(4):317–330

DeVlieg R, Sitton K, Feikert E et al (2002) ONCE (ONe-sided Cell End effector) robotic drilling system. SAE Technical Paper 2002-01-2626. doi:10.4271/2002-01-2626

Firestone L, Cook K, Culp K et al (1991) Comparison of autofocus methods for automated microscopy. Cytometry Part-A 12(3):195–206. doi:10.1002/cyto.990120302

Geusebroek J-M, Cornelissen F, Smeulders AWM et al (2000) Robust autofocusing in microscopy. Cytometry Part-A 39(1):1–9. doi:10.1002/(SICI)1097-0320(20000101)39:1<1: AID-CYTO2>3.0.CO;2-J

Handa DP, Foxa MDT, Harana FM, Petersb C, Morganc SA, McLeanc MA et al (2000) Optical focus control system for laser welding and direct casting. Opt Lasers Eng 34(4–6):415–427

Jin S, Cho J, Kwon K et al (2010) A dedicated hardware architecture for real-time auto-focusing using an FPGA. Mach Vis Appl 21(5):727–734. doi:10.1007/s00138-009-0190-2

Kehtarnavaz N, Oh HJ (2003) Development and real-time implementation of a rule-based auto-focus algorithm. Real-Time Imaging 9(3):197–203. doi:10.1016/S1077-2014(03)00037-8

Krotkov E (1988) Focusing. Int J Comput Vision 1(3):223–237. doi:10.1007/BF00127822

Liu XY, Wang WH, Sun Y (2007) Dynamic evaluation of autofocusing for automated microscopic analysis of blood smear and pap smear. J Microsc 227(1):15–23. doi:10.1111/j.1365-2818. 2007.01779.x

Mateos-Pérez JM, Redondo R, Nava R et al (2012) Comparative evaluation of autofocus algorithms for a real-time system for automatic detection of mycobacterium tuberculosis. Cytometry Part-A 81A(3):213–221. doi:10.1002/cyto.a.22020

Olsson T, Haage M, Kihlman H et al (2010) Cost-efficient drilling using industrial robots with high-bandwidth force feedback. Robot Comput Integr Manuf 26(1):24–38. doi:10.1016/j.rcim. 2009.01.002

Osibote OA, Dendere R, Krishnan S et al (2010) Automated focusing in bright-field microscopy for tuberculosis detection. J Microsc 240(2):155–163. doi:10.1111/j.1365-2818.2010.03389.x

Pal NR, Pal SK (1991) Entropy: a new definition and its applications. IEEE Trans Syst Man Cybern 21(5):1260–1270

Razlighi QR, Kehtarnavaz N (2009) A comparison study of image spatial entropy. In: Proceedings of SPIE-IS&T electronic imaging, visual communications and image processing, San Jose, CA, pp 72571X-1–72571X-10. doi:10.1117/12.814439

Santos A, De Sol Ortiz, rzano C, Vaquero JJ et al (1997) Evaluation of autofocus functions in molecular cytogenetic analysis. J Microsc 188(3):264–272. doi:10.1046/j.1365-2818.1997. 2630819.x

Shannon CE (2001) A mathematical theory of communication. ACM SIGMOBILE Mob Comput Commun Rev 5(1):3–55. doi:10.1145/584091.584093

Shih L (2007) Autofocus survey: a comparison of algorithms. In: Proceedings of SPIE-IS&T electronic imaging, digital photography III, San Jose, California, pp 65020B1–65020B11

Summers M (2005) Robot capability test and development of industrial robot positioning system for the aerospace industry. SAE Technical Papers 2005-01-3336. doi:10.4271/2005-01-3336

Tsai DM, Chou CC (2003) A fast focus measure for video display inspection. Mach Vis Appl 14 (3):192–196. doi:10.1007/s00138-003-0126-1

Zhan Q, Wang X (2012) Hand–eye calibration and positioning for a robot drilling system. Int J Adv Manuf Technol 61(5–8):691–701. doi:10.1007/s00170-011-3741-4

Zhang S, Liu J-H, Li S et al (2011) The research of mixed programming auto-focus based on image processing. In: Zhu R, Zhang Y, Liu B, Liu C (eds) Information computing and applications: Communications in computer and information science. Springer, Berlin, pp 217–225

Zhu W, Qu W, Cao L et al (2013) An off-line programming system for robotic drilling in aerospace manufacturing. Int J Adv Manuf Technol 68(9–12):2535–2545. doi:10.1007/s00170-013-4873-5

A New Scene Segmentation Method Based on Color Information for Mobile Robot in Indoor Environment

Xu-dong Zhang, Qi-Jie Zhao, Qing-XU Meng, Da-Wei Tu and Jin-Gang Yi

Abstract Scene segmentation is the basis of autonomous robots environmental understanding. For scene objects show different color characteristics aggregation in the mobile service robot operating indoor environment, this thesis proposes a scene segmentation method based on color layering and Multi-Size filtering. This paper slices the scene by constructing the color layering model and then designs the multi-size filter, which is designed according to the results of detecting the numbers and size of the connected domains in one layer, to segment the target. This paper also builds a robot operating system and the experiments of global environment and local scene are constructed with an average accuracy of scene segmentation and hierarchical, respectively, reaching 96.2 and 92.5 %. The results show that the method can effectively segment the scenes with salient color features.

Keywords Service robot · Scene segmentation · Color layering · Multi-size filtering

1 Introduction

The scene understanding research is widely concerned because that scene understanding, environmental information detection and perception are helpful for autonomous operation of service robot, intelligent driverless cars, etc. (Uckermann

X. Zhang · Q.-J. Zhao (✉) · Q.-X. Meng · D.-W. Tu · J.-G. Yi
School of Mechatronic Engineering and Automation, Shanghai University,
Shanghai 200072, China
e-mail: zqj@shu.edu.cn

Q.-J. Zhao · D.-W. Tu
Shanghai Key Laboratory of Intelligent Manufacturing and Robotics,
Shanghai 200072, China

J.-G. Yi
Department of Mechanical and Aerospace Engineering, Rutgers University,
Piscataway, NJ 08854, USA

© Zhejiang University Press and Springer Science+Business Media Singapore 2017   353
C. Yang et al. (eds.), *Wearable Sensors and Robots*, Lecture Notes in Electrical
Engineering 399, DOI 10.1007/978-981-10-2404-7_28

et al. 2014; Borges and Moghadam 2014; Li et al. 2014). In recent years, with the development of the aging society, autonomous service robot research is paid more attention. The robot's understanding of the work environment is critical among the problems of autonomous service robots practical application.

There are many methods to obtain environment information, among which the most frequently used as follows: Zhuang et al. (2011) complete identification of indoor scene frame through extracting plane feature of the indoor environment and determine the spatial relationship by the method of 3D laser ranging; Martinelli (2015) establishes a set of indoor global positioning system using RFID technology, which measures the current position through the reader fixed to robot accepting conversion signal RSSI and UHF-RFID phase. In addition, the method of helping robot obtaining scene information through interactive has been widely used. For example, Muthugala makes robot judge and understand environmental information by detecting distance and obtaining fuzzy language information of the operator (Muthugala et al. 2015); What is more, the methods of scene information acquisition based on vision are most widely used. Machine vision can not only achieve the detection but also recognize the objects in the scene after adding recognition algorithm.

The basis of scene understanding is to segment target objects in the scene. Some scholars had carried out researches about robot's work environment scene segmentation. For example, the indoor scene segmentation is summarized as three kinds: floors, walls and clutter objects (Posada et al. 2013), thus it can help the robot to identify indoor structure; moreover, some researchers do semantic segmentation for robot's cluttered working environment (Hoof et al. 2013; Potapova et al. 2014). As the scene depth information acquiring more accessible, the segmentation methods based on RGB-D information are hotspots. Karpathy et al. (2013) use PCL, Kinect and Fusion to distinguish objects in indoor 3D grid environment; Silberman provides a mark indoor scenarios' method which based on RGB-D data and the support relations of objects. This work is to emphasize more attachment relationship between each other than the accurate segmentation (Silberman et al. 2012). However, limited by its reasoning and decision-making ability, it is difficult for robot completely independent to make correct understanding about the scene. Some scholars put forward a way of human–computer interaction to segment scene (Johnson and Agah 2009; Mansur et al. 2008; Ren et al. 2012). According to the characteristics of service robot's semi-structure working environment, the environment scene segmentation and understanding by the way of human–computer interaction will be a feasible and practical method.

According to the characteristics of service robots working indoor and human–computer interaction collaboration, there are two major scenes which robot often meets: global environment and local operation scene. Because of objects in the room keeping static in a certain period of time, different targets can be distinguished through color image. In addition, different objects perform the aggregation of different size and color in various color layers. Based on the above analysis, this paper presents the method of layering colors into several layers which different targets include and designing multi-size filter to segment objects of the scene in the human–computer interaction working condition.

2 Segmentation Method

A segmentation method based on color layering and multi-size filtering is presented for the global environment and local operation scene of mobile robot indoor operation process. Considering the characteristics of the three channels [Hue (H), saturation (S), Value (V)] in HSV color space change independently, the information is easy to be extracted and analyzed from it. As shown in Fig. 1, we obtain scene source images and histograms of components, construct color layer model according to the aggregation of color, split layers with diverse color information from the original scene. The color layers which have been split are elaborated and including variety of objects with similar Hue in the scene can be layered further based on the H channel using color layered model. Detect the connected domains of the split layers and design the multi-size filter according to the quantity and size of them, so we segment objects from different color layers.

According to the distribution in three component (hue H, saturation S, value V) histograms of scene color image, and the result of calculating color degree of aggregation C_i, scene image is divided to bizarre layers including only black or white information and non-bizarre layers including various colors. The formulas (1–2) show Layered Model:

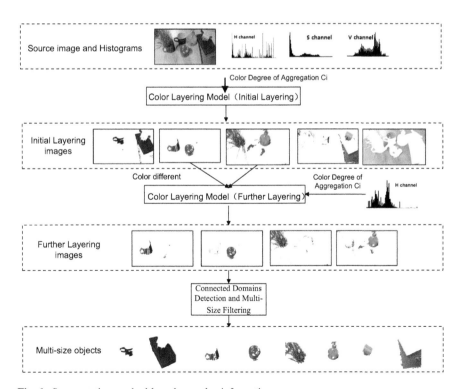

Fig. 1 Segmentation method based on color information

$$C_i = \{ \alpha[H_{di}H_{ui}], \beta[S_{di}S_{ui}], \gamma[V_{di}V_{ui}] \} \tag{1}$$

$$L_i(C_i) = \begin{cases} L_0 = \{C_0 | \alpha = 0, \beta = \gamma = 0.5\}, \text{white singular layer} \\ L_1 = \{C_1 | \alpha = \beta = 0, \gamma = 1\}, \text{black singular layer} \\ L_i = \{C_i | \alpha = 1, \beta = \gamma = 0, i > 1\}, \text{color nonsingular layers} \end{cases} \tag{2}$$

In the formulas, α, β, γ represent the weight of three components in layered colors; $[H_{di}\ H_{ui}]$, $[S_{di}\ S_{ui}]$, $[V_{di}\ V_{ui}]$ represent the range of each component determined from the histogram; $L_i\ (C_i)$ indicates the segmented ith layer segmented whose color aggregation is C_i.

Detect the size and number of connected domains in the segmented layers by statistical information. Multi-size filter is designed mainly by the size of connected domains. To filter out noise and segment target objects by the filter. Formula 3 show multi-size filter Model

$$MF_i[R_s(k)] = \begin{cases} \text{Noise}, R_s(k) < R_{s0} \\ Obj[k], R_{sd}(k) < R_s(k) < R_{su}(k) \end{cases} \tag{3}$$

In the formulas, $MF_i[R_s(k)]$ is the filter of valued $R_s(k)$ size in ith layer, $k = 1, 2, 3, \ldots, R_n$. R_n is the number of connected domains and R_s0 represents minimum resolvable target size. $Obj[k]$ is segmented target in the kth scale. $R_{sd}(k)$ and $R_{su}(k)$ are the upper and lower cutoff sizes for segmenting the target of valued $R_s(k)$.

3 Algorithm Implement

Specific algorithm steps of scene segmentation method based on color layering and Multi-Size filtering is shown in Fig. 2:

3.1 Scene Layering Algorithm

3.1.1 Primary Layering

According to color layered model (1), (2) and the ratio of three components (H, S, V) in the color diagram, we initially identify the model's coefficients to segment original scene into several layers:

(1) bizarre layers

white $L_0 = \{0.5 * (S \in [S_{di}S_{ui}]) + 0.5 * (V \in [V_{di}V_{ui}]) | S_{ui} < 0.15, V_{di} > 0.6\}$
black $L_1 = \{V \in [V_{di}V_{ui}] | V_{ui} < 0.15\}$;

Fig. 2 The flow chart of
segmentation method

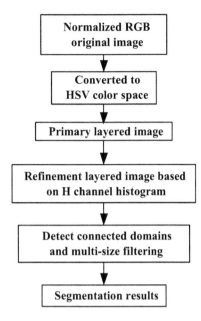

(2) non-bizarre layers:

$$L_i = \{H \in [H_{di} H_{ui}]\},$$

H_{di} and H_{ui} are the value of adjacent troughs in H histogram.

3.1.2 Refinement Layering

Draw H component histogram of the singular layer, respectively, then split the layer into a new layers based on the histogram distribution. $f(h)$, the longitudinal axis of H Component histogram, stands for the frequency of H value tonal point, the horizontal axis H stands for hue scale (Fig. 3). Take three H values of the wave

Fig. 3 H channel histogram
statistical information

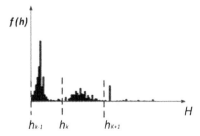

trough h_{k-1}, h_k, h_{k+1} as threshold value and re-split layer into $L_k[h_{k-1},h_k]$, $L_{k+1}[h_k, h_{k+1}]$. The method of identifying wave trough is as follows:

Find initial value. $P_0 = \{k, h(k) \mid h(k) <= h(k-1)$ & $h(k) \leq h(k+1)$, min $\leq k \leq$ max$\}$, P_0 is a set of initial values of wave trough.

Find obvious value. $P_1 = \{p_k, h(p_k) \mid h(p_k) \leq h(p_{k-1})$ & $h(p_i) \leq h(p_{k+1})$, $p_k \in P_0\}$, P_1 is a set of obvious values of wave trough.

Remove value beyond the threshold. When the frequency of value k (H_k) is beyond $0.5* H_{k,max}$, remove this value from P_1 and finally generates a new set A.

Remove adjacent value. If two values in set A are close enough, the two values are considered as in similar color region. Therefore, when the distance between the two values is less than $T(T < 15)$, remove the larger one and regenerates a set B.

If set B consists of M elements$\{p_1, p_2,..., p_M\}$, every two adjacent elements divides into a new layer, thus the coarse segmentation layers are subdivide and get the whole layers L_i.

3.2 Multi-size Filtering Algorithm

According to the multi-scale filter model, detected connected domains on all layers, and designed a filter based on the number and scale of connected domains, concrete algorithm steps are as follows:

(1) Detected based on Seed-Filling method and counted the number R_n of connected domain on each layer;

(2) Counted scale size $R_s(j)$, $j = 1,2, ..., R_n$ of each connected domain RC_j and sorted;

(3) Calculated the area sum of all the connected domains on each layer $S = \sum_{j=1}^{R_n} R_s(j)$;

(4) Counted the Proportion which is the area $R_s(j)$ take up the area sum of all the connected domains $(RC_j) = R_s(j)/S$;

(5) According to the ratio of the connected domain areas, determined the filter connected domain's scale $R_{s0}(0)$, $R_{sd}(k)$ and $R_{su}(k)$, k is the scale series, divided different scales target objects Obj[k] by filter;

$$
\text{Obj}[k] = \begin{cases} \text{Noise}, R_s(0) < R_{s0}(0), f(\text{RC0}) < 0.1 \\ \text{Obj[small]}, R_{s0}(k) < R_s(k) < R_{s1}(k), f(s0) > 0.1 \\ \text{Obj[mid]}, R_{s1}(k) < R_s(k) < R_{s2}(k), f(s1) > 0.3 \\ \text{Obj[big]}, R_s(k) > R_{s2}(k), f(s2) > 0.5 \end{cases}
$$

4   Experiments

In order to verify feasibility and effect of the proposed method, we construct a HCI and cooperative system which composed with a mobile robot, HCI interface, four degrees of freedom mechanical arm and color camera. HCI interface is shown in Fig. 4. Users give commands of task execution to the robot through the interface. Mobile robot completes the task indoor according to the task information, such as grasping object. During the movement, the robot uses color camera to acquire real-time global environment and the local operation scene and executes scene segmentation in grasping object stage (Fig. 5). Robot feeds back segmentation results and presents them on HCI interface, making HCI collaboration convenient.

The main purpose of this experiment is to verify the effect of proposed scene segmentation method, not to evaluate the result of task performance. Experimenters issue to the robot task instruction through the interface, and control the movement and operation of the robot. For different stages of the process of the mobile robot operations, Experimenters collected a representative of 22 scene images including global environment and local operation scenes, which have 174 separable regions or objects with obvious feature. Conduct segmentation experiments by building color layering models and filters. Whether stratification and filtering results is correct or not is judged by the experimenters according to segmentation results presented on the interface. Through the experiment, the average individual scene image processing time is 2.1 s. Statistical analysis and results are shown in Table 1. Figure 6 is a partial segmentation results.

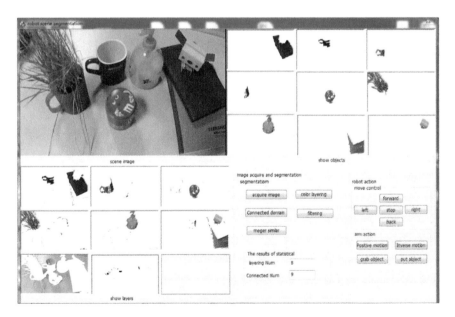

Fig. 4 The human–computer interaction interface

(a)

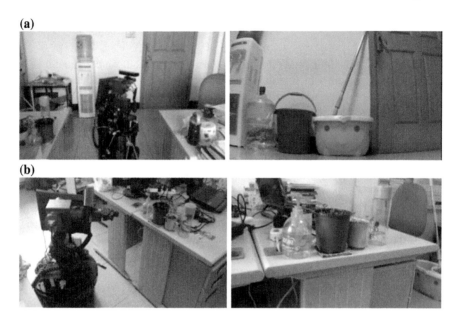

(b)

Fig. 5 The experimental environment. **a** The global environment scene. **b** The local operation scene

Table 1 Results and statistics

	Layering accuracy (%)	Segmentation accuracy (%)
Global environment scene	94.8	90.7
local operation scene	97.6	94.3
Average	96.2	92.5

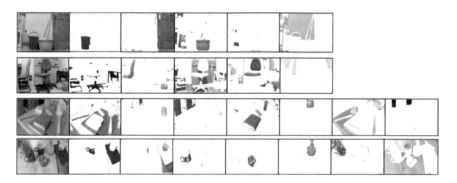

Fig. 6 Part results of color layering experiments

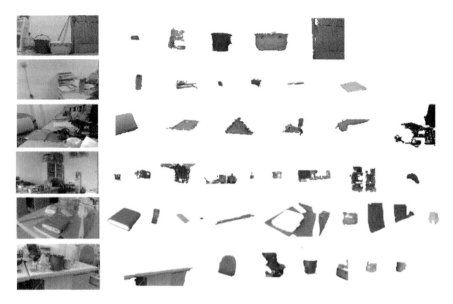

Fig. 7 Scene segmentation result

From Fig. 6 and Table 1, it can be seen from the layering results that target objects in each scene are not same because of color histogram statistics aggregation was different. In the local operation scene, the object hue difference is obviously higher than the global environmental scene, and the layering correct effect reach 97.6 %. As seen from segmentation results in Table 1 and Fig. 7, the target areas with a clear difference in hue channel have gotten a good effect, and the scale of the target object in local operation scene has a bigger proportion of the whole scene. Segmentation correct rate reaches 94.3 % after filtered for targets with different scales. Because there are more target objects and great difference in scale and Hue, besides, some small-scale target is treated as noise and filtered, segmentation correct rate in global environmental scene is 90.7 %.

According to the experimental analysis, global environmental scene includes objects of various sizes and color information and occlusions between objects often occur because of uneven illumination and reflection which will greatly affect the layered effect. As shown in Fig. 8, the difference of color between the wall and the bucket is too similar to be layered in the same layer. As the diffuse reflection of the door, the ground shows similar color with the door and be divided into different layers. In the below picture, the chair is shielded and having the shadows that they are divided into different layers. Because lighting conditions are relatively stable in local operation scene, the layering and segmenting process have good results. The average of layering and segmentation accuracy of the proposed method gets to 96.2 and 92.5 %. It shows that this method based on color layering and multi-size filtering is effective for regions or objects which have obvious characteristics of segmentation in the scene.

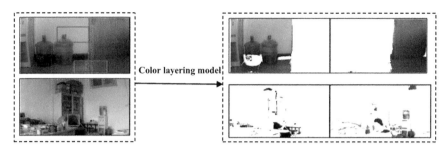

Fig. 8 Layering wrong affected by the light

5 Conclusion

Based on the problem of mobile service robots' operating environment scene understanding problem, we focus on scene segmentation method which has an important supporting role for environment understanding. This thesis proposes a scene segmentation method based on color layering and multi-size filtering: first, splitting layers by proposed color layer model to get different color layers because of environment scene can be treated as multi-color layers superimposed. Secondly, using proposed multi-size filtering to filter segment all the layers in order to get different size object areas. Finally, through HCI experiment, the result has verified this method can segment scene image effectively. Our experiment has only found and segmented connected domains and not judged whether two adjacent domains belong to the same object. Therefore, our approach will segment objects which contain many colors into severer domains. For understanding robot working environment, it is important to identify one whole object. Our further work will focus on test domains' properties and problems associated with each other.

Acknowledgments Project supported by the National Natural Science Foundation of China (No. 61101177).

References

Borges PVK, Moghadam P (2014) Combining motion and appearance for scene segmentation. In: 2014 IEEE international conference on robotics and automation (ICRA), pp 1028–1035
Hoof HV, Kroemer O, Peters J (2013) Probabilistic interactive segmentation for anthropomorphic robots in cluttered environments. IEEE-RAS Int Conf Humanoid Rob 169–176
Johnson DO, Agah A (2009) Human robot interaction through semantic integration of multiple modalities, dialog management, and contexts. Int J Social Robot 1(4):283–305
Karpathy A, Miller S, Li FF (2013) Object discovery in 3D scenes via shape analysis. In: 2013 IEEE international conference on robotics and automation (ICRA), IEEE, pp 2088–2095
Li J, Jin L, Fei S et al (2014) Robust urban road image segmentation. In: 11th IEEE world congress on intelligent control and automation (WCICA), pp 2923–2928

Mansur A, Sakata K, Rukhsana T et al (2008) Human robot interaction through simple expressions for object recognition. In: The 17th IEEE international symposium on robot and human interactive communication, pp 647–652

Martinelli F (2015) A robot localization system combining RSSI and phase shift in UHF-RFID signals. IEEE Control Syst Technol 99:1

Muthugala MAVJ, Jayasekara ABP et al (2015) Interpreting fuzzy linguistic information in user commands by analyzing movement restrictions in the surrounding environment. In: IEEE Moratuwa engineering research conference (MERCon), pp 124–129

Posada LF, Narayanan KK, Hoffmann F et al (2013) Semantic classification of scenes and places with omnidirectional vision. In: IEEE European conference on mobile robots, pp 113–118

Potapova E, Varadarajan KM, Richtsfeld A et al (2014) Attention-driven object detection and segmentation of cluttered table scenes using 2.5D symmetry. In: IEEE international conference on robotics and automation, pp 4946–4952

Ren X, Bo L, Fox D (2012) Rgb-(d) scene labeling: features and algorithms. In: 2012 IEEE conference on computer vision and pattern recognition (CVPR), pp 2759–2766

Silberman N, Hoiem D, Kohli P et al (2012) Indoor segmentation and support inference from RGBD images. In: Proceedings of the 12th European conference on computer vision—Volume Part V. Springer, Berlin, pp 746–760

Uckermann A, Eibrechter C, Haschke R et al (2014) Real-time hierarchical scene segmentation and classification. In: 14th IEEE-RAS international conference on humanoid robots (humanoids), pp 225–231

Zhuang Y, Lu X, Li Y (2011) Mobile robot indoor scene cognition using 3D laser scanning. Acta Automatica Sinica 37(10):1232–1240

Dynamic Hopping Height Control
of Single-Legged Hopping Robot

Zhi-wei Chen, Bo Jin, Shi-qiang Zhu, Yun-tian Pang and Gang Chen

Abstract In order to control the vertical hopping height of two degree-of-freedom articulated single-legged hopping robot, a hopping height control scheme based on the energy conservation in the course of hopping was proposed. The kinematic model of the legged hopping robot and the hybrid dynamic model on flight and stance phase according to the different constraint condition was established, and the ground contact model based on the impact collisions between end-effector and ground was analyzed. On the one hand when robot was controlled to hop higher, energy was imparted into the robot system by increasing the pre-compressed length of the virtual spring. On the other hand, when robot was controlled to hop lower, the redundant energy of the robot system was dissipated by the inelastic collisions when the end-effector touched down the ground. The control scheme is demonstrated by the simulation experiment of hopping height control implemented in MATLAB/Simulink. The robot's hopping height increases from 0.55 to 0.8 m and then decreases from 0.8 to 0.55 m with 0.05 m intervals. The phase plot shows that the dynamic hopping with different height approaches respective periodic and stable orbit. And the time domain hopping plot shows that the control scheme has fast dynamic response and nearly no steady-state error. The experiment result shows great efficiency of the control scheme proposed here.

Keywords Single-legged hopping robot · Dynamic hopping height control · Energy compensation algorithm

Z. Chen · B. Jin (✉) · S. Zhu · Y. Pang · G. Chen
State Key Laboratory of Fluid Power and Mechatronic Systems, Zhejiang University, Hangzhou 310027, China
e-mail: bjin@zju.edu.cn

© Zhejiang University Press and Springer Science+Business Media Singapore 2017   365
C. Yang et al. (eds.), *Wearable Sensors and Robots*, Lecture Notes in Electrical
Engineering 399, DOI 10.1007/978-981-10-2404-7_29

1 Introduction

Dynamic gait of legged robot that uses a ballistic flight phase does not need a continuous path of support, a broad base of support, nor closely spaced footholds; so that the dynamic robot system should be able to traverse more difficult terrain than static system (Hodgins and Raibert 1991). MIT cheetah robot which is inspired biologically is able to run with the speed of 22 km/h and even jump over hurdles with dynamic gait (Bosworth et al. 2014). Some kinds of mammals have the ability to navigate uneven terrain by hopping accurately from one safe placement spot to the next. For a robot to possess the similar capability, three problems are required solved at least (Bhatti et al. 2013). The first and fundamental problem that should be solved is a low-level controller which executes a series of hops from one stepping point to another. Raibert and his colleagues (Raibert 1986), who made some pioneer work on their single-legged robot, decoupled the problem into a three-part controller: (1) hopping, (2) forward speed, (3) posture. Meanwhile, they extend the single-legged robot controller to the bipeds. Jessica et al (Hodgins and Raibert 1991) explored three methods for controlling step length by adjusting one of the three parameters which are forward running speed, running height, and duration of ground contact. The legged robot is therefore able to travel on rough terrain by the feet placed on the available footholds. From the previous research, it is concluded that the vertical motion of a legged robot is the first step for dynamic gait on.

Much work has been carried out since the hopping height control is realized as one of the most important method to improve legged robot's ability to navigate steep and mountainous terrain. Spring loaded inverted pendulum (SLIP) (Yu et al. 2012; Council et al. 2014) is the simplest and fundamental template to analyze dynamical locomotion of robots from biomechanics and robotics perspectives. Assuming as a conservative system, several simulation analysis of hopping height control of a robot is done based on SLIP. Similarly more steady-state hopping height control strategies are realized on the telescopic single-legged robot (Prosser and Moshe 1993; Naik et al. 2005; Cherouvim and Evangelos 2009; Azahar et al. 2013). In order to break the fundamental limitation of SLIP model assuming no energy loss during impact, a more accurate representation of hopping procedure is proposed by considering inelastic impact of robot's end-effector with the ground. The control schemes of the maximum jumping height control were then developed on the hopping robot with two mass model and four-link model respectively (Bhatti et al. 2012; Mathis and Mukherjee 2013). A modified SLIP model with impact compensation (SLIPic) (Hutter et al. 2010) is utilized onto a segmented robotic leg. Continuous hopping with desired hopping height on uneven ground is achieved with a deadbeat controller by adjusting the angle of attack. An instantaneous control method over the step-time (Bhatti et al. 2012, 2013) is developed to meet the rapid changes in the desired height for a vertically hopping robotic leg. The robotic leg is modified from the hydraulically actuated leg of the quadruped robot HyQ. In order to theoretically research the feasibility to navigate steep and mountainous terrain for a legged robot, this paper mainly contributes to the hopping height control of the one-legged hopping robot.

In this paper, the hopping height of the single-legged hopping robot is controlled by an energy compensation algorithm to compensate the energy loss in the course of the phase of robot hopping. The robot system, which is based on the kinematic model and dynamic model of the single-legged hopping robot, takes into account the inelastic impact collision with the ground.

2 Model of the Single-Legged Robot

The articulated single-legged robot is composed of three rigid bodies: the upper body (consisting of hip), the thigh, and the shank with mass and their inertia with respect to the corresponding centers of mass as shown in Fig. 1. The thigh is attached to the main body by the hip joint, whereas the shank is linked to the thigh by the knee joint. The hip and knee joint are able to rotate actively within 120° in the sagittal plane. Considering the weight of some kinds of actuator such as hydraulic cylinder, the center of mass (CoM) of the thigh and shank link deviate from the centroid of the link. The single-legged robot's upper body is constrained on a vertical slider, so that it is free to slide vertically on the slider in order to perform a vertical jumping motion. The slide also limits hip joint motion on sagittal plane. Legged robots belong to the class of floating base systems, which can move anywhere in space using the ground as support (Hutter 2013). We form the floating base coordinate to describe the kinematic and dynamic model.

Fig. 1 Single-legged robot constrained by a vertical slider

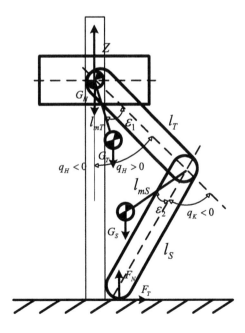

2.1 Kinematic Model

The generalized coordinate is used to mathematically describe the motion of floating base robot. The body fixed frame B, which can be arbitrarily displaced with respect to the inertial frame I, is displaced on hip. Robot kinematic are described as a function of the vector,

$$\mathbf{q} = (\mathbf{q}_b \quad \mathbf{q}_r)^{\mathrm{T}} \tag{1}$$

which includes the vector $\mathbf{q}_b$ describing the unactuated floating base coordinate and the actuated joint coordinate $\mathbf{q}_r$. As the x axis is fixed by the vertical slider, the information of x axis is not required here. So that the vector $\mathbf{q}_b$ has only the vertical height z. The vector $\mathbf{q}_r$ has the element of hip and knee joint angle $\mathbf{q}_r = (q_H \quad q_K)^{\mathrm{T}}$. q_H and q_K are hip and knee joint angle, respectively, the symbol of which is as shown in Fig. 1. In summary, we get the minimal coordinates

$$\mathbf{q} = (z \quad q_H \quad q_K)^{\mathrm{T}} \tag{2}$$

According to the Kinematics of the robot, we get the CoG(Center of Gravity) of thigh $\mathbf{r}_{mT}$ and shank $\mathbf{r}_{mS}$, here the abscissa value of the floating base coordinate is regarded as 0.

$$\mathbf{r}_{mH} = (0 \quad z)^{\mathrm{T}} \tag{3}$$

$$\mathbf{r}_{mT} = \begin{bmatrix} x_{mT} \\ z_{mT} \end{bmatrix} = \begin{bmatrix} l_{mT}\sin(q_H + \varepsilon_1) \\ z - l_{mT}\cos(q_H + \varepsilon_1) \end{bmatrix} \tag{4}$$

$$\mathbf{r}_{mS} = \begin{bmatrix} x_{mS} \\ z_{mS} \end{bmatrix}$$
$$= \begin{bmatrix} l_T\sin q_H + l_{mS}\sin(q_H + q_K + \varepsilon_2) \\ z - l_T\cos q_H - l_{mS}\cos(q_H + q_K + \varepsilon_2) \end{bmatrix} \tag{5}$$

the CoG of legged robot,

$$\mathbf{r}_{CoG} = \frac{m_H\mathbf{r}_{mH} + m_T\mathbf{r}_{mT} + m_S\mathbf{r}_{mS}}{m_H + m_T + m_S} \tag{6}$$

and the end-effector position of the legged robot,

$$\mathbf{r}_{ee} = \begin{bmatrix} x_{ee} \\ z_{ee} \end{bmatrix}$$
$$= \begin{bmatrix} l_T\sin q_H + l_S\sin(q_H + q_K) \\ z - l_T\cos q_H - l_{mS}\cos(q_H + q_K) \end{bmatrix} \tag{7}$$

where m_H, m_T, m_S denote CoM of hip, thigh, and shank. l_{mT} denotes length between hip joint and CoM of upper limb; l_{mS} denotes the length between knee joint and CoM of lower limb; ε_1 denotes shift angle to the CoM of upper limb; ε_2 denotes shift angle to the CoM of lower limb; l_T, l_S denote length of upper limb and lower limb, respectively. When the upper is heavy enough and for simplicity, the hip (CoG of upper leg) is always considered as the CoG of leg.

The leg Jacobian matrix $\mathbf{J}$ is significant in both kinematics and dynamics. It relates the velocity of the robot end-effector as a function of the hip and knee joint angle velocity. The transpose of Jacobian matrix relates the joint torque with the ground force which will be used in the next section of this paper. The Jacobian matrix is obtained by directly differentiating the forward kinematic equation.

$$\mathbf{J} = \frac{\partial \mathbf{r}_{ee}}{\partial \mathbf{q}_r} \tag{8}$$

2.2 Dynamic Model

To derive the equations of motion, we must first analyze the dynamic motion of a cyclic hopping process. The hopping motion is divided into two phases, flight phase and stance phase. On flight phase, the robot has ascending, apex, and descending process in sequence; on stance phase, the robot has compression, bottom, and thrust in sequence. Two transient states, lift-off and touch-down, separate flight phase and stance phase and influence the hopping motion dynamically. Lift-off dynamically transits hopping from stance phase to flight phase, and touch-down transits vice versa. According to the different constraint condition, non-holonomic constraint of flight and holonomic constraint of stance phase, we establish different dynamic models (Peng et al. 2012; Hutter 2013).

2.2.1 Flight Phase

On flight phase, the single-legged robot with floating base is exerted only by the gravity, and the freefall trajectory is generated due to the horizontal constraint by the vertical slider. The dynamics of the single-legged robot on flight phase is modeled as follows:

$$\mathbf{M}(\mathbf{q})\ddot{\mathbf{q}} + \mathbf{B}(\mathbf{q}, \dot{\mathbf{q}}) + \mathbf{g}(\mathbf{q}) = \mathbf{S}^T \boldsymbol{\tau} \tag{9}$$

where the vector $\mathbf{q} \in R^{3 \times 1}$ contains the generalized coordinates as (2) shows. $\dot{\mathbf{q}}$ and $\ddot{\mathbf{q}}$ are the first and twice derivative of $\mathbf{q}$, respectively. The terms $\mathbf{M}(\mathbf{q}) \in R^{3 \times 3}$, $\mathbf{B}(\mathbf{q}) \in R^{3 \times 3}$, $\mathbf{g}(\mathbf{q}) \in R^{3 \times 1}$ are the inertia matrix, the matrix containing the centrifugal and Coriolis terms, and the gravitational torque vector. The vector $\boldsymbol{\tau}$ is the actuator torque $\tau = (\tau_H \quad \tau_K)^T$. The term $S \in [0_{1 \times 1} \quad I_{2 \times 2}]$ is the selected matrix which separates the actuated joint coordinates from the unactuated floating base coordinates.

2.2.2 Stance Phase

When the single-legged robot contacts the ground, there is ground reaction force exerted on the end-effector of the robot. The hybrid dynamic equation of motion with contact model with respect to the inertial frame is written as

$$M(q)\ddot{q} + B(q, \dot{q}) + g(q) = S^T \tau + J^T \lambda \tag{10}$$

with the ground reaction force λ, the other terms are the same with (9).

By constraints, we mean locations of the end-effector of the robot on the ground, where external forces or torques are applied. No motion is observed with respect to the inertial frame, which means that the end-effector sticks to the ground and has no any slipping. So that the contact constraint reduces to

$$\dot{r}_{ee} = J\dot{q} = 0 \tag{11}$$

$$\ddot{r}_{ee} = J\ddot{q} + \dot{J}\dot{q} = 0 \tag{12}$$

Combining the contact constraint (11), (12) with robot dynamics yields a closed from solution for the ground reaction force

$$\lambda = \left(JM^{-1}J^T\right)^{-1}\{JM^{-1}(B + g - S^T\tau) - \dot{J}\dot{q}\} \tag{13}$$

The result shows that we receive the ground reaction force based on the description of the robot dynamics without any further contact force sensor information. Inserting (13) into (10), we get the equation of constrained dynamics,

$$M(q)\ddot{q} + N^T(B(q, \dot{q}) + g(q)) + J^T \Lambda \dot{J}\dot{q} = (SN)^T \tau \tag{14}$$

where $\Lambda = \left(JM^{-1}J^T\right)^{-1}$, $N = [I - M^{-1}J^T \Lambda J]$. The result is independent of the ground reaction force λ.

2.2.3 The Ground Contact Model Based on the Impact Collisions

The impact of the robot with ground is inelastic collision and the infinitesimal impact time results in high peak impact force with a rapid dissipation of energy. During the infinitesimally short time of collisions, the joint positions of a robotic system remain unchanged since joint angular velocities are finite whose integrals over an infinitesimally short time interval are zero (Zheng and Hemami 1985; Hutter 2013). To resolve the contact impulse, one may integrate both sides of Eq. (14) in an infinitesimal time.

$$
\lim_{\Delta t \to 0} \int_{t_0}^{t_0 + \Delta t} (\mathbf{M}(\mathbf{q})\ddot{\mathbf{q}})\mathrm{d}t
$$

$$
+ \lim_{\Delta t \to 0} \int_{t_0}^{t_0 + \Delta t} \left(\mathbf{B}(\mathbf{q}, \dot{\mathbf{q}}) + \mathbf{g}(\mathbf{q}) - \mathbf{S}^{\mathrm{T}}\boldsymbol{\tau}\right)\mathrm{d}t \tag{15}
$$

$$
= \lim_{\Delta t \to 0} \int_{t_0}^{t_0 + \Delta t} \left(\mathbf{J}^{\mathrm{T}}\boldsymbol{\lambda}\right)\mathrm{d}t
$$

where t_0 denotes the initial time of the impact. Because of the infinitesimally short time of collisions $\Delta t \to 0$, the second term on the left of the Eq. (15) equals 0. And it is simplified as

$$
\mathbf{M}(\mathbf{q})[\dot{\mathbf{q}}(t_0 + \Delta t) - \dot{\mathbf{q}}(t_0)]
$$

$$
= \lim_{\Delta t \to 0} \int_{t_0}^{t_0 + \Delta t} \left(\mathbf{J}^{\mathrm{T}}\boldsymbol{\lambda}\right)\mathrm{d}t \tag{16}
$$

We may further denote F as the generalized impulsive force,

$$
F = \lim_{\Delta t \to 0} \int_{t_0}^{t_0 + \Delta t} \boldsymbol{\lambda}\mathrm{d}t \tag{17}
$$

Considering the contact constraint of the impact,

$$
\dot{\mathbf{r}}_{\mathrm{ee}}(t_0 + \Delta t) = \mathbf{J}\dot{\mathbf{q}}(t_0 + \Delta t) = \mathbf{0} \tag{18}
$$

we get the generalized impact force,

$$
F = \left(\mathbf{J}\mathbf{M}^{-1}\mathbf{J}^{\mathrm{T}}\right)^{-1}\mathbf{J}\dot{\mathbf{q}}(t_0) = \boldsymbol{\Lambda}\mathbf{J}\dot{\mathbf{q}}(t_0) \tag{19}
$$

Substituting (19) into (16) yields the generalized velocity after impact,

$$
\dot{\mathbf{q}}(t_0 + \Delta t) = \left[\mathbf{I} - \mathbf{M}^{-1}\mathbf{J}^{\mathrm{T}}\boldsymbol{\Lambda}\mathbf{J}\right]\dot{\mathbf{q}}(t_0)
$$

$$
= \mathbf{N}\dot{\mathbf{q}}(t_0) \tag{20}
$$

2.2.4   Contact Switches

The dynamic equation of the hopping process of the robot is subdivided into two intervals by the contact model, the flight phase and stance phase. Corresponding to the different dynamic equation, the control scheme is also subdivided into two intervals. The switching judgement between touch-down and lift-off on time is necessary for the hopping robot control.

Apparently the contact event of touch-down happens when the end-effector of the robot touches the ground, the kinematics of the robot satisfies the following relation:

$$z - l_T \cos q_H - l_S \cos(q_H + q_K) = 0 \tag{21}$$

On stance, the normal constraint force of the ground reaction is always upward. The instant the normal constraint force changes to zero or negative, it represents that the end-effector of the robot lifts off from the ground. Consequently the contact event of lift-off is the instance of the zero crossing of the normal constraint force λ_z of the end-effector.

$$\lambda_z = 0 \tag{22}$$

3   Hopping Height Control Scheme

Most of the energy of the hopping system dissipates during hopping because of friction, damp of the components and loss of impact collisions with the ground. The hopping system is always unstable, the height of which is gradually decreasing if there is no energy imparted into the robot system. The intuitive way to keep the height control constant in the course of hopping is quantitatively compensating the loss of energy in some kind of ways. The hopping height control can only be implemented on stance phase as the robot system has internal force only on flight. This paper proposes a hopping height control scheme based on the virtual model control with energy conservation when the robot is on stance.

3.1   Virtual Model Control on Stance

Virtual model control uses imagined mechanical components, which produces a straight-forward means of controlling joint torques to produce a desired robot motion behavior. The virtual component can be springs, dampers, and so on. It is an intuitive control scheme and right now widely applied though the legged robot joint torques of real actuators (Hutter et al. 2011; Cunha 2013). The advantage of virtual model control is that no dynamic model of the robot is required. They can be placed at strategic locations, where at least one should be attached to the robot. For a legged robot, the frequently applied two locations are hip joint and end-effector. The great benefit of virtual model control is intuitive and requires relatively small amounts of computation. Furthermore, a high-level controller can be implemented as a state machine that simply changes virtual component connections or parameters at the state transitions (Pratt et al. 2001).

The virtual model control validates the instant the legged robot touches down the ground and invalidates when the leg lifts off the ground. During the stance time, the spring virtually retracts and extends along with the fluctuation of the hip and energy is interchanging between the robot system and the spring as Fig. 1 shows. On compression section, kinetic energy of the robot system transforms to the spring and on thrusting, potential energy of spring transforms to the robot system.

Using the kinematics of the legged robot, the force of the virtual spring can be transformed into the joint torque with the transpose of the Jacobian matrix.

$$\tau = \mathbf{J}_{FH}^{T}\mathbf{F}_{v} \tag{23}$$

$$\mathbf{J}_{FH} = \frac{\partial \mathbf{r}_{mH} - \partial \mathbf{r}_{ee}}{\partial \mathbf{q}_{r}} \tag{24}$$

While both, the matrix $\mathbf{J}_{FH}$ is the relative Jacobian of the robot, which is solely a function of joint coordinate. The virtual force $\mathbf{F}_{v}$, which is simply calculated as a linear spring element between the end-effector of robot and hip joint, was decomposed into a vertical and a horizontal component. A modified implementation is presented by Hutter et al. (2011). The virtual force in vertical direction provides the springy behavior,

$$F_{v}^{y} = k_{s}(y_{0} - (y_{mH} - y_{ee})) \tag{25}$$

where k_{s} denotes the spring stiffness; y_{0} denotes the original vertical length between the hip joint and the end-effector when the end-effector contacts the ground; y_{mH} and y_{ee} denote the absolute ordinate value of hip joint and end-effector, respectively. The calculation requires only relative position of the hip against the end-effector, so that can be implemented with only joint sensors.

The forward motion is limited by the vertical slider and the end-effector sticks to the ground, the horizontal virtual force is smaller than the static friction force. Here we set the horizontal force as a constant, and the value of which is equal to static friction force.

$$F_{v}^{x} = \mu F_{v}^{y} \tag{26}$$

Here, μ represents the coefficient of static friction between the end-effector of robot and the ground.

3.2   Energy Compensation Algorithm

In order to compensate the loss energy in the collision in the instant the robot touches down the ground, the energy injection algorithm is implemented during the stance phase. The intuitive way to inject energy into a robot system with virtual mode control on stance is by virtual component (Hutter et al. 2010). We select spring as the virtual component. On the compression section of stance phase, the kinetic energy of the robot system is transformed to the potential energy of the spring. On the thrusting section, the potential energy of the spring is transformed to the kinetic energy. The spring is pre-compressed with impact energy compensation in the instant the robot touches the ground, i.e., the original length of the leg spring is changed to $y_0' = y_0 + \Delta y$, with the pre-compressed spring length Δy, and the quantity of system energy increased according to

$$\Delta E = \frac{1}{2} k_s \Delta y^2 \qquad (27)$$

When the energy of (27) matches the total loss of energy, the single-legged robot would hop in a stable state, i.e., hopping periodically with the same height. While in the actual robot, the cause of energy loss is not only the collision, but also the friction and damper. In some other situation, the energy is also consumed by the auxiliary test bench, such as the vertical slider in this paper. Therefore the loss energy cannot always be obtained precisely by calculating the kinetic energy of the CoG point before and after impact. The total energy of the robot system equals to the potential energy on apex where the leg finished adjusting to the desired posture. From the viewpoint of potential energy, the loss of energy is visible for the controller as it is easy for the slider with position sensor to obtain the apex height

$$E_H = mgH \qquad (28)$$

where E_H denotes the total energy on apex; m denotes the total mass of the robot; g denotes the gravity acceleration; H denotes the apex height. By comparing the energy on apex height, the extra energy that should be injected into or removed from the robot system is known.

When the hopping height is lower than the desired height, the algorithm is (29).

$$\Delta E_{cm}^k = K_p \left(E_{dH} - E_H^{k-1} \right)$$
$$+ K_i \sum_{j=2}^{k} \left(E_{dH} - E_H^{j-1} \right) \qquad (29)$$

ΔE_{cm}^k denotes the potential energy required compensation for the kth $(k \geq 2)$ hopping of the robot system; E_{dH} denotes the potential energy of the desired hopping apex height; E_H^k denotes the potential energy of apex height for the

kth$(k \geq 2)$ hopping; K_p, K_i are the constant parameter tuned by experience. For the first hopping, i.e. $k = 1$, there's no virtually pre-compressed spring and no energy compensated to the robot system. The pre-compressed length of the virtual spring is

$$\Delta y = \begin{cases} \text{sign}\left(\Delta E_{cm}^k\right) \frac{\sqrt{2|\Delta E_{cm}^k|}}{k_s}, & k \geq 2 \\ 0, & k = 1 \end{cases} \tag{30}$$

When the hopping height is higher than the desired height which implies that the energy of the robot system is larger than that with the desired apex height, energy dissipates naturally by the inelastic impact collisions. More details should be focused on the pre-compressed length in this situation. On the one hand, the value of Δy due to (30) may become negative which denotes that the virtual spring becomes a pre-stretched spring, making no contributions to dissipate energy of the robot system, therefore the result of Δy has the lower bound as (31) shows. On the other hand, in order to quickly get out of the negative value of the integrator when $\Delta y < 0$, the value of the integrator resets to 0 arbitrarily. The energy dissipation depends on the impact collisions, friction, and damper.

$$\Delta y = \begin{cases} \Delta y, & \Delta y > 0 \\ 0, & \Delta y \leq 0 \end{cases} \tag{31}$$

3.3 The Spring Stiffness

The mammals are able to adjust the stiffness of their legs by itself in accordance with the ground stiffness, the forward motion and the ruggedness of the ground. The spring stiffness in the virtual model control, which is inspired by biology, also determines the dynamic performance of hopping of the single-legged robot. Surprisingly, relative individual-leg stiffness is similar in trotters, runners, and hoppers using from one to four legs in stance (Holmes et al. 2006). The research result by Daniel Z (Koditschek et al. 2004) indicates that the relative individual-leg stiffness satisfies $k_{rel,ind} \approx 10$. The relative individual-leg stiffness $k_{rel,ind}$ can be estimated by dividing the relative SLIP stiffness by this number.

$$k_{rel,ind} = \frac{k_{rel}}{n} \tag{32}$$

Here, n equals the number of legs supporting body weight (e.g., for an insect $n = 3$, and for a trotting quadruped or a hopper such as kangaroo $n = 2$), $n = 1$ in this paper. k_{rel} denotes the relative stiffness and can be calculated

$$k_{\text{rel}} = \frac{F_{\text{vert}}/mg}{\Delta l/l} \tag{33}$$

where F_{vert} denotes the vertical whole-body ground reaction force; Δl denotes the compression of the whole-body leg spring; l denotes the hip height.

So that the absolute spring stiffness can be calculated from (32), (33)

$$k_{\text{s}} = \frac{F_{\text{vert}}}{\Delta l} = n\frac{mg}{l}k_{\text{rel,ind}} \tag{34}$$

4 Simulation Experiment and Result

Figure 2 shows the virtual prototype of the single-legged robot constrained by vertical slider with CAD software. By the CAD software, mass and inertia of each segment are estimated with respect to the corresponding joint axis. The main specifications of the robot are shown in Table 1. The kinematics of the robot and the dynamic equations of motion on flight and stance phase presented on Sect. 2 are established in Simulink. As this paper mainly focuses on the dynamic model of hopping motion and the height control scheme, the hydraulic actuator dynamic model is neglected. The block diagram of the control scheme of single-legged robot

Fig. 2 CAD model of the single-legged robot with vertical slider

Table 1 Main specifications of the single-legged robot

Specification	Value
Leg segment length (m)	0.35
Hip joint range of motion (°)	[−50, 70]
Knee joint range of motion (°)	[−140, −20]
Hip mass (kg)	10
Upper limb mass (kg)	5.359
Upper limb inertia (kg m²)	0.33
Lower limb mass (kg)	1.64
Lower limb inertia (kg m²)	0.0497

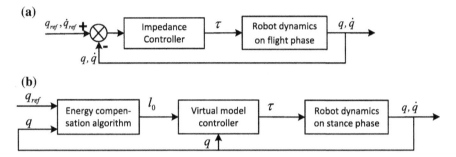

Fig. 3 Block diagram of the control scheme of hopping when the single-legged robot on flight phase (**a**) and on stance phase (**b**)

is divided into flight (Fig. 3 a) and stance (Fig. 3 b) control scheme. The instant the end-effector lift-off, the robotic leg retracts to the desired posture (e.g., the hip and knee joint is 45° and 90°, respectively). An impedance controller is used to command the legged robot to the desired posture. The joint toque is calculated as

$$\begin{bmatrix} \tau_H \\ \tau_K \end{bmatrix} = K_P \begin{pmatrix} q_{Hd} - q_H \\ q_{Kd} - q_H \end{pmatrix} + K_D \begin{pmatrix} -\dot{q}_H \\ -\dot{q}_K \end{pmatrix} \tag{35}$$

K_P, K_D denote the impedance controller parameters, and the value are 500, 35, respectively. q_{Hd}, q_{Kd} denote the desired hip and knee joint angle. $\dot{q}_H$, $\dot{q}_K$ denote the hip and knee joint angle velocity. On stance phase, the control scheme is switched to the virtual model controller with energy conservation. As the Sect. 2.2 represents, the control scheme switching is driven by the contact event of flight and stance phase, touch-down and lift-off. The virtual spring stiffness is determined $k_s = 3400\,\text{N/m}$ by substituting specification value of Table 1 to (35). Considering the energy output capacity and the hopping resolution, the hopping height range is

(a)

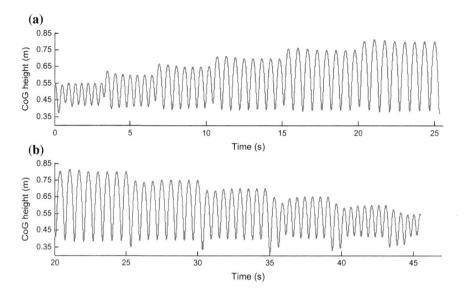

(b)

Fig. 4 Continuous hopping experiment with CoG dynamic hopping height versus time, hopping height from 0.55 to 0.8 m (**a**) and from 0.8 to 0.55 m (**b**)

[0.055, 0.305] m. The hopping height here denotes the altitude difference between the CoG apex height and CoG height in the instant the robot touches down the ground. When the values of the controller gains are set to $K_p = 1.1$, $K_i = 1.6$, the hopping of single-legged robot has better dynamic response and static error on the global hopping height range.

In this work, a periodic hopping experiment with dynamic height control is presented. The desired hopping height can be dynamically controlled and randomly set in an appropriate range, in order to testify the validity of the height control algorithm.

The desired hopping height first increases from 0.55 to 0.8 m (Fig. 4 a) and then decreases from 0.8 to 0.55 m (Fig. 4 b) with 0.05 m height interval. The single-legged robot continuously hops eight times on the desired hopping height before it is controlled to the next desired height. The initial condition of the experiment is that the robot is free-falling with no velocity from height of 0.55 m.

Figure 4 is the time domain results of dynamic hopping height control experiment. It shows that the hopping height quickly converges to the desired apex height. Three transition hops are enough to converge to the desired height when the desired height is higher than the previous desired height. During the transition hops, the contracting length of leg on stance is distinctive with the stable hops because of the adjustment of the spring pre-compressed length. The contracting length of the leg adjusts more when the height is lower than the previous desired height. The largest adjustment happens when the height transits from 0.7 to 0.65 m.

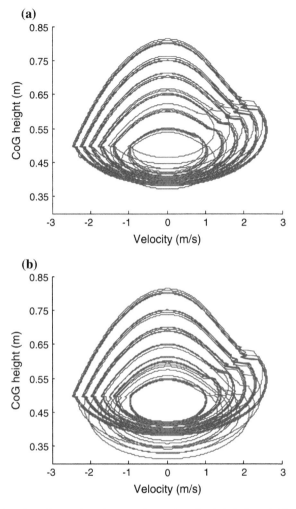

Fig. 5 Phase plot of continuous hopping experiment, hopping height from 0.55 to 0.8 m (**a**) and from 0.8 to 0.55 m (**b**)

The limit cycle behavior for different amplitude of apex height and hopping transition are investigated. Figure 5 is the phase plot of continuous hopping experiment showing the vertical position versus vertical velocity of CoG. It shows that the dynamic hopping with the same height approaches to a periodic orbit. After the transition hops, the periodic orbit converges to stable (Fig. 6).

Fig. 6 The pre-compressed
length of spring on stance on
the continuous hopping
experiment

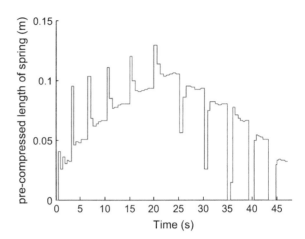

5   Conclusions

In this research, an articulated single-legged robot with two degree of freedom is presented. The robot's upper body is constrained on a vertical slider, so that it is free to slide vertically on the slider. We first established the kinematic model of the legged hopping robot and the hybrid dynamic model on flight and stance phase. Meanwhile, the ground contact model based on the impact collisions is established.

On the basis of the articulated single-legged robot model, we have developed a hopping height control scheme to control the vertical hopping height of a single-legged hopping robot dynamically. The control scheme is based on the virtual model control with the energy conservation when the robot is on stance. An imagined spring is fixed between the hip and the end-effector on the virtual model. The spring stiffness is determined by the previous research of biology. When the hopping height is lower than the desired height, the extra energy which is conserved on the pre-compressed spring is injected into the robot system on the thrusting section of stance phase. The pre-compressed length of the virtual spring is determined according to the required potential energy. When the hopping height is higher than the desired height, energy dissipates by the contact impulse. In this situation, the negative pre-compressed length of the spring, which denotes that the virtual spring become a pre-stretched spring, makes no contributions to dissipate energy of the robot system. The phase plot shows that the dynamic hopping with different height approaches respective periodic and stable orbit. And the time domain hopping plot shows that the control scheme has fast dynamic response and nearly no steady-state error. The experiment result shows great efficiency of the control scheme proposed here.

Acknowledgments Project supported by Science Fund for Creative Research Groups of National Natural Science Foundation of China (No. 51521064), National Natural Science Foundation of China (N0. 41506116), and Zhejiang Provincial Natural Science Foundation of China (No: LY13E050001).

References

Azahar AH, Horng CS, Kassim AM (2013) Vertical motion control of a one legged hopping robot by using central pattern generator (CPG). In: IEEE symposium on industrial electronics and applications, pp 7–12. doi:10.1109/ISIEA.2013.6738958

Bhatti J, Plummer AR, Sahinkaya MN, Iravan P (2012) Fast and adaptive hopping height control of single-legged robot. In: 11th Biennial conference on engineering systems design and analysis, pp 303–309. doi:10.1115/ESDA2012-82564

Bhatti J, Iravani P, Plummer A, Sahinkaya MN (2013) Instantaneous control of a vertically hopping leg's total step-time. In: IEEE international conference on robotics and automation, pp 1–6. doi:10.1109/ICRA.2013.66305.48

Bosworth W, Kim S, Hogan N (2014) The effect of leg impedance on stability and efficiency in quadrupedal trotting. In: 2014 IEEE/RSJ international conference on intelligent robots and systems, pp 4895–4900. doi:10.1109/IROS.2014.6943258

Cherouvim N, Evangelos P (2009) Control of hopping speed and height over unknown rough terrain using a single actuator. In: IEEE international conference on robotics and automation, pp 2743–2748. doi:10.1109/sROBOT.2009.5152232

Council G, Yang S, Revzen S (2014) Deadbeat control with (almost) no sensing in a hybrid model of legged locomotion. In: International conference on advanced mechatronic systems, Kumamoto, pp 475–480. doi:10.1109/ICAMechS.2014.6911592

Cunha TB (2013) Hydraulic compliance control of the quadruped robot HyQ. Italy, University of Genoa. PhD thesis, University of Genoa, Italy

Hodgins JK, Raibert MH (1991) Adjusting step length for rough terrain locomotion. IEEE Trans Robot Autom 7(3):289–298. doi:10.1109/70.88138

Holmes P, Full RJ, Koditschek D, Guckenheimer J (2006) The dynamics of legged locomotion: models, analyses, and challenges. SIAM Rev 48(2):207–304. doi:10.1137/S0036144504445133

Hutter M (2013) StarlETH & Co.—design and control of legged robots with compliant actuation. Switzerland, ETH Zurich. PhD thesis, ETH Zurich, Switzerland

Hutter M, Remy CD, Höpflinger MA, Siegwart R (2010) SLIP running with an articulated robotic leg. In: IEEE/RSJ international conference on intelligent robots and systems, pp 4934–4939. doi:10.1109/IROS.2010.5651461

Hutter M, Remy CD, Hoepflinger MA, Siegwart R (2011). ScarlETH: design and control of a planar running robot. In: IEEE/RSJ international conference on intelligent robots and systems, pp 562–567. doi:10.1109/IROS.2011.6094504

Koditschek DE, Full RJ, Buehler M (2004) Mechanical aspects of legged locomotion control. Arthropod Struct Dev 33(3):251–272. doi:10.1016/j.asd.06.003

Mathis FB, Mukherjee R (2013) Apex height control of a two-mass hopping robot. In: IEEE international conference on robotics and automation, pp 4770–4775. doi:10.1109/ICRA.2013.6631259

Naik KG, Naik KG, Mehrandezh M, Mehrandezh M (2005) Control of a one-legged hopping robot using an inverse dynamic-based PID controller. In: IEEE CCECE/CCGEI, pp 770–773. doi:10.1109/CCECE.2005.15570.42

Peng X, Bin G, Kunxiu D, Shenghai H (2012) Establishment and experimental study of the floating basis dynamics model of a hopping robot. Appl Sci Technol 39(4):31–36. doi:10.3969/j.issn.1009-671X.2012.04.007

Pratt J, Chew C, Torres A, Dilworth P, Pratt G (2001) Virtual model control: an intuitive approach for bipedal locomotion. Int J Robot Res 20(2):129–143. doi:10.1177/02783640122067309

Prosser J, Moshe K (1993) Control of hopping height for a one-legged hopping machine. In: 32nd conference on decision and control, pp 2688–2693. doi:10.1117/12.14382.8

Raibert MH (1986) Legged robots that balance. The MIT Press, London, pp 16–18

Yu H, Li M, Guo W, Cai H (2012) Stance control of the SLIP hopper with adjustable stiffness of leg spring. In: IEEE international conference on mechatronics and automation, pp 2007–2012. doi:10.1109/ICM.A.2012.6285130

Zheng Y, Hemami H (1985) Mathematical modeling of a robot collision with its environment. J Robot Syst 2(3):289–307. doi:10.1002/rob.4620020307

The Stability Analysis of Quadrotor Unmanned Aerial Vechicles

Yun-ping Liu, Xian-ying Li, Tian-miao Wang, Yong-hong Zhang and Ping Mei

Abstract The problems of dynamic stability of the quadrotor Unmanned Aerial Vehicles (UAV), such as: cornering, wear, and explosion of oar take place due to the aerodynamic force and gyroscopic effect during takeoff and landing process; the vibration; reduction of instruction tracking accuracy; and out of control are prone to take place due to the influence of atmospheric turbulence and motion coupling during yawing. However, the optimized structural parameters of the aircraft is very important for improving the stability of the motion control and the energy saving. Therefore, the relationship of quantification between structural parameter of quadrotor UAV and dynamic stability is built with the method of Lyapunov exponent starting from structure design of mechanical, which guides the mechanical-structural design and provides important basis for optimizing the control system. This relationship lays a basic foundation for enhancing the reliability and stability for the flight mission. Compared with the direct method of Lyapunov, the method of Lyapunov exponent is easier to build, and the calculation process is simpler.

Keywords Quadrotor unmanned aerial vehicles · Take-off · Landing · Yawing · Dynamic stability · Lyapunov exponent

Y. Liu (✉) · X. Li · Y. Zhang · P. Mei
Jiangsu Collaborative Innovation Center on Atmospheric Environment
and Equipment Technology, Nanjing University of Information Science
and Technology, Nanjing 210044, China

Y. Liu · T. Wang
College of Mechanical Engineering and Automation,
Beijing University of Aeronautics and Astronautics Robot Research Institute,
Beijing 100191, China

© Zhejiang University Press and Springer Science+Business Media Singapore 2017   383
C. Yang et al. (eds.), *Wearable Sensors and Robots*, Lecture Notes in Electrical
Engineering 399, DOI 10.1007/978-981-10-2404-7_30

1 Introduction

As is well known, the quadrotor Unmanned Aerial Vehicles (UAV) offers superior properties, such as small volume, high mobility, and vertical take-off; landing over a wide range of operating conditions, makes it an attractive option in executing flight tasks, such as: dropping the bomb, detecting the atmospheric pressure, carrying goods, and so on (Bai et al. 2012). However, the problems of dynamic stability of the quadrotor UAV, such as cornering, wear, and explosion of oar are prone to take place due to the aerodynamic force and gyroscopic effect. At the stage of takeoff and landing, the probability of the flight accident is greatly increased and the proportion of the total accident rate is as high as 60–75 % due to the complex air flow of the ground (Xu 2011). At the stage of yawing, the instability of quadrotor UAV leads to the increased probability of the flight accident as a result of the influence of air turbulences and Dutch rolls (Xiao 2014). Research on the dynamic stability should be done for the quadrotor UAV because of the high value of devices and the danger for the ground of the crash.

The dynamic stability is the natural system of returning to the correct track when the system deviates from its trajectory due to the external disturbance, which is affected by the change of mechanical structure parameters or control torque into the dynamic characteristics of the whole system (Kumon et al. 2011; Dong 2010; Tan 2013). Meanwhile, the value of control torque is closely related to the structural parameters of the system. Therefore, the optimization of the mechanical structure parameters of the aircraft is significant for improving the dynamic stability and saving energy. In theory, there are two main methods for the traditional dynamic stability, that is, the direct method for solving dynamic equations and the direct method of Lyapunov (Pflimlin et al. 2007). However, the quadrotor UAV is a kind of nonholonomic system with multiple characteristics, such as nonlinearity, multivariate, and coupling highly. Therefore, the mathematical model used to analyze the dynamic stability of this system is too hard for traditional methods to be established with the direct method of Lyapunov (Li and Song 2013; Sum and Wu 2012). In particular, the quantitative relationship between structural parameter and dynamic stability of the system is hard to build. Therefore, the establishment of the quantitative relationship between the mechanical structure parameters of the quadrotor UAV and the dynamic stability at the stage of takeoff, landing, and yawing is provided with important scientific significance and application value (Nguyen and Vu 2013). The method of Lyapunov exponent can be used to describe the extent of the convergence or divergence exponentially of the two tracks, that is, the original initial value and the initial value as a result of the disturbing influence, which provides an effective solution to the problem of dynamic stability. Ding et al. applied the method of Lyapunov exponent into the field of biomechanics. The dynamic stability of the upper body motions is studied when walking (Dingwell and Marin 2006). Wu et al. studied the dynamic stability of biped robot in motion using the method of Lyapunov exponent (Yang and Wu 2006). As a result, the method of Lyapunov exponent can be used as a quantitative analysis method for the dynamic stability of the quadrotor UAV at the stage of takeoff, landing, and yawing. Compared with the direct method of Lyapunov, this method of Lyapunov exponent can be constructed. Particularly, the quantitative relationship between the structural parameters and the dynamic stability can be established and analyzed (Stephen et al. 2012).

2 Dynamic Modeling

Figure 1 illustrates the coordinate system for the quadrotor UAV $E(X, Y, Z)$ and the geodetic coordinate system $B(x, y, z)$. There are two hypotheses, that is, the quadrotor UAV is rigid body and four axes of propeller are perpendicular to the plane of the body. When four forces are equal ($F_1 = F_2 = F_3 = F_4$), the quadrotor UAV is at the state of rising, falling, or hovering. The system is at the state of pitching, when $F_2 = F_4$ and $F_1 \neq F_3$. The system is at the state of rolling, when $F_1 = F_3$ and $F_2 \neq F_4$. When $F_1 = F_3 \neq F_2 = F_4$, the system is at the state of yawing.

The dynamic model of the quadrotor UAV based on Newton Euler equation is represented by:

$$\dot{\mathbf{q}} = \mathbf{V}(\mathbf{q})\mathbf{p} \tag{1}$$

$$\mathbf{M}(\mathbf{q})\dot{\mathbf{p}} + \mathbf{C}(\mathbf{q}, \mathbf{p})\mathbf{p} + \mathbf{F}(\mathbf{p}, \mathbf{q}, \mathbf{u}) = 0 \tag{2}$$

where $\mathbf{V}(\mathbf{q})$ is Kinematics matrix. $\mathbf{M}(\mathbf{q})$ is inertial matrix. $\mathbf{C}(\mathbf{q}, \mathbf{p})$ is gyroscopic matrix. The force function $\mathbf{F}(\mathbf{p}, \mathbf{q}, \mathbf{u})$ includes all of the aerodynamic, engine, and gravitational forces and moments.

$$\mathbf{p} = (p, q, r, u, v, w)^{\mathrm{T}}$$
$$\mathbf{q} = (\phi, \theta, \psi, x, y, z)^{\mathrm{T}}$$

Fig. 1 Schematic diagram of system

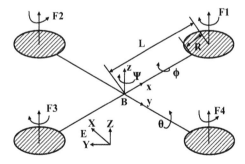

$$
\mathbf{V}(\mathbf{q}) =
\begin{bmatrix}
1 & 0 & 0 & 0 & 0 & 0 \\
S_\phi T_\theta & C_\phi & S_\phi C_\theta^{-1} & 0 & 0 & 0 \\
C_\phi T_\theta & -S_\phi & C_\phi C_\theta^{-1} & 0 & 0 & 0 \\
0 & 0 & 0 & C_\theta C_\psi & C_\theta S_\psi & -S_\theta \\
0 & 0 & 0 & S_\phi S_\theta C_\psi - C_\phi S_\psi & S_\phi S_\theta S_\psi + C_\phi C_\psi & S_\phi C_\theta \\
0 & 0 & 0 & C_\phi S_\theta C_\psi + S_\phi S_\psi & C_\phi S_\theta S_\psi - S_\phi C_\psi & C_\phi C_\theta
\end{bmatrix}
$$

$$
\mathbf{M}(\mathbf{q}) =
\begin{bmatrix}
I_x & 0 & 0 & 0 & 0 & 0 \\
0 & I_y & 0 & 0 & 0 & 0 \\
0 & 0 & I_z & 0 & 0 & 0 \\
0 & 0 & 0 & m & 0 & 0 \\
0 & 0 & 0 & 0 & m & 0 \\
0 & 0 & 0 & 0 & 0 & m
\end{bmatrix}
$$

$$
\mathbf{C}(\mathbf{q},\mathbf{p}) =
\begin{bmatrix}
0 & -I_y r & I_z q & 0 & -mw & mv \\
I_x r & 0 & -I_z p & mw & 0 & -mu \\
-I_x q & I_y p & 0 & -mv & mu & 0 \\
0 & 0 & 0 & 0 & -mr & mq \\
0 & 0 & 0 & mr & 0 & -mp \\
0 & 0 & 0 & -mq & mp & 0
\end{bmatrix}
$$

$$
\mathbf{F}(\mathbf{p},\mathbf{q},\mathbf{u}) =
\begin{bmatrix}
-LU_2 \\
-LU_3 \\
-U_4 \\
mgS_\theta - U_1\left(C_\phi C_\psi S_\theta + S_\phi S_\psi\right) \\
-mgC_\theta S_\phi - U_1\left(-S_\phi C_\psi + C_\phi S_\theta S_\psi\right) \\
-mgC_\theta C_\phi - U_1 C_\phi C_\theta
\end{bmatrix}
$$

$$
\begin{cases}
U_1 = F_1 + F_2 + F_3 + F_4 \\
U_2 = F_2 + F_4 \\
U_3 = F_1 + F_3 \\
U_4 = K(F_1 + F_2 + F_3 + F_4)
\end{cases}
$$

$$
F = \frac{1}{2}\rho A C_T R^2 \Omega^2
$$

The I_x, I_y, and I_z are the rotational inertia of the system axis x, y, z. $A = \pi R^2$ is the area of rotating wing per rotation. ρ is the air density. $K = C_Q/C_T$ is the ratio of the rotor torque coefficient (C_Q) to the Rotor lift coefficient (C_T). Ω is the rotor speed. U_1 is the control value of vertical velocity. U_2 is the input value of roll control. U_3 is the input value of pitch control. U_4 is the control value of yawing. F is the pulling force of every rotor. m is the quality of system. g is the gravity acceleration. φ is the rolling angle around the x axis. θ is the pitching angle around the y axis. ψ is the yawing angle around the z axis. Meanwhile, the trigonometric functions are assumed, that is, $S_\varphi = \sin\varphi$, $C_\varphi = \cos\varphi$, $S_\theta = \sin\theta$, $C_\theta = \cos\theta$, $C_\theta^{-1} = \sec\theta$, $T_\theta = \tan\theta$, $S_\psi = \sin\psi$, $C_\psi = \cos\psi$.

From Eq. (1) and Eq. (2), we obtain the generalized coordinate vector $\mathbf{q}$ and the quasi-velocity vector $\mathbf{p}$.

$$\dot{\mathbf{q}} = \begin{bmatrix} p + rC_\phi T_\theta + qS_\phi T_\theta \\ qC_\phi - rS_\phi \\ rC_\phi C_\theta^{-1} + qS_\phi C_\theta^{-1} \\ uC_\theta C_\psi + v\left(C_\psi S_\theta S_\phi - C_\phi S_\psi\right) + w\left(C_\phi C_\psi S_\theta + S_\phi S_\psi\right) \\ uC_\theta S_\psi + v\left(C_\phi C_\psi + S_\theta S_\phi S_\psi\right) + w\left(-C_\psi S_\phi + C_\phi S_\theta S_\psi\right) \\ wC_\theta C_\phi - uS_\theta + vC_\theta S_\phi \end{bmatrix} \tag{3}$$

$$\dot{\mathbf{p}} = \begin{bmatrix} -(-I_y qr + I_z qr - LU_2)/I_x \\ -(I_x pr - I_z pr - LU_3)/I_y \\ -(-I_x pq + I_y pq - U_4)/I_z \\ -\left[\left(-mrv + mqw + mgS_\theta - U_1(C_\phi C_\psi S_\theta + S_\phi S_\psi)\right)\right]/m \\ -\left[\left(mru - mpw - mgC_\theta S_\phi - U_1(-C_\psi S_\phi + C_\phi S_\theta S_\psi)\right)\right]/m \\ -\left[\left(-mqu + mpv - mgC_\theta C_\phi - U_1 C_\theta C_\phi\right)\right]/m \end{bmatrix} \tag{4}$$

Equations (3) and (4) are transformed into the form of the system–state equation, that is, Eqs. (5):

$$\dot{X} = f(X) \tag{5}$$

where $X = [\mathbf{q} \ \mathbf{p}]^{\mathrm{T}} = (\phi, \theta, \psi, x, y, z, p, q, r, u, v, w)^{\mathrm{T}}$ is the state quantity.

3 Lyapunov Exponent Calculating

Lyapunov exponent is the ratio of average index of the system when the extent of the convergence or divergence exponential of the two tracks, that is, the original initial value and the initial value as a result of the disturbing influence. In the direction of less than zero of the Lyapunov index, the phase volume begins to shrink and the system is stable and the system and the quadrotor UAV is not sensitive to the initial conditions. On the contrary, the system will lose its steadiness (Zhang et al. 2013).

The calculation formula of Lyapunov exponent is written as:

$$\lambda = \lim_{n\to\infty} \frac{1}{n} \sum_{i=0}^{n-1} \ln\left|\frac{df(X)}{dX}\right|_{X_i} \tag{6}$$

The value of Lyapunov exponent is determined by the Jacobi matrix $|df/dX|_{X_i}$. Figure 2 presents the main structural parameters can be obtained from Eqs. (6), which affects the value of Lyapunov exponent. The main structural parameters conclude centroid vector L, System quality m, and the rotational inertia of system I_x, I_y, and I_z.

The tangent vector W of the system track is obtained from Eqs. (5), which can be presented by

$$\frac{\mathrm{d}W}{\mathrm{d}t} = J[X(t)]W \tag{7}$$

where $X(t)$ is the solution of Eqs. (5). $J[X]$ is the Jacobi matrix of this equation.

Figure 3 shows the calculation process of Lyapunov exponent.

The sampling instant T is fixed at 0.1 s and the iterative value K is fixed at 100 in order to calculate the Lyapunov exponent, that is, $\lambda_1, \lambda_2 \ldots \lambda_6$. The initial condition of Eqs. (7) is set as $\left\{ u_1^{(k-1)}, u_2^{(k-1)}, \ldots u_6^{(k-1)} \right\}$ for the iteration of $k(k = 1, 2, \ldots N)$. The initial condition $\left\{ u_1^{(k-1)}, u_2^{(k-1)}, \ldots u_6^{(k-1)} \right\}$ is transformed into $\left\{ w_1^{(k-1)}, w_2^{(k-1)}, \ldots w_6^{(k-1)} \right\}$ through the T_s integral transformation. Then, $\left\{ w_1^{(k-1)}, w_2^{(k-1)}, \ldots w_6^{(k-1)} \right\}$ is transformed by GramSchm orthogonal and turns into the vector $\left\{ v_1^{(k-1)}, v_2^{(k-1)}, \ldots v_6^{(k-1)} \right\}$. Finally, the ultimate vector is obtained, that is, $\left\{ u_1^k, u_2^k, \ldots u_6^k \right\}$. Repeat this process until the Lyapunov exponent reaches the maximum number of iterations N. The exponent $\lambda_1, \lambda_2 \ldots \lambda_6$ obtained forms Lyapunov exponent spectrum.

The simulation of the quadrotor UAV is analyzed by the Lyapunov exponent spectrum calculated base on the dynamics model during takeoff, landing, and yawing, and all the calculation is completed with Mathematica. The measured parameters of rotational inertia are presented in Table 1. Table 2 shows the structure parameters of the system.

Fig. 2 The main parameters affected the value of Lyapunov exponent

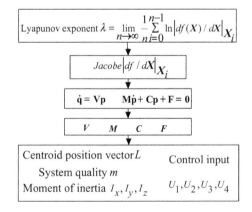

Fig. 3 The calculation process of Lyapunov exponent

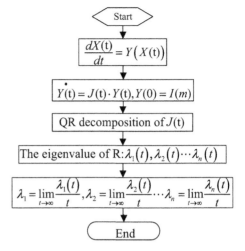

Table 1 Measured parameters of rotational inertia

	r (m)	a (m)	$T(s^{-1})$	w (rad/s)	I (kg m^2)
X	0.125	0.49	1.1716	5.3631	9.5065E−003
Y	0.105	0.46	1.3860	4.5334	1.00E−002
Z	0.185	0.58	1.1374	5.5244	1.658E−002

Table 2 Structure parameters of the system

Types	Contents
M	0.875 kg
G	9.8 m/s^2
L	0.225 m
C_T	1.0792E−005
C_Q	1.8992E−007
R	0.125 m
P	11.69 kg/m^3
I_x	9.5065E−003 kg m^2
I_y	1.00E−002 kg m^2
I_z	1.658E−002 kg m^2
I_{rz}	6.00E−005 kg m^2
K	1.7598E−002

4 Stability Analysis

The simulation result reveals that the convergence speed of Lyapunov exponent spectrum to zero at the state of takeoff is faster than that of landing according to Figs. 4 and 5. Compared with the stability of system during landing, the system at the state of takeoff has better stability.

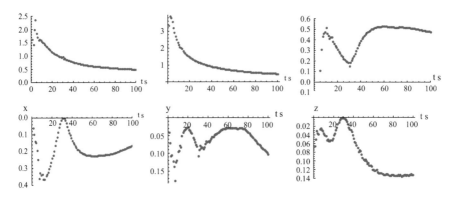

Fig. 4 Lyapunov exponent spectrum during takeoff

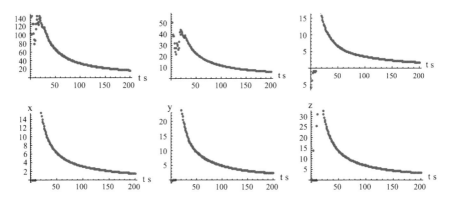

Fig. 5 Lyapunov exponent spectrum during landing

Under the condition of changing vector L, the convergence speed of Lyapunov exponent spectrum to zero is faster with the increase of the value of vector L observed in Fig. 6. Therefore, the stability of the quadrotor UAV can be improved by revising the structural parameters of the system.

Seen from Fig. 7, when the motor speed of No. 1 and 3 are fixed at 3500r and the motor speed of No. 2 and 4 are fixed at 1000r, the system of quadrotor UAV displays the state of yawing in the situation one.

Seen from Fig. 7, when the motor speed of No. 1 and 3 are fixed at 3000r and the motor speed of No. 2 and 4 are fixed at 1500r, the system of quadrotor UAV also displays the state of yawing in the situation two.

Figures 9 and 10 show the simulation results when the value of vector L is changed to 0.5 and 1 under the same condition of other parameters.

Seen from Figs. 8, 9 and 10, the convergence speed of Lyapunov exponent spectrum to zero is faster with the increase of the value of vector L, which can obtain better stability of system at the state of yawing. Therefore, the stability of the quadrotor UAV can be improved by revising the value of vector L.

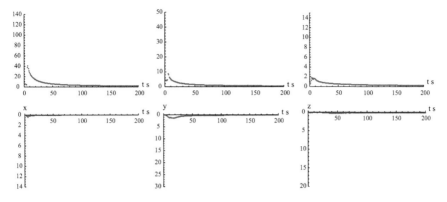

Fig. 6 Modified Lyapunov exponent spectrum during landing

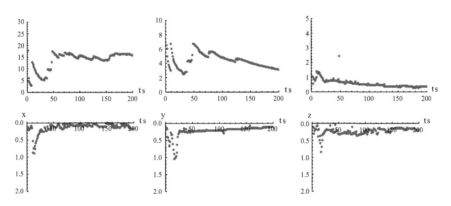

Fig. 7 Lyapunov exponent spectrum during yawing

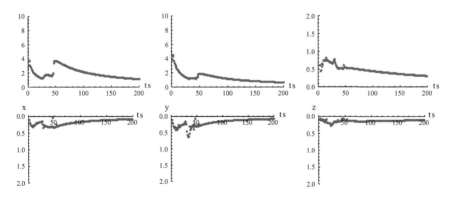

Fig. 8 Lyapunov exponent spectrum during yawing

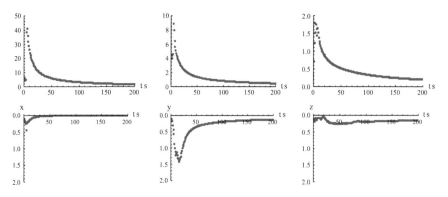

Fig. 9 Modified Lyapunov exponent spectrum

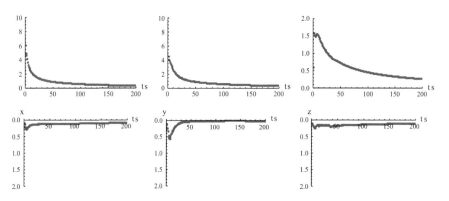

Fig. 10 Modified Lyapunov exponent spectrum

Under the state of situation two, the weight of system is given different values, that is, 2 and 0.5 kg. The results of simulation are displayed in Figs. 11 and 12, respectively.

Seen from Figs. 8, 11 and 12, the results obtained reveal that the convergence speed of Lyapunov exponent spectrum to zero with the lighter weight of system than that of heavy ones. Therefore, the lighter weight of system corresponds to the better stability of the quadrotor UAV during yawing.

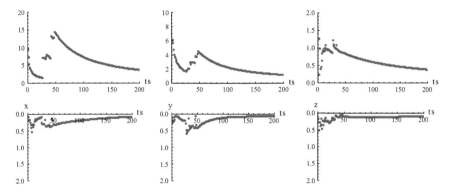

Fig. 11 Lyapunov exponent spectrum during yawing with the system weight of 2 kg

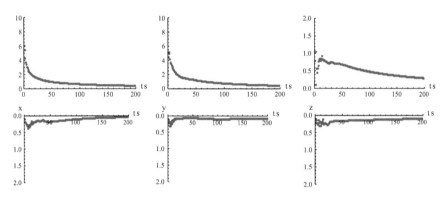

Fig. 12 Lyapunov exponent spectrum during yawing with the system weight of 0.5 kg

5 Conclusion

In this paper, the problems of dynamic stability of the quadrotor UAV are investigated during takeoff, landing, and yawing process. From the optimized structural parameters point of view, the relationship of quantification between structural parameter of quadrotor UAV and dynamic stability is built with the method of Lyapunov exponent, which will guide the mechanical-structural design, and will provide important basis for optimizing the control system. Compared with the direct method of Lyapunov, the method of Lyapunov exponent is easier to build, and the calculation process is simpler. Based on the results obtained from the simulation analysis, the following conclusions could be drawn:

(1) Based on the comprehensive comparison, the quantized relationship between input torque and the dynamic stability of system, the results obtained reveal that the dynamic stability during yawing is better with the smaller difference between input values for any two sets of motors.

(2) The relationship of quantification between structural parameter of quadrotor UAV and dynamic stability is built, which provides important basis for optimizing the control system.

Acknowledgments This research is supported by the Natural Science Foundation of Jiangsu province (BK20130999), the National Natural Science Foundation of China (51405243, 51575283), the graduate student research innovation project of the ordinary universities in Jiangsu province (SJLX_0385).

References

Bai YQ, Liu H, Shi ZY et al (2012) Robust flight control of quadrotor unmanned air vehicles. J Robot 34(5):519–524

Dingwell JB, Marin LC (2006) Kinematic variability and local dynamic stability of upper body motions when walking at different speeds. J Biomech 39:444–452

Dong MM (2010) Design and dynamic analysis of a helicopter landing gear parameters. Nanjing University of Aeronautics and Astronautics, Nanjing

Kumon M, Katupitiya J, Mizumoto I (2011) Robust attitude control of vectored thrust aerial vehicles. In: 18th IFAC world congress, Milano, vol 28(8), pp 2607–2613

Li YB, Song SX (2013) Hovering control for quadrotor unmanned helicopter based on fuzzy self-tuning PID algorithm. J Control Eng China 20(5):910–914

Nguyen HD, Vu HL (2013) VOLKER M. Robust stability of differential-algebraic equations. J Surv Differ-Algebraic Equ 63–95

Pflimlin JM, Soueres P, Hamel T (2007) Position control of a ducted fan VTOL UAV in crosswind. Int J Control 80(5):666–683

Stephen S, Rachel H, Cheryl H (2012) Quantifying stability using frequency domain data from wireless inertial measurement units. J Systemics Cybern Inform 10(4):1–4

Sum Y, Wu CQ (2012) Stability analysis via the concept of Lyapunov exponents: a case study in optimal controlled biped standing. Int J Control 85(12):1952–1966

Tan YG (2013) The dynamic analysis of a controllable under actuated robot. Wuhan University of Technology, Wuhan

Xiao W (2014) Research on the control technology of lateral coupling for hypersonic vehicle. Nanjing University of Aeronautics and Astronautics, Nanjing

Xu YL (2011–2015) Crash source Investigation on the safety of take-off and landing. DB/OL. Xinhua

Yang C, Wu Q (2006) On stabilization of bipedal robots during disturbed standing using the concept of Lyapunov exponents. J Robotica 24:621–624

Zhang WC, Tan SC, Gao PZ (2013) Chaotic forecasting of natural circulation flow instabilities under rolling motion based on Lyapunov exponents. J Acta Phys Sin 6

Hand Exoskeleton Control for Cerebrum Plasticity Training Based on Brain–Computer Interface

Qian Bi, Canjun Yang, Wei Yang, Jinchang Fan and Hansong Wang

Abstract Rehabilitation therapy with exoskeleton robots has been widely adopted to realize normal traction training of muscles, but the plasticity training of cerebrum is usually ignored during rehabilitation with exoskeletons. This paper presents a new exoskeleton aided hand rehabilitation method for post-stroke patient to validate the feasibility and reliability of cerebrum plasticity training. The approach is based on the Brain–Computer Interface (BCI) technology with which the EEG can be acquired and processed to obtain the patient's hand motion intention by applying Independent Component Analysis (ICA) algorithm. The hand exoskeleton system is motivated and controlled by the motion intention to assist the hand movement. Experiments of hand exoskeleton motion control and force control based on BCI validated the feasibility and reliability of the system. Despite the 1.8–2.9 s time delay of response during experiment, the subject's hand motion intention was well acquired by BCI and the corresponding hand motion was executed by hand exoskeleton.

Keywords Rehabilitation · Hand exoskeleton · BCI · Cerebrum plasticity training

Q. Bi · C. Yang (✉) · W. Yang · J. Fan · H. Wang
The State Key Lab. of Fluid Power Transmission and Control, Zhejiang University,
Hangzhou 310027, People's Republic of China
e-mail: ycj@zju.edu.cn

Q. Bi
e-mail: biqianmyself@zju.edu.cn

W. Yang
e-mail: zjuaway@zju.edu.cn

J. Fan
e-mail: jcfan@zju.edu.cn

H. Wang
e-mail: zjuwhs@zju.edu.cn

© Zhejiang University Press and Springer Science+Business Media Singapore 2017
C. Yang et al. (eds.), *Wearable Sensors and Robots*, Lecture Notes in Electrical
Engineering 399, DOI 10.1007/978-981-10-2404-7_31

1 Introduction

Exoskeleton robots have been developed to replace the traditional therapist-dependent post-stroke rehabilitation. However, the plasticity training of cerebrum cannot be ignored besides normal traction training of muscles for the patients. Doyle et al. (1987) and Carr and Shepherd (1987) find that with specific task training and enough repeating times, the patient's brain cortex will store the corresponding motion mode. The traditional exoskeleton for rehabilitation is mostly used to realize normal traction training without the patient's forwardly participation. Since the patient's brain nerve cannot participate into control his/her limb movement during training with exoskeleton, the plasticity of cerebrum will not be utilized or improved.

Although the post-stroke patient cannot control the limb movement with brain cortex through normal nerve conduction, the corresponding brain cortex activities still exist. Thus the Brain–Computer Interface (BCI) technology can be used to acquire the potential signals of brain cortex during motion imagination. With the potential signals, the motion imagination intention can be acquired and used to control the exoskeleton to assist the limb movement.

Currently, research on exoskeleton system based on BCI technology is still in the initial stage. Progress has been made in upper limb exoskeleton control. Gomez et al. (2011) presented exoskeleton shared control based on BCI. The BCI signals are processed by support vector machine to determine the wearer's intention which controls the seven degree of freedoms upper limb exoskeleton for rehabilitation. Antonio et al. (2012) designed Gaze-BCI-Driven upper limb exoskeleton system. The task target in workspace is focused on by eye tracking technology and the acquired BCI signals are processed by support vector machine to obtain the signal features. The results of subject experiment with this system showed positive effects. Because of low BCI signal recognition rate for hand, few progresses have been made in hand exoskeleton based on BCI. Although the electroencephalograph (EEG) or magnetoencephalograph (MEG) of hand motion can be acquired (Reinkensmeyer et al. 2000; Georgopoulos et al. 2005), few studies have been conducted focused on single finger motion intention recognition applying noninvasive EEG acquisition (Waldert et al. 2008). Sam et al. (2011) adopted BCI to control the hand orthosis for post-stroke rehabilitation. The subject's motion intention is obtained by comparing frequency spectrum between imagination of hand movement and calmness state.

In this research, to realize the plasticity training of cerebrum and stimulate the patient with feedback signals from EEG during rehabilitation therapy, the BCI is applied to the hand exoskeleton control system as decision-making layer. Figure 1 shows principle of hand exoskeleton system based on BCI. Outside the patient body, another auxiliary control loop is established consisting of BCI system and exoskeleton system. The BCI system is the auxiliary decision-making layer and the exoskeleton system is the auxiliary executive layer for the patient. With these two systems, the patient can participate in the hand rehabilitation training with his/her motion intention.

Fig. 1 Principle of
exoskeleton based on BCI

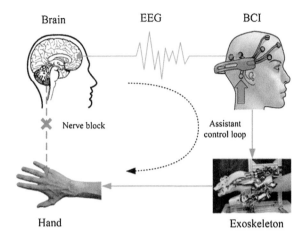

In a word, this paper solves two problems, the hand motion imagination intention acquisition based on BCI technology and hand exoskeleton control by the motion intention. Sections 2 and 3 introduce EEG acquisition and processing including description of EEG for hand motion imagination and the Independent Component Analysis (ICA) method for EEG signal processing. Section 4 is hand exoskeleton control experiments which introduce experiment test-bed, experiment process, and the results. Finally, conclusions are made in Sect. 5 and direction of further research is pointed out.

2 EEG from Hand Motion Imagination

Human hand is the most dexterous part of the human body and the various complex motions are controlled by nervous system. Among the functional partition of brain cortex, the human hand projection area covers most of the whole motor map in precentral gyrus with an exact percentage of thirty percents (Quandt et al. 2012) as shown in Fig. 2. The multiple joints combinations of each finger make it difficult to obtain the characteristic information in finger motion from potential change of brain cortex.

Furthermore, the influence of physiological electrical signal is not negligible during EEG extraction. The physiological electrical signal mainly consists of blinking or eye movement artifact signal, electromyography artifact signal, and head movement artifact signal.

Blinking or eye movement artifact signal: when blinking or eye movement happens, the eye movement electrical signal will arise, whose frequency and amplitude is related to those of blinking or eye movement. This artifact signal can almost influence all channels, which can be seen from the waveforms.

Fig. 2 The brain cortex
functional partition

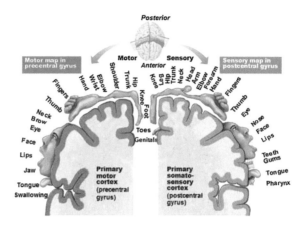

Electromyography artifact signal: when the temporal parts' muscles contract caused by swallowing or biting motions, the electromyography artifact signal will arise obviously. Both the frequency and amplitude are high which can be seen from the channels near the temporal parts.

Head movement artifact signal: when the subject's head moves during the experiment process, the head movement artifact signal will arise, which leads to drifting of the data from all the channels.

However, the human EEG own rhythms which are similar to sine waves. The range of frequency is 1–60 Hz and the frequency of adults' EEG is mostly 10 Hz. Classification of the EEG basic rhythms are shown in Fig. 3. Among these EEG basic rhythms, α rhythm is the basic EEG frequency for normal person that arises when the subject calms down or closes the eyes quietly. Its frequency is 8–13 Hz and its amplitude is 20–100 μV. When the subject is stimulated, α rhythm will immediately replaced by β rhythm which represents the excitability improvement of brain cortex.

In general, the human EEG belongs to nonstationary random signal. Xiao and Ding (2013) study EEG power spectrum during hand motion imagination and find that the EEG power concentrate upon α rhythm and power concentration appears when frequency is 10 Hz. Figure 4 shows this phenomenon intuitively.

In this research we use EPOC (Emotiv Inc) EEG acquisition helmet as shown in Fig. 5. The electrode placement of EPOC is in accordance with international 10/20 electrode placement standard. Fourteen channels are chose to measure the EEG of subject with a sampling frequency of 2048 Hz.

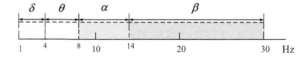

Fig. 3 Electroencephalogram (EEG) basic rhythms

Fig. 4 Power spectrum of
EEG in finger movements

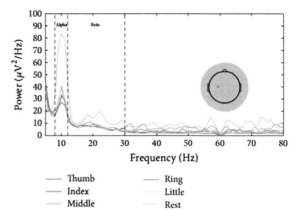

—— Thumb	—— Ring
—— Index	···· Little
—— Middle	···· Rest

Fig. 5 Emotiv EPOC

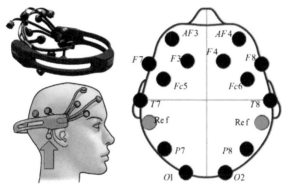

In conclusion, the motion imagination characteristic signal cannot be observed directly. Meanwhile, how the potential signals are mixed on brain cortex is unknown. So it is a typical blind source separation problem. We adopted typical blind source separation method ICA method to process the EEG.

3   EEG Processing Based on ICA

3.1   The ICA Processing Method

Figure 6 shows basic principle of ICA (Comon 1994; Amari and Gardoso 1997). The source signal, mixed matrix, and noise signal are unknown. Thus observed signal is adopted to estimate decomposition matrix to acquire independent decomposition signals.

However, the following assumptions should be made:

(1) The mixed system is modeled as a linear instantaneous mixed model;

Fig. 6 Principle of ICA

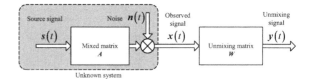

(2) The mean values of both source signal and observed signal are zero.

Hence, the observed signal can be expressed by Eq. (1).

$$\mathbf{x}(t) = A\mathbf{s}(t) = \sum_{i=1}^{n}\sum_{j=1}^{n} a_{ij}s_j \tag{1}$$

where $\boldsymbol{x}(t) \in \mathbf{R}^n$ is the observed signal, $A \in \mathbf{R}^{n \times n}$ is the mixed matrix and $s(t) \in \mathbf{R}^n$ represents source signal.

The aim of ICA process is to obtain the linear transformation matrix W, which makes the output variables independent. Equation (2) shows process of ICA.

$$\boldsymbol{y}(t) = W\mathbf{x}(t) = WA\mathbf{s}(t) = G\mathbf{s}(t) \tag{2}$$

where $\boldsymbol{y}(t) \in \boldsymbol{R}^n$ is the decomposition signal, $W \in \boldsymbol{R}^{n \times n}$ is the decomposition matrix and $\boldsymbol{G} \in \boldsymbol{R}^{n \times n}$ represents the system matrix.

The ICA decomposition process consists of whiting processing and objective function optimization. First, the observed signal $x(t)$ is whiting processed to remove the relevance of each observed signal. Then, the optimization object function $J(W)$ is utilized with the variable W. This object function determines independent degree of the output decomposition signals. Therefore, the object decomposition matrix W will be solved while $J(W)$ reaches to the extreme. That is to say, the ICA algorithm is an optimization process which aims to make the decomposition signals approach to the source signals.

To improve the convergence rate of ICA algorithm, we use the FastICA algorithm proposed by Hyva ̈rinen and Oja (1997) and Hyva ̈rinen et al. (1999). Based on batch processing of sampling data from observed variables, this algorithm adopts maximum negative entropy as the object function. The ICA processing speed is then improved by isolating an independent component from observed signals each time. The procedure of FastICA includes the following four steps:

Step 1: Data standardization and whiting processing;
Step 2: Object function iteration by adjusting the weighting vector shown in Eq. (3)

$$\begin{aligned}\omega_i(k+1) &= E\{\boldsymbol{v}G[\omega_i^T(k)\boldsymbol{v}]\}\\ &\quad - E\{\dot{G}[\omega_i^T(k)v]\}\omega_i(k)\end{aligned} \tag{3}$$

where k is the iteration number, ω_i is the ith line of decomposition matrix $\boldsymbol{W}$, $\boldsymbol{v}$ is the source signal matrix and $G(\cdot)$ stands for an non-quadratic function shown in Eq. (4).

$$\begin{cases} G(u) = \frac{1}{a}\log_2 \cosh(au) \\ \dot{G}(u) = \tanh(au) \end{cases} \quad 1 \le a \le 2 \tag{4}$$

Step 3: Normalization of weighting vector with Eq. (5).

$$\omega_i(k+1) = \frac{\omega_i(k+1)}{\|\omega_i(k+1)\|} \tag{5}$$

Step 4: Source signal decomposition with Eq. (6).

$$y = \boldsymbol{W}\boldsymbol{v} \tag{6}$$

3.2 EEG Analysis Experiment

Applying seven channels of EPOC, the EEG was acquired during the subject hand gripping motion imagination with visual induction. With the help of FastICA algorithm, the decomposition signals were obtained and the corresponding decomposition matrix was solved.

The test subject was a healthy adult male from whom informed consent was obtained. At the beginning, the subject saw the picture which showed the hand exoskeleton gripping motion and imagined this motion for 10 s. Then the picture changed to a black one and the subject got rest for 10 s. Reciprocating three times, one group of experiment was finished. The subject was asked to conduct three groups of experiment. Figure 7 shows observed signals and decomposition signals. All the observed signals were almost disorderly while the first decomposition signal showed obvious independence.

For further analysis of relationship between observed signals and motion imagination, the first decomposition signal was filtered by Butterworth filter as shown in Fig. 8. When the subject imagined the hand gripping motion, the filtered signal declined from the peak to the trough. However, when the subject calmed down, the filtered signal rose from the trough back to the peak. All the three cycles showed the same phenomena. Figure 9 shows the other two groups of experiment. Similarly, during motion imagination, the filtered signal declined from the peak to the trough while it rose from the trough back to the peak when the subject calmed

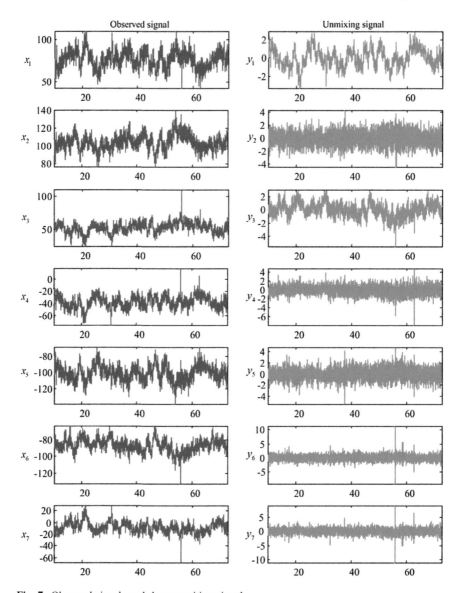

Fig. 7 Observed signals and decomposition signals

down. However, during the third cycle of group three, the filtered signal did not rise regularly. One reasonable explanation was that the subject could not always focus attention during repeated experiments.

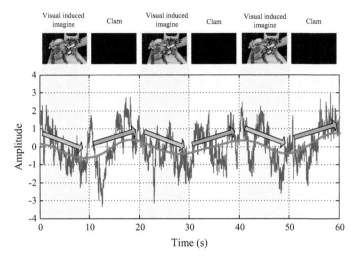

Fig. 8 Data analysis of group one

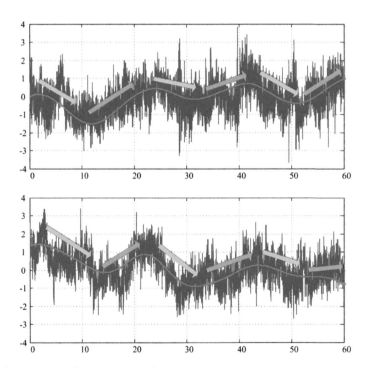

Fig. 9 Data analysis of group two and three

4 Hand Exoskeleton Control Experiment

Figure 10 shows sketch map of hand exoskeleton system based on BCI control. The system consists of EEG acquisition subsystem, feature recognition algorithm, data exchange subsystem, finger control subsystem, and motor-driven subsystem. The objectives of the experimental study are (1) to validate feasibility and reliability of BCI control in hand exoskeleton system, (2) to investigate the response speed of BCI control.

4.1 Experimental Test-Bed

Since control method of one finger is the sufficient condition for hand exoskeleton control, we study on the index finger exoskeleton control in this paper. The index

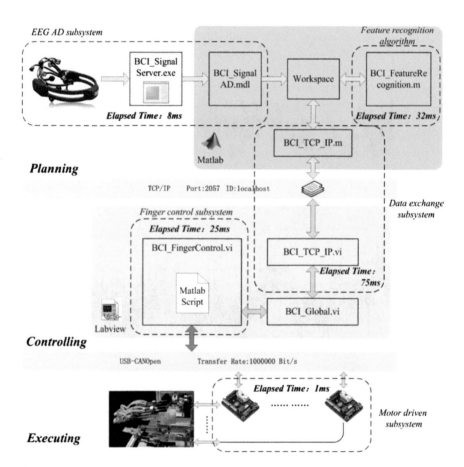

Fig. 10 Hand exoskeleton system based on BCI control

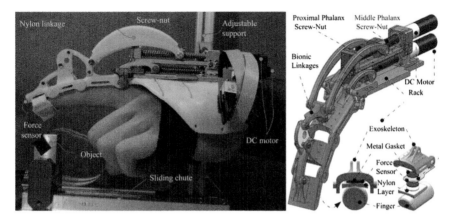

Fig. 11 Index exoskeleton system (Qian et al. 2014)

finger exoskeleton device is designed in accordance with the shape of each phalanx, shown in Fig. 11. The linkages are designed arc so that they have better mechanical performance, bigger pressure angles, and larger motion space than common linkages. Both linkages and exoskeleton phalanxes are manufactured in nylon material through rapid prototyping, as their appearance shapes are too complicated to manufactured in common way. Meanwhile, the nonmetal parts have low inertia, which is helpful for control performance. The force sensor is a pressure sensitive resistance between two metal gaskets.

First, the subject EEG was acquired by EPOC and sent to upper computer (personal computer). Then the source signals were successively filtered, whiting processed, and unmixed applying the FastICA algorithm. Finally, the obtained motion intention signals from the subject were sent to the hand exoskeleton system as control instruction. As shown in Fig. 10, it takes time during data acquisition, transmission, and exchange. The elapsed time of each subsystem is listed in Table 1. The TCP/IP protocol data exchange accounted for 65.8 % of all the elapsed time which implied that real-time control of the hand exoskeleton system with EEG was almost impossible. However, the motion intention signals were meaningful for rehabilitation considering the low-speed training. With the intention of the subject, the hand exoskeleton was able to help realize the subject's hand motion.

4.2 Results and Discussion

In this study, the hand exoskeleton motion control and force control based on BCI were conducted to validate the feasibility and reliability of BCI control. Meanwhile, the response characteristic was investigated. The hand exoskeleton motion control is a classical position control method, and the force control here is the model reference adaptive impedance control (MRAIC) which is introduced by Qian and Can-jun (2013).

Table 1 Elapsed time of subsystems

Subsystem	Elapsed Time（ms）
EEG AD Subsystem	8
Feature Recognition Algorithm	5
Data Exchange Subsystem	75
Finger Control Subsystem	25
Motor Driven Subsystem	1

A. *Hand exoskeleton motion control based on BCI*

Figure 12 is experiment result of hand exoskeleton motion control based on BCI. The black-dotted line stands for visual induced (VID) signal. When the amplitude is high level, it means that the subject is watching the picture of the hand

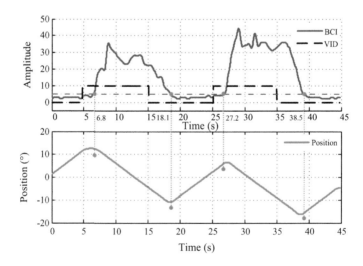

Fig. 12 Hand exoskeleton motion control based on BCI

exoskeleton gripping motion and imagining the gripping motion. Otherwise, the subject stops imagining and gets relaxed. The blue thick line is motion intention from decomposition signal and blue thin dotted line is threshold of motion intention which is set to be 5. When the motion intention exceeds the threshold, it means that the subject has the gripping intention. Then, the hand exoskeleton will drive the subject's finger to flex. The subject was under visual induction when t was 5 s while the motion intention was acquired when t was 6.8 s. There was a delay of 1.8 s. Furthermore, the motion intention disappeared 3.1 s later after visual induction stopped. The results implied that the brain activity was able to respond quickly when stimulated while it needed more time to calm down when visual induction disappeared. This phenomenon leaded to increase of flexion angle after two cycles as shown in Fig. 12.

B. *Hand exoskeleton force control based on BCI*

Similarly, the hand exoskeleton force control experiments were conducted. In this experiment, once the motion intention is acquired, the hand exoskeleton will be controlled to grip the object with 0.5 N interaction forces. When the motion intention disappears, the interaction force will be 0 N. Figures 13, 14 and 15 show results of three group experiments. In Fig. 13, the subject was under visual induction when t was 3 s and the motion intention was acquired 2 s later. The motion intention disappeared 2.7 s later after visual induction stopped. The interaction force control results showed good tracking characteristics which validated that hand exoskeleton force control based on BCI was reliable.

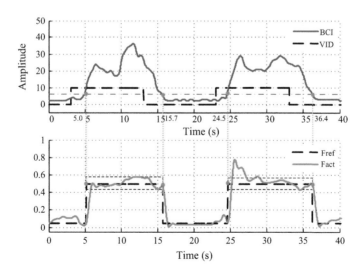

Fig. 13 Hand exoskeleton force control based on BCI (group one)

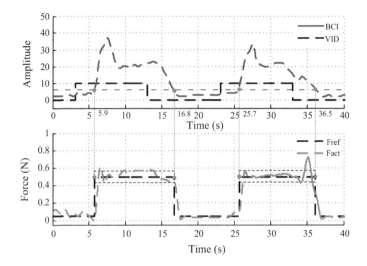

Fig. 14 Hand exoskeleton force control based on BCI (group two)

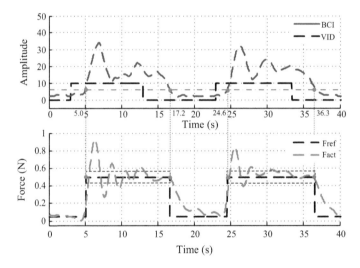

Fig. 15 Hand exoskeleton force control based on BCI (group three)

5 Conclusions

This paper studied the relationship between hand gripping motion and EEG to provide a new cerebrum plasticity training-combined hand rehabilitation method for post-stroke patient. The subject's EEG was acquired and processed by FastICA algorithm. The decomposition signals were filtered and then the motion intention of subject was extracted. By applying BCI technology, the hand exoskeleton motion control and force control experiments were conducted. The results showed feasibility and validity of hand exoskeleton control based on BCI. The key conclusions from this research can be summarized as follows

(1) EEG processing was a typical blind source separation problem and the FastICA algorithm was an effective method to solve this problem.
(2) The EEG analysis experiment validated the relevance of acquired motion intention and visual induction.
(3) The hand exoskeleton motion and force control experiments validated the feasibility and reliability of EEG control. The elapsed time of data exchange implied difficulty of real-time BCI control. However, the extracted motion intention could well represent the subject's intention which could be applied for hand rehabilitation.

Further work will involve improving the precision of intention acquisition by optimization of hardware system and ICA method.

Acknowledgments This work was supported in part by Science Fund for Creative Research Groups of National Natural Science Foundation of China (No.:51221004).

References

Amari S, Gardoso JF (1997) Blind source separation-semiparametric statistical approach. IEEE Trans Signal Process 45(11):2692–2700
Bi Q, Yang C-j (2014) Human-machine interaction force control using model reference adaptive impedance control for index finger exoskeleton. J Zhejiang Univ Sci C 15(4):275–283
Carr JH, Shepherd RB (1987) A motor relearning program for stroke, 2nd edn. Aspen Publisher, New York, pp 151–154
Comon P (1994) Independent component analysis, a new concept. Sig Process 36:287–314
Doyle JC, Lenz K, Packard A (1987) Design examples using μ-synthesis: space shuttle lateral axis FCS during reentry. NATO ASI Ser Modell Robustness Sensitivity Reduction Control Syst 34:127–154
Frisoli A, Loconsole C, Leonardis D, Banno F et al (2012) A new gaze-BCI-driven control of an upper limb exoskeleton for rehabilitation in real-world tasks. IEEE Trans Syst 42(6):1169–1180
Georgopoulos AP, Langheim FJP, Leuthold AC, Merkle AN (2005) Magnetoencephalographic signals predict movement trajectory in space. Exp Brain Res 167(1):132–135
Gomez-Rodriguez M, Grosse-Wentrup M, Hill J, Gharabaghi A et al (2011) Towards brain-robot interfaces in stroke rehabilitation. In: IEEE international conference on rehabilitation robotics. Switzerland, 29 Jun–1 July, 2011, pp 1–6
Hyva rinen A (1999) Fast and robust fixed-point algorithm for independent component analysis. IEEE Trans Neural Netw 10(3):626–634
Hyva rinen A, Oja E (1997) A fast fixed-point algorithm for independent component analysis. Neural Comput 9(7):1483–1492
Quandt F, Reichert C, Hinrichs H et al (2012) Single trial discrimination of individual finger movements on one hand: a combined MEG and EEG study. NeuroImage 59:3316–3324

Reinkensmeyer DJ, Kahn LE, Averbuch M et al (2000) Understanding and teaching arm movement impairment after chronic brain injury: progress with the arm guide. J Rehabil Res Dev 37(6):653–662

Waldert S, Preissl H, Demandt E, Braun C et al (2008) Hand movement direction decoded from MEG and EEG. J Neurosci 28(4):1000–1008

Xiao R, Ding L (2013) Evaluation of EEG Features in decoding individual finger movements from one hand. Comput Math Methods Med 2013:1–10

Lateral Balance Recovery of Quadruped Robot on Rough Terrains

Guo-liang Yuan, Shao-yuan Li, He-sheng Wang and Dan Huang

Abstract In recent years, the proportion of nuclear power use in China has gradually risen; however, of the numerous nuclear power plant accidents warn us to consider aspects related to the relief provided at the scene of the accident. Therefore, the development of an emergency rescue robot for the nuclear power plant becomes the necessity of the hour in addition to being a world-level problem in the field of international nuclear disaster relief. At the accident scene, due to the unexpected situation that may exist, the emergency rescue robot may be hampered during its motion. Considering this situation, this article discusses the situation wherein a quadruped robot achieves the lateral balance recovery while moving on rough terrains. Therefore, we propose a flywheel pendulum model and the capture point theory to derive relevant equations. Finally, in the simulation, we exert a lateral force on our quadruped robot, the performance of which will prove our methodology's validity.

Keywords Quadruped robot · Capture point · Lateral balance recovery · Rough terrain

1 Introduction

With the rapid development of science and technology since the 1960s, the field of research and application of robotics has entered a critical period of vigorous development. Inspired by natural biological structures, through bionics design,

G. Yuan (✉) · H. Wang · D. Huang
Institution of Electrical and Automation, Shanghai Jiao Tongi University,
Shanghai 310027, China
e-mail: 1130329124@sjtu.edu.cn

S. Li · H. Wang · D. Huang
Department of Automation, Shangha Jiao Tongi University, Shanghai 310027, China

© Zhejiang University Press and Springer Science+Business Media Singapore 2017   411
C. Yang et al. (eds.), *Wearable Sensors and Robots*, Lecture Notes in Electrical
Engineering 399, DOI 10.1007/978-981-10-2404-7_32

people have produced large amounts of artificial mechanical structures, and with these man-made mechanical structures, have manufactured a wide variety of robots (Pratt et al. 2006). These robots are already used in the industrial, medical, and scientific research areas, facilitating greatly increased productivity; however, the traditional mode of production has also been affected and has undergone unprecedented changes. People around the world have recognized the important role of the robot: it can serve as an attractive tool for future development, and countries have invested heavily in the field of robotics research, hoping to achieve new breakthroughs and achievements. At the same time, with the promotion of applications using nuclear power, and considering the occurrence of accidents at nuclear power plants, the international community has paid increasing attention to application of robots in nuclear power (Maki and McIlroy 1977). Although nuclear accidents have not occurred in China, precautions must be taken to prevent them. Therefore, the study of robots that can be used at the scene of nuclear disaster has become a key focus of domestic research on robots.

In recent years, domestic and foreign scholars in the field of bionics research have presented various equivalent simplified models, by rationally reducing the order to achieve effective control of the speed and dynamic motion of the quadruped robot, and have achieved gratifying results (Raibert et al. 2008). Their research involved robots mostly undisturbed during high-speed motion; there is still lack of research into dynamic and complex situations whereby strong external disturbances may be encountered, endangering stability control. Considering the robot relief process at the nuclear disaster scene, there may be a subsequent accident caused by external shocks, which may lead to robot rollover. This will not only affect the robot's work results but could also cause damage or result in a non-functional robot (Atkeson and stephens 2007). Therefore, for the robot to achieve balance in the midst of external shocks has become one area of research focus. Therefore, to overcome the complexity of the work environment, it is essential to consider the rough road.

Based on these considerations, we propose to solve the quadruped robot's walk on uneven surfaces using the capture point theory. By determining the point off the foot and the posture compensation angle, we can let the quadruped robot achieve complete restoration of the balance in rough terrain against an external force (Arslan and Saranli 2012). Finally, in the simulation, we exert a lateral force on the quadruped robot, the performance thereafter proving our method valid.

2 Balance Recovery Algorithm: Capture Point Theory

The theory used in this article is adapted from the "catch point" theory proposed by Pratt et al. in 2006. The capture point theory proposed a humanoid robot in the beginning in order to solve the external equilibrium recovery state.

In this article, we use a simplified model of the support foot model for the capture point theory, namely, a flywheel pendulum model, shown in Fig. 1.

It supports legs abstracted into a scalable and massless pendulum, which is abstracted into a flywheel body (Yu et al. 2012).

The model gives the solutions for the equations of conservation of motion energy of columns:

Fig. 1 Flywheel pendulum model

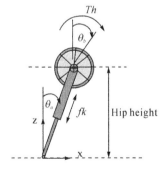

$$m\ddot{x} = f_k \sin\theta_a - \frac{\text{Th}}{l}\cos\theta_a \tag{1}$$

$$m\ddot{z} = -mg + f_k \cos\theta_a + \frac{\text{Th}}{l}\sin\theta_a \tag{2}$$

$$J\ddot{\theta}_b = \text{Th} \tag{3}$$

wherein m represents the mass of the flywheel, J represents the moment of inertia of the flywheel, g stands for acceleration on free fall, x and z are the coordinates of the center of gravity used for representation, l is the distance from the toe to the center, θ_a and θ_b represent an angle relative to the vertical leg of the flywheel, Th represents the flywheel torque, and f_k is the driving force of the legs (Maufroy et al. 2012).

Because the supporting leg is following swinging leg motion, the speed in the vertical direction is zero. So, $\dot{z} = 0$, $z = z_0$. From Eq. (2), we can derive the following:

$$f_k = \frac{mg}{\cos\theta_a} - \frac{1}{l}\frac{\sin\theta_a}{\cos\theta_a}\text{Th} \tag{4}$$

In this case, $\cos\theta_a = \frac{z}{l}$, $\sin\theta_a = \frac{x}{l}$; thus, we get

$$f_k = \frac{mg}{z_0}l - \frac{1}{l}\frac{x}{z_0}\text{Th} \tag{5}$$

Substituting for f_k into Eq. (1), we obtain that the following:

$$\ddot{x} = \frac{g}{z_0}x - \frac{1}{mz_0}\text{Th} \tag{6}$$

$$\ddot{\theta}_b = \frac{1}{J}\text{Th} \tag{7}$$

When the flywheel torque is zero, we know

$$\ddot{x} = \frac{g}{z_0}x \tag{8}$$

Analysis of its orbital energy model then yields the following equation (Buchli et al. 2011):

$$E_{\text{LIP}} = \frac{1}{2}\dot{x}^2 - \frac{g}{2z_0}x^2 \tag{9}$$

Assume that the center of gravity is toward the direction of movement of the support foot; when $E_{\text{LIP}} > 0$, the center has enough support across the foot and forward movement; when $E_{\text{LIP}} < 0$, the focus will fail to support the foot before reaching the reverse movement; when $E_{\text{LIP}} = 0$, the center of gravity is sufficient to reach and maintain the support above (Xuewen 2012).

When $E_{\text{LIP}} = 0$, we get two feature vectors:

$$\dot{x} = \pm x \sqrt{\frac{g}{z_0}} \tag{10}$$

The formula represents a robot's stable and unstable states. When x and $\dot{x}$ have the same symbol, it represents the center of gravity acts toward the movement of the support foot; the state is the steady state, but it is unstable.

If the swinging leg caused no change in position of the support foot, then the process of the pendulum motion results in conservation of orbital energy of the robot. Assuming full exchange support and no delay in time, there is no any energy loss in the middle; so given the speed of the previous time, one can track the energy needed to arrive at the next foothold. From Eq. (10), we can calculate the following (Rubin et al. 2013):

$$x_{\text{capture}} = \dot{x} \sqrt{\frac{z_0}{g}} \tag{11}$$

Therefore, given a moment's speed, we can calculate only the point of capture, which allows the robot to restore balance off the points of the foot.

3   Balanced Gait Recovery

To simplify gait planning, for calculating the robot movement and lateral movement separately, researchers generally assume that before each robot motion, the lateral motion of the coupling is small enough to ignore. In this article, we ignore the first movement of the robot and the lateral movement of the coupling; we just consider the recovery movement for attaining balance after the lateral force, assuming that the initial transverse velocity is zero. The flow chart is shown in Fig. 2.

3.1   Turn State Prediction

Next, we analyze the state of the robot using the most widely used zero moment point (ZMP) stability criterion.

Assuming that the robot mass is concentrated at the center of gravity, the coordinates of the center of gravity are (x_g, y_g, z_g). On the flat ground, the robot's center of gravity at ground level is a constant, and the ZMP location is $(x_{\text{zmp}}, y_{\text{zmp}}, 0)$, represented by the following equation:

Fig. 2 Flow chart

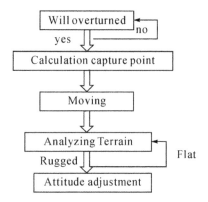

$$\begin{pmatrix} x_{zmp} \\ y_{zmp} \end{pmatrix} = \begin{pmatrix} x \\ y \end{pmatrix} - \frac{z}{g} \begin{pmatrix} \ddot{x} \\ \ddot{y} \end{pmatrix} \tag{12}$$

When the ZMP point is inside the robot foot and ground forming a support polygon, the robot is in steady state; when the ZMP point is outside the robot foot and ground and forms a support polygon, the robot is going to flip (Rebula et al. 2007).

Thus, by the ZMP stability criterion, we can determine the angle needed to restore balance to the robot gait and reach steady state.

3.2 Gait Designed to Restore Balance

Because the capture point theory was proposed to solve the case of humanoid robot in balance by the recovery in case of an external force, it applies to to the problem of the theoretical quadruped robot affected by external shocks to restore balance; we thus propose the concept of "virtual leg" (Kim et al. 2010).

As shown in Figs. 3 and 4, we will place the quadruped robot between the pieces of mechanical legs called "restless legs." We assume that only one side of the front foot of the robot is used as a machine to achieve balance. Once the gait balance required by the swing of the proposed restless leg of the robot is completed, the robot follows the same side of the front foot to move the legs in synchronization in a complete gait. Therefore, just as humanoid robots in general, only the feet of the quadruped robot have similar gait (Havoutis et al. 2013).

In order to avoid collision accident between the swinging leg and the ground obstacles, we set the foot end of the swinging leg to take a relatively simple way to the end of the track after swinging two points, the equivalent of the foot end of the swinging leg walking three sides of a rectangle; thus the robot can well avoid obstacles (Fig. 5).

Fig. 3 Side view of virtual leg

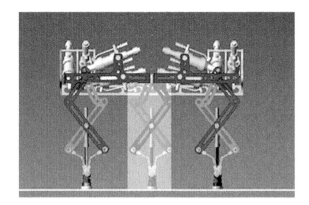

Fig. 4 Rear view of virtual leg

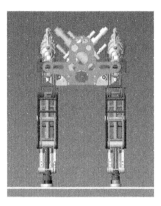

Fig. 5 Walking gait

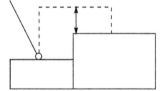

3.3 Balance Recovery

When the ground is flat, the robot receives the signal in the foot, and we can detect when the robot's lateral velocity is altered by external shocks. Using the capture point theory, we can calculate the position of the point of capture, as well as the current location of the robot; the robot can then be manipulated to achieve the balanced lateral displacement.

When the desired lateral displacement is small, i.e., within the maximum swing angle of the robot, the robot can be expected to achieve balance in a single step. If

Fig. 6 The reference frames
of the quadruped robot

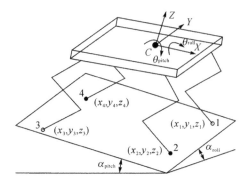

the desired lateral displacement is not small, the robot cannot accomplish restoration of balance; so the robot needs to calculate the maximum angle of the swing again after the capture point location, and therefore requires more steps to complete the balance restoration maneuver (Haowei et al. 2011).

In the rough natural environment, unexpected changes in terrain caused by the strong impact of foot on the ground is the main cause of instability in the robot. We study the three-dimensional extent of the rugged terrain, the real-time adjustment of trunk space, robot attitude, change point of the foot in contact with the ground state, so that the robot is stabilized in the rough terrain (Pongas et al. 2007). The schematic diagram of the quadruped robot walking on uneven ground is shown in Fig. 6.

According to the spatial location of the support foot, we calculate the current tilt angle in actual terrain (Auke 2008):

$$\alpha_{\text{pitch}} = \arctan \frac{Z_f - Z_h}{X_f - X_h} \tag{13}$$

$$\alpha_{\text{roll}} = \arctan \frac{Z_1 - Z_r}{Y_1 - Y_r} \tag{14}$$

where α is the terrain inclination.

$$\begin{cases} Z_f = \frac{z_1 = z_2}{2}, & Z_h = \frac{z_3 = z_4}{2} \\ X_f = \frac{x_1 = x_2}{2}, & X_h = \frac{x_3 = x_4}{2} \\ Z_f = \frac{z_1 = z_1}{2}, & Z_h = \frac{z_1 = z_1}{2} \\ Y_f = \frac{y_1 = y_4}{2}, & Y_h = \frac{y_2 = y_3}{2} \end{cases} \tag{15}$$

In the formula, X, Y, and Z represent the coordinates of the point of first full contact with the ground.

Setting terrain inclination threshold in the X and Y directions, when α_{pitch} or α_{roll} are larger than the threshold, the land is identified as rugged terrain. We introduce preplanned centroid trajectory for the ground-to-torso angle, as well as attitude control parameters for adjustment of the body, so that the body posture is able to meet the current terrain (Park and Park 2012).

$$\begin{cases} \theta^*_{i\_des} = \theta_{i\_des} + k_i\alpha_i \\ k_i = 0, \ \alpha_i \in [-A, A] \end{cases} \tag{16}$$

In the equation, θ_{i_des} represents the pitch or roll direction of the control angle of the trunk attitude, A represents threshold for the terrain inclination, and k_i represents the torso posture adjusting gain angle (Barasuol et al. 2013).

When the terrain is not flat, there will be a difference in distance of the point drop between the current and the original positions.

$$\Delta x_{capture} = \dot{x} \frac{A}{\sqrt{zg} + \sqrt{(z+A)g}} \tag{17}$$

The difference in distance makes the quadruped robot possess the kinetic energy in the y-direction. Therefore, when there is less rugged local potential energy, the kinetic energy is not enough to make the robot move laterally. Otherwise, the robot cannot remain stable.

4 Quadruped Robot Motion Simulation

We use the simulation software RecurDyn V7R5 to accomplish this simulation.

The parameters used are as follows: Quadruped walking robot cycle: $T = 21$ s; Moving distance: $L = 275$ mm; Lift height: $H = 100$; Rough terrain height: $z = A\cos(2\pi x) + A\sin(2\pi y)$, $A = 0.1$ m.

The quadruped robot traversing the rough terrain is shown below:

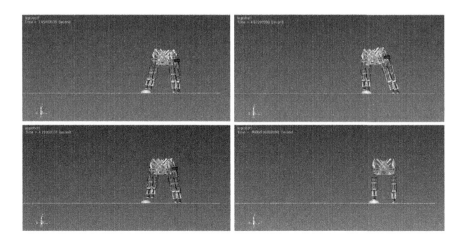

Simulation results show that by determining and adjusting points off the foot of the body posture angle, effectively reducing the impact of the foot, we can achieve balance recovery in the large rugged terrain covered by the robot.

5 Conclusion

At the nuclear disaster site, due to the presence of complex interfering factors and uncertain terrain, the robot may experience external shocks during walking.

Gait Planning is proposed to consider the four-legged robot at the accident scene in the face of external shocks and the presence of two rugged terrain interferences. In this emergency situation, we drop enough points of the foot by completing the selection and adjustment of attitude, so that the robot can be enabled to restore balance in the case of such interference, thereby continuing its relief work.

In this way, we can not only make the robot successfully complete its assigned work, and to avoid accidental damage to the robot due to the ensuing aftershocks, but also extend the service life of the robot.

References

Arslan O, Saranli U (2012) Reactive planning and control of planar spring-mass running on rough terrain. IEEE Trans Rob 28(3):577–579

Atkeson CG, Stephens B (2007) Multiple balance strategies from one optimization criterion. In: 7th IEEE-RAS international conference on humanoid robots. IEEE, Piscataway, NJ, USA, pp 65–72

Auke JI (2008) Central pattern generators for locomotion control in animals and robots: a review. Neural Networks 21(4):642–653

Barasuol V, Buchli J, Semini C, Frigerio M, De Pieri E, Caldwell DG (2013) A reactive controller framework for quadrupedal locomotion on challenging terrain. In: Proceedings of the IEEE international conference on robotics and automation, Karlsruhe, pp 2539–2545

Buchli J, Stulp F, Theodorou E et al (2011) Learning variable impedance control. Int J Robot Res 30(7):831–849

Haowei W, Pengfei W, Xin W et al (2011) Touchdown angle's impact on bounding gait of the quadrupeds. In: Proceedings of the 2011 IEEE 5th international conference on robotics, automation and mechatronics (RAM). Qingdao, China, IEEE, Piscataway USA, pp 178–183

Havoutis I, Semini C, Buchli J et al (2013) Quadrupedal trotting with active compliance. In: 2013 IEEE international conference on mechatronics (ICM), Vicenza, Italy, IEEE, Piscataway, NJ, USA, pp 610–616

Kim TJ, So B, Kwon O, Park S (2010) The energy minimization algorithm using foot rotation for hydraulic actuated quadruped walking robot with redundancy. In: Proceedings of the 6th German conference on robotics, Munich, pp 786–791

Maki B, McIlroy W (1977) The role of limb movements in maintaining upright stance: the "charge-in-support" strategy. Phys Ther 77(5):488507

Maufroy C, Kimura H, Nishikawa T (2012) Stable dynamic walking of the quadruped "Kotetsu" using phase modulations based on leg loading/unloading against lateral perturbations. In: 2012

IEEE international conference on robotics and automation (ICRA). Saint Paul MN USA, Piscataway, NJ, USA. IEEE, pp 1883–1888

Park J, Park JH (2012) Impedance control of quadruped robot and its impedance characteristic modulation for trotting on irregular terrain. In: 2012 IEEE/RSJ international conference on intelligent robots and systems (IROS2012), Vilamoura, Portugal, IEEE, Piscataway, NJ, USA, pp 175–180

Pongas D, Mistry M, Schaal S (2007) A robust quadruped walking gait for traversing rough terrain. In: Proceedings of IEEE international conference on robotics and automation, Roma, Italy, pp 1474–1479

Pratt J, Carff J, Drakunov S et al (2006) Capture point: a step toward humanoid push recovery. In: 6th IEEE-RAS international conference on humanoid robots, Piscataway, NJ USA: IEEE, pp 200–207

Raibert M, Blankespoor K, Nelson G et al (2008) Bigdog, the rough-terrain quadruped robot. In: World congress of international federation of automation control. Springer, Heidelberg

Rebula JR, Neuhaus PD, Bonnlander BV et al (2007) A controller for the LittleDog quadruped walking on rough terrain. In: 2007 IEEE international conference on robotics and automation, Roma, Italy. IEEE, Piscataway, USA, pp 1467–1473

Rubin C, Yangzheng C, Lin L et al (2013) Inverse kinematics of a new quadruped robot control method. Int J Adv Robot Syst 10:1–8

Xuewen R, Yibin L, Jiuhong R et al (2012) Inverse kinematics of a new quadruped robot. J Mech Sci Technol 26(4):1171–1177

Yu HT, Li MT, Wang PF, Cai HG (2012) Approximate perturbation stance map of the SLIP runner and application to locomotion control. J Bionic Eng 9(4):411–422

Research on the Application of NoSQL Database in Intelligent Manufacturing

Chuan-hong Zhou, Kun Yao, Zhen-yang Jiang and Wu-xia Bai

Abstract Intelligent manufacturing is a new trend that is a deep integration of information and industrialization. As to the information, database is very important. For the past 20 years, relational databases have been widely used in a lot of fields because of their rich feature, such as query capabilities and transaction management and so on. However, they do not have the capacity of storing and processing a large number of data effectively, and at the same time, they are not very efficient to make transactions and join operations. In order to adapt to the new demand, some of new databases have been invented which are not in accordance with relational model. These databases are known as NoSQL. The underlying data and transaction model of the NoSQL are different from relational database. Some of organizations have shown much interest in NoSQL and adopted this new technology, which promote further research on the NoSQL. In this thesis, we research NoSQL database about their origin and characteristics. Then, we compare between one of the NoSQL database, MongoDB, to the standard relational database, SQL Server. We contrast their performance from different aspects. Results show that MongoDB performs better than the relational database.

Keywords NoSQL databases · MongoDB · SQL server

1 Introduction

The concept of "Industry 4.0" indicates the development trends of manufacturing in the future: Intelligent manufacturing (Shu 2014). However, database plays an important role in intelligent manufacturing. It is not only the core of various information systems such as Management information system, Decision support

C. Zhou · K. Yao (✉) · Z. Jiang · W. Bai
School of Mechatronic Engineering and Automation,
Shanghai Key Laboratory of Intelligent Manufacturing and Robotics,
Shanghai University, Shanghai 200072, China
e-mail: 18321733305@163.com

system, but is also acting as a major tool to scientific research and decision-making management. With the data processing in large scale and rapid growth of big data, traditional relational database exposed some problems that were insurmountable, which in turn gives the opportunity for new database model (Abramova and Bernardino 2013), so NoSQL (Vijaykumr and Saravanakumar 2010) database emerged and developed rapidly. Some famous technology companies such as Google, Facebook have put NoSQL database into practice, thus, it is a new attempt to apply NoSQL database to manufacturing field.

2 Research Status

The word of NoSQL appeared for the first time in 1998, which refer to the database with the characteristics of light weight, open source, none SQL function that created by Carlo Strozzi. In 2009, Johan Oskarsson launched a debate about the distributed-open resource database (Pramod and Martin 2013), and Eric Evans put forward the concept of NoSQL again. At this time, NoSQL mainly refer to the design patterns of database with non-relational, distributed, no ACID. A seminar about "no:sql(east)" which was hosted at Atlanta in 2009 was milestone, presenting its slogan "select fun, profit from real_world where relational=false;". Thus, the most popular explanation on NoSQL is that non-relational, emphasizing Key-Value Stores and advantage of document database, instead of simply opposing RDBMS.

Comparing to RDBMS, NoSQL database has its own advantages (Zheng2013):

- Flexible extensibility
  With the increasing amount of data, there is a difficulty that developers have to face: How to extend database? Vertical expansion has its limitation on computing power and disadvantage on cost. However, horizontal expansion is easy to extend and not expensive.
- Mass data and high performance
  NoSQL database has a very high reading and writing performance with low latency, especially when it is related to big data. All of these benefits from the structure of database which are simple and non-relational.
- Flexible data model
  There is no need to build data model in advance. You can update data storage mode at any time, so, it is convenient to add field to database table.

Currently, NoSQL database has been applied to some international technology companies, such as Google, Yahoo, Twitter, Amazon, Facebook. At the same time, some Chinese Internet companies have put NoSQL database into practice or intend to turn to NoSQL database, such as Sina, Alibab, Visual China, Jingdong Mall (Chen et al. 2011). Although NoSQL develop fast, but it is not mature, it needs to be improved in practice (Stonebraker and Cattell 2011).

3 NoSQL Database

3.1 BASE Versus ACID

CAP theory (Han et al. 2011) and weak order consistency model BASE is the academic cornerstone of NoSQL database. According to CAP theory, a distributed system cannot meet the consistency, availability, partition-tolerance at the same time. It just can meet two attribute at most (Brewer 2010).

BASE (Roe 2012; Tudorica and Bucur 2011) is the abbreviated of Basically Available, Soft-state, Eventually Consistency which is against ACID model. It not only allows the data to meet only the Eventually Consistency but also can be connected flexibly (Soft-state, which allows the data to be inconsistent in a period of time).

The traditional relational databases use transaction to realize the data access with the characteristics of atomic, consistency, isolation, and persistence (ACID) (Roe 2012), which guarantees the data integrity and accuracy. However, with the development of Internet and big data, especially in the Web 2.0 era, high availability and high scalability is more important than consistency. Therefore, NoSQL database can fulfill the eventual consistency on the basis of BASE model rather than strong consistency.

3.2 Types of NoSQL Databases

Because of the model of bivariate table is simple, logic clarity and easy to be achieved, RDBMS adopt bivariate table to store data. However, bivariate table is not flexible when it comes to big data. Thus, from the perspective of data model, NoSQL proposes new concept to copy with the new demands.

With high adherence to NoSQL databases, different databases have been developed. All of those are based on same principles but own some different characteristics. Typically can be defined four categories (Indrawan-Santiago 2012):

Key-value Store. It is the simplest NoSQL store. Each key corresponds to an arbitrary value. Key-value store does not provide specific operation for attribute value of the data. It usually only provides simple operation such as set, get, and delete. Examples of Key-value Store databases are: Volemort, Berkeley DB.

Document Store. Document store is a structured document. It is usually converted into the structure of JSON or BSON to store data. The document can store list, key-value, and complex structure document. Thus, document store is very flexible. The typical Document Store databases are: MongoDB, CouchDB.

Column-family. The data model's characteristic is columnar storage, that is each row of data is stored in different column, and the collection of these columns is called the cluster. Each data which is stored in column contains a timestamp attribute, so that the multiple versions of the same data item in column can be saved. Examples of Column-family databases: HBase, Cassandra.

Graph database. The storage structure of graph is very different from the previous storage mode, such as data model, data query method, and so on. The guiding ideology of storage structure of graph is that data is not equal, relational data on storage or key-value on storage may not be the best way of storage. Examples of Graph databases: Neo4J, HyperGraphDB.

3.3 Reliability

The perfect state of reliability is that database will make all of the write operation written to the persistent storage device, and at the same time, copy multiple copies to different nodes which locate in different places to prevent data loss. The requirements on reliability affect performance, thus, different NoSQL system adopt different strategy on reliability.

3.3.1 Single-Machine Reliability

The requirement on single-machine reliability is that write operation can work regularly even if database encounter power outages or restarting machine. Generally, the guarantee of the single-machine reliability is accomplished by writing the data to the disk, but this often causes the disk IO to become the bottleneck of the system. By minimizing the random write, instead of the sequential write of each disk device, it can reduce the performance of the single-machine reliability. HBase, Cassandra, Redis and Riak can guarantee the trouble recovery by flushing data on low frequency to the disk, and realize the high performance of the reliability by recording the operation to the log file. It can also improve the throughput efficiency by aggregating write operation, for example, the group commit mechanism of the Cassandra will put the concurrent write operations on a short period of time together and flush to disk (Hewitt 2010).

3.3.2 Multi-machine Reliability

Because the hardware can sometimes cause the damage that cannot recover, the reliability of the single machine is not enough. As for some important data, multi-machine backup is the essential reliability measures. For example, HBase provides multi-machine reliability by synchronizing each write operation to more than two nodes (Cbodorow 2014).

3.4 Distributed Extension

For the growing application load, a simple solution is to upgrade the machine by increasing the memory and adding hard drive to cope with the load. But with the increase of the amount of data, the cost of upgrading the hardware cannot bring the corresponding improvement on performance, then, it is needed to let the data to be stored on different machines and share access pressure by the means of distributed extension. Because the most of the NoSQL systems and query condition are based on the key-value model, data partition (Lu 2013) is usually performed according to the data key, that is, the key attributes decide which machine will store the data. According to the different operation on key, data partition is divided into two ways: range partition, according to the value of key itself to allocate data; hash partition, according to key hash value to allocate data.

4 Application of MongoDB in Information Acquisition System

For information acquisition system, we choose respectively MongoDB which is the most typical product of the NoSQL database and SQL Server which is the representative of relational database and then compare the two databases from many aspects.

4.1 Introduction to System

This article is mainly about the temperature and humidity acquisition system of a machine tool factory with one small warehouse in Shanghai city. The stereoscopic warehouse is currently installed with 18 temperature and humidity sensors which are divided into two rows of symmetrical layout and run 24 h a day. The system reads the data by the means of Labview which are initially obtained by the temperature and humidity sensor and then lets the data store into the database.

4.1.1 Storage Requirement

The data items that need to be recorded include storage time, temperature, and humidity.

4.1.2 Query Requirement

As for the information acquisition system, the demand for query is not complex, but it needs to ensure the rate of query when it comes to the large amount of information. A common query is to return the information based on a query condition, and sort the results in chronological order.

4.1.3 Expanding Requirement

As data size collected by this system is very large (the amount of temperature and humidity information stored in one day more than 50 thousands) and increase rapidly, there is a high requirement for scalability. Thus, it needs to expand database in a convenient and low cost way. Comparing to MongoDB database, SQL Server database has the limitation on distributed expansion when the amount of the data is increasing. And when there is a need to add new kinds of the acquisition information, such as dust, gas concentration, and so on, SQL Server database needs to design the form again and, MongoDB, due to the various types of data structure, implements the expansion of the data structure easily.

The following are SQL Server and MongoDB database solutions, including design data model, query fulfillment, scalability, and analyze the pros and cons and compare them.

4.2 SQL Server Solutions

4.2.1 Storage Scheme

For the information acquisition system, SQL Server designs two tables, Temperature information table td_temp and Humidity information table td_hdt (Tables 1, 2).

4.2.2 Query Scheme

A common query is to return the information according to a query condition, and sort the results in chronological order. For example, querying the temperature and humidity information which were collected by temperature and humidity sensor 1
select td_tmp.Date,td_tmp.Temp1, td_hdt.Hdt1 *from* td_tmp *inner join* td_hdt *on* td_tmp.Date=td_hdt.Date *where* tmp.Date between 'starttime' and 'endtime' *order by* td_tmp.Date desc

Table 1 td_temp

Field name	Data type	Instruction
Date	datatime	date
Temp1	float	sensor1
Temp2	float	sensor2
…………	………	……………………
Temp18	float	sensor18

Table 2 td_hdt

Field name	Data type	Instruction
Date	datetime	date
Hdt1	float	sensor 1
Hdt2	float	sensor 2
……….	………	……………………
Hdt18	float	sensor 18

This query uses table joining between two tables. It is not efficiency when it comes to a large amount of data (Parker et al. 2013).

4.2.3 Expansibility

Any machine has physical limitations: memory capacity, hard disk capacity, processor speed, etc. With the increasing amount of data, only improving the hardware configuration is not only expensive, but also unable to keep pace with the growth of data, so SQL Server has poor scalability (Huang 2010). Besides, when the type of information collected by the sensor is increased, the new form of information is needed to redesign, this is a little trouble.

4.3 MongoDB Solutions

4.3.1 Storage Scheme

The data structures of MongoDB are similar to JSON, which is called BSON (Banker 2012). The document storage of MongoDB can achieve an array of types and embedded documents (Boicea et al. 2012).

Collective structure
```
{
  {date: storage-time datetime}
  {sensor1:{{Temp1:Temperature float},{Hdt1: Humidity float }}}
  {sensor2:{{Temp2:Temperature float},{Htd2: Humidity float }}}
  ……
  {sensor18:{{Temp18:Temperature float},{Htd18: Humidity float }}}
}
```

4.3.2 Query Scheme

A common query is to return the information according to a query condition, and sort the results in chronological order. For example, querying the temperature and humidity information which were collected by temperature and humidity sensor 1. MongoDB shell query statements as follows:

db.posts.find("date": {"$gt" :start time, "$lt": end time},{"sensor1":1}).*sort* ({date:-1})

4.3.3 Expansibility

MongoDB provides automatic slicing mechanism for distributed extension; when the amount of data of acquisition system grows to a degree that the existing slicing system cannot meet the needs, it can be added the distributed Mongod server (Pokorny 2011).

When the sensor adds new information, because the data structure of MongoDB is varied, it is easy to design new data structure. Such as adding two information: dust and odor concentration,

Collective structure
{

  {date: storage-time datetime}
  {sensor1:{{Temp1: Temperature float},{Hdt1: Humidity float },{Dust1: Dust float },{odor1: Odor float }}}
  {sensor2:{{Temp2: Temperature float },{Htd2: Humidity float },{Dust2: Dust float },{odor2: Odor float }}}
  ……
  {sensor18:{{Temp18: Temperature float },{Htd18: Humidity float },{Dust18: Dust float },{odor18: Odor float }}}

}

4.4 Database Performance Test

This paper runs SQL Server and MongoDB instances on the same configuration of the computer and then tests the performance of MongoDB and SQL Server from three aspects: INSERT, QUERY, DELETE.

Computer configuration is as follows: single-core CPU 2.1 GHz, 2 GB Memory, Win7 operating system, MongoDB 2.4.12, SQL Server 2008. Using SQL Server 2008, query analyzer and mongodb administrator shell to get time-consuming. Here is the performance comparison

First we conduct a test about insert operation. We inserted some objects which have the same fields and information into the two databases.

In order to get an intuitively comparison, we insert the data into database with different magnitude and then we calculate how much time it took and finally make a chart to show their performance.

As is shown in the Table 3, MongoDB database is more efficient when it come terms to big data.

The Fig. 1 intuitively shows the comparison of the two databases.

Second, we conduct a test about query operation. This paper tests a common query: To find the temperature and humidity information of Sensor1 in a certain period of time. On the condition of different amount of data, such as 100, 1000, 10000, 100000, 1000000, we computed how much it took, and then compare SQL Server with MongoDB on performance (Table 4).

From the results we know that MongoDB has better performance than SQL Server on query.

Fig. 2 intuitively shows the comparison of the two databases.

Third, we conduct a test about delete operation. Usually the database will save the data in the last period of time, at the same time, delete the data that are too old, so that it can help the staff to know the latest dynamics. Thus, we delete the data from the database with different magnitude and then we calculate how much time it took and finally make a chart to show their performance.

Table 3 Insert time (m sec)

No. of records	MongoDB	SQL Server
100	4	27
1000	34	876
10000	562	7220
100000	3150	70287
1000000	45871	732078

Fig. 1 Insert time (m sec)

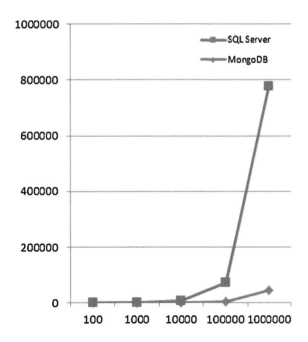

Table 4 Query time (m sec)

No. of records	MongoDB	SQL Server
100	5	42
1000	47	1362
10000	856	9261
100000	4967	93810
1000000	58751	982078

Fig. 2 Query time (m sec)

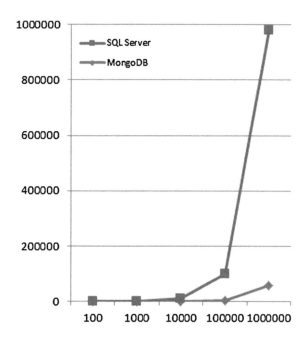

As is shown in Table 5, MongoDB database has better performance on delete operation. With the expansion of database records, the time spent by SQL Server deleting the records is increasing rapidly than MongoDB.

Fig. 3 intuitively shows the comparison of the two databases.

Table 5 Delete time (m sec)

No. of records	MongoDB	SQL Server
100	2	24
1000	10	71
10000	37	159
100000	66	1032
1000000	107	28079

Fig. 3 Delete time (m sec)

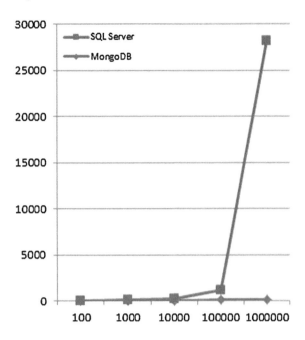

5   Conclusions

Through studying the characteristics of NoSQL database, comparing the performance of SQL Server 2008 database and MongoDB database from three aspects, this paper concludes that MongoDB database has better performance. The experiment achieved good results and made an attempt in the field of intelligent manufacturing.

Acknowledgments This work is supported by subject of Science and Technology Commission of Shanghai Municipality in key technology development and demonstration application of the unmanned factory for industrial robot production (No. 14DZ1100700). We thank Shanghai Key Laboratory of Intelligent Manufacturing and Robotics for assistance with advice.

References

Abramova V, Bernardino J (2013) NoSQL databases: MongoDB vs Cassandra. In: Proceedings of the international C* conference on computer science and software engineering, pp 14–22. doi:10.1145/2494444.2494447

Banker K (2012) MongoDB in action. Manning Publications Co, Shelter Island, pp 5–8

Boicea A, Radulescu F, Agapin LI (2012) MongoDB vs Oracle-database comparison. In: 2012 third international conference on emerging intelligent data and web technologies, pp 330–335. doi:10.1109/EIDWT.2012.32

Brewer E (2010) CAP twelve years later: how the "rules" have changed. Computer 45(2):23–29. doi:10.1109/MC.2012.37

Cbodorow K (2014) MongoDB: the definitive guide. Post & Telecom Press, Beijing, pp 204–206

Chen F, Yongqiang Z, Zhiwei X (2011) CCIndex for Cassandra: a novel schema for multi-dimensional range queries in Cassandra. In: 2011 IEEE 7th international conference on semantics knowledge and grid (SKG), pp 130–136. doi:10.1109/SKG.2011.28

Han J, Haihong E, Le G, Du J (2011) Survey on NoSQL database. In: 2011 6th international on pervasive computing and applications (ICPCA), pp 363–366. doi:10.1109/ICPCA.2011.6106531

Hewitt E (2010) Cassandra: the definitive guide. China Machine Press, BeiJing, pp 123–128

Huang XL (2010) The development and application of non-relational database NoSQL. Fujian Comput 26(07):30. doi:10.3969/j.issn.1673-2782.2010.07.018

Indrawan-Santiago M (2012) Database research: are we at a crossroad? Reflection on NoSQL. In: 2012 15th international conference on network-based information systems (NBiS), pp 45–51. doi:10.1109/NBiS.2012.95

Lu J-H (2013) Big data challenge and NoSQL database technology. Publishing House of Electronics Industry, Beijing, pp 58–60

Parker Z, Poe S, Vrbsky SV (2013) Comparing NoSQL MongoDB to an SQL DB. In: Proceedings of the 51st ACM southeast conference, Article No 5. doi:10.1145/2498328.2500047

Pokorny J (2011) NoSQL databases: a step to database scalability in web environment. In: iiWAS2011—13th international conference on information integration and web-based applications and services, pp 278–283. doi:10.1145/2095536.2095583

Pramod J, Martin F (2013) NoSQL distilled. China Machine Press, BeiJing, p 10

Roe C (2012) ACID vs. BASE: the shifting pH of database transaction procesing. http://www.dataversity.net/acid-vs-base-the-shifting-ph-of-database-transaction-processing/

Shu Z (2014) Industrial 4 and intelligent manufacturing. Mach Des Manuf Eng 43(8):1–4. doi:10.3969/j.issn.2095-509X.2014.08.001

Stonebraker M, Cattell R (2011) 10 rules for scalable performance in 'simple operation' datastores. Commun ACM 54(6):72–80. doi:10.1145/1953122.1953144

Tudorica BG, Bucur C (2011) A comparison between several NoSQL databases with comments and notes. In: Roedunet international conference (RoEduNet), pp 1–5. doi:10.1109/RoEduNet.2011.5993686

Vijaykumr S, Saravanakumar S (2010) Implementation of NoSQL for robotics. In: 2010 international conference on Emerging trends in robotics and communication technologies (INTERACT), pp 195–200. doi:10.1109/INTERACT.2010.5706225

Zheng G-L (2013) Research and implementation of campus energy data collection system based on MongoDB. MS Thesis, South China University of Technology, Guangzhou, China

Design and Application of Auditory Evoked EEG Processing Platform Based on Matlab

Rong Yang, Hui-qun Fu, Xiu-feng Zhang, Li Wang, Ning Zhang and Feng-ling Ma

Abstract To improve the processing speed and efficiency of imaging EEG signals, a processing system based on Matlab GUI and Virtual Reality sink which integrates various algorithms such as Independent Component Analysis, Hilbert–Huang Transform, Support Vector Machine, etc., is designed in this paper. The EEG data used is collected based on a novel auditory evoked method proposed in this paper. The experimental results indicate that the novel evoked method has preferable recognition that can reach to 87.2 %, and the classification results can control the movement of the designed virtual arm accurately. This research provides a certain theory and experiment basis for the research of the stroke rehabilitation robot based on the imaging movement EEG signal of auditory evoking.

Keywords Graphical user interface · Virtual reality · Hilbert–Huang transform · Support vector machine · Auditory EEG · Stroke rehabilitation

1 Introduction

Electroencephalogram (EEG) is a kind of signal that contains abundant physiological information and can reflect different thinking activities. A lot of studies have proved that EEG has a connection with the real movement and the imagine movement (Shen 2011). At present, lots have been done on the research of brain–computer interface based on visual stimulus which has already been widely used in the rehabilitation of stroke patients (Gan 2011). However, the use of brain–computer interface based on visual stimulus for stroke rehabilitation is limited because

R. Yang (✉) · X. Zhang · L. Wang · N. Zhang · F. Ma
National Research Center for Rehabilitation Technical Aids,
Beijing 100176, China
e-mail: wwqq346@163.com

H. Fu
101 Institute of the Ministry of Civil Affairs, Beijing 100070, China

© Zhejiang University Press and Springer Science+Business Media Singapore 2017
C. Yang et al. (eds.), *Wearable Sensors and Robots*, Lecture Notes in Electrical
Engineering 399, DOI 10.1007/978-981-10-2404-7_34

stroke not only destroys the motor function of the patients, but also the advanced cognitive function related to it (Liang et al. 2012). The study of brain–computer interface based on auditory stimuli becomes very important and necessary. It has been shown that the choice of stimulus has a great influence on the recognition of the EEG signal (Johnson et al. 2006). The traditional auditory evoked method is too complicated for the patients to memorize during the EEG signal acquisition (Xu et al. 2011). This leads to a high error rate and is unfavorable for the feature extraction and pattern recognition of the EEG signal. According to the characteristics of stroke patients, this paper proposes a novel experimental method of imaging left-right hand movement based on clear auditory instructions and the application result proves the feasibility and advancement.

Graphical User Interface (GUI) and Virtual Reality Sink (VR Sink) are two important extended functions in the Matlab software. These functions can not only use the powerful data processing and analysis ability of Matlab, but also have advantages in interactive and integrative platform design and application. The platform developed by GUI and VR sink is powerful, easy used and can also be run independently of the Matlab software (Chen 2014). Therefore, based on GUI and VR Sink, this paper designs a processing and test platform which integrates several algorithms to test the availability of the proposed novel experiment method and improve the processing speed of the imaging EEG signals. The experimental result shows that the designed platform satisfies the requirement of high speed, high efficiency, and high stability in the processing of EEG signals.

2 Theory and Methods

It has been proved that when imaging the hand movement, the energy between 8 and 12 Hz in the brain kinaesthesia area which is called mu rhythm, is changed differently on the ipsolateral side and contralateral side (Shen 2011). Based on this theory, in this paper, after the data preprocessing period, Hilbert–Huang Transform (HHT) is used to extract the instantaneous energy of the mu rhythm in the EEG signals as the feature and Support Vector Machine (SVM) algorithm is used to recognize the movement of the left-right hand. Figure 1 shows the flow chart of the data processing.

2.1 Data Preprocessing

Ocular Artifact Reduction and Filtering are two central and important parts in this period (Cui 2013). These steps effectively improve the Signal Noise Ratio (SNR) and the detection efficiency of EEG data, which improve the effectiveness of the data finally. In this paper, for Ocular Artifact Reduction, Independent Component Analysis (ICA) is used to eliminate the electrooculogram

Fig. 1 The flow chart of the data processing

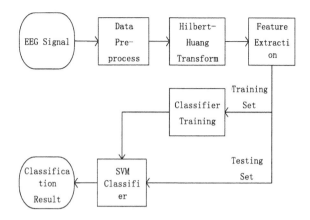

(EOG) signals produced during the movement imagery period. The characteristic of the ICA is that the multichannel's observed signals are decomposed to some mutual independent component according to statistical independent fundamental and optimization algorithm, consequently helps us to analyze signals and enhance the SNR (Zhou et al. 2005).

Prior to extraction of the feature, the referenced EEG signal is filtered in an 8–30 Hz band. The filter band is chose because it encompasses the mu frequency band, which has been shown to be most important for movement classification. Furthermore, in a recent study, it is shown that a broad frequency band gives better classification results compared to narrow bands (Müller-Gerking et al. 1999).

2.2 Feature Extraction Based on HHT

HHT is a new kind of processing algorithm for unsteady signal proposed by N.E. Huang in 1998. It consists of two steps: Empirical Mode Decomposition (EMD) and Hilbert transform (Shen 2011). EMD is the core of HHT. It separates the signal into a finite number of Intrinsic Mode Function (IMF) with different features scale based on the internal characteristics of the signal. The Hilbert spectrum constructed by the instantaneous frequency of the IMFs solved by the Hilbert transform can reflect the energy distribution of the signal in time and frequency accurately. HHT algorithm, which has higher time-frequency resolution, is very suitable for the process of EEG signals (Yuan et al. 2010).

2.2.1 EMD

The IMFs decomposed by EMD must satisfy two characteristics to make the instantaneous frequency of EEG signal stable and meaningful (Yan 2012):

(1) IMFs must satisfy the definition of narrow zone signal for Gaussian distribution: the number of all the local extremums and the zero-crossings must be equal or the difference between them is 1 at the most.
(2) IMFs must satisfy the definition of the local symmetric: the mean value determined by the local extremums must be zero at any time.

The EMD is ended until the remaining component becomes a monotone function or less than a predetermined value. By analysis of EMD, the raw EEG can be expressed as

$$x(t) = \sum_{i=1}^{n} c_i(t) + r_n(t) \tag{1}$$

where $\sum_{i=1}^{n} c_i(t)$ is the sum of IMFs and $r_n(t)$ is the remaining component.

2.2.2 Hilbert Transform

The Hilbert transformation of the IMFs is given as

$$y_i(t) = \frac{1}{\pi} P \int_{-\infty}^{\infty} \frac{c_i(\tau)}{t - \tau} d\tau \tag{2}$$

With the instantaneous amplitude

$$a_i(t) = \sqrt{y_i(t)^2 + c_i(t)^2} \tag{3}$$

and the instantaneous frequency

$$\omega_i(t) = \frac{d}{dt} \left(\arctan \frac{y_i(t)}{c_i(t)} \right) \tag{4}$$

the Hilbert spectrum in which the amplitude is changed according to time and instantaneous frequency, is given as

$$H(\omega, t) = \sum_{i=1}^{n} a_i(t) \exp\left(j \int \omega_i(t) \, dt \right) \tag{5}$$

To get the marginal spectrum, integrations are performed on the Hilbert spectrum due to the calculation of

$$h(\omega) = \int_0^T H(\omega, t)\, dt \tag{6}$$

The marginal spectrum reflects the amplitude accumulation on each and every frequency point, which is also the total energy of each point.

2.3 Recognition Based on SVM

In this experiment, SVM is adopted to recognize and classify the feature of EEG signals extracted by HHT. SVM is a kind of new machine learning method based on statistical learning theory proposed by Vapnik (Wu 2007). Through nonlinear mapping, the sample space is mapped into a high-dimensional or an infinite-dimensional feature space to transfer the nonlinear separable problem into a linear separable problem (Sun 2010). Based on structural risk minimization rules, SVM has the optimal classification ability and extended ability which is very suitable for the classification of small sample. An SVM classifier based on the Radial Basis Function (RBF) is applied in this paper (Huang et al. 2014). The RBF is shown as

$$K(\| x, y \|) = \exp\left\{ - \| x - y \|^2 \big/ (2 * \sigma)^2 \right\} \tag{7}$$

where $K(\| x, y \|)$ is the monotonic function of the Euclidean distance between x and y, and σ is the width parameter of the function.

3 Platform Design Based on Matlab

3.1 GUI Design

There are four main models designed based on Matlab GUI to realize the processing and classification of the EEG signal: data import model, feature extraction model, classification model, and menu and toolbar design model.

(1) Data import model: All the EEG signals are sampled at 250 Hz. The data import model has three main functions: importing, modifying, and deleting

EEG data. In the experimental data, the 4 s to the 10 s is the pure motive imagery period. Data in this period is used to analysis in this paper.

(2) Feature extraction model: The functions of the HHT algorithm, the spectrum display and the feature matrix display are all integrated in this model. This model processes the EEG signals from C3 and C4 electrodes with the HHT, displays the drawn Hilbert spectrum and marginal spectrum on the figures, and then displays the extracted feature vector on the table as a 1×2 matrix, where 2 is the number of channels in this experiment.

(3) Classification model: In this model, the functions of the SVM algorithm are integrated. Users classify the extracted feature vector based on the optional type of core function, penalty coefficient, and width of core function. The classification result is expressed as "Right hand" or "Left hand" on the figure.

(4) Menu and toolbar design model: Users can use the file menu to save the drawn marginal spectrum, and can also derive the extracted feature matrix of the EEG data as Excel format. The zoom in, zoom out, and drag tools on the toolbar can be used to modify the spectrum and axes so that users can observe and analysis the power spectrum easily and clearly.

3.2   Virtual Reality Design

A 3D virtual character is designed based on the VR Sink model to simulate the movement of the human arm as Fig. 2 shows. The trajectory of the arm on the designed virtual character is changed according to the classification result of the EEG signal.

Fig. 2 3D virtual character

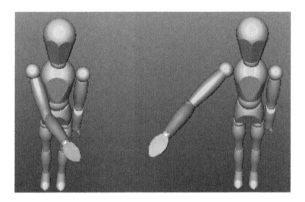

4 Experiment Test

4.1 Data Acquisition

A new experiment method based on clear auditory instructions is proposed in this paper using the E-prime software. The flow of the experiment is shown in Fig. 3. Two kinds of clear voice left-right hand movement instructions are used as the auditory stimulus in the experiment. At the beginning, the subject must clear their minds and stay relaxed. At the 2 s, when hearing the "start" instruction, the subject must be focused and prepare to do the imaging tasks. From the 4 s, the subject is instructed to imagine a movement of the left or the right hand depending on the auditory instruction they heard from the computer. The auditory instruction is consisted by the name of left or right hand spoken in Chinese and is randomized and cycled. One whole imaging task is stopped by a short warning tone ("beep") at 10 s.

EEG signal is recorded by Nuero-scan equipment in this experiment. The electrode mounted on the right ear is the reference electrode and the electrodes named C3 and C4 mounted on the brain kinaesthesia area are the recording electrodes. The actual recording potential is the difference between the electrode potential and the reference potential.

One male and one female right-handed subject took part in the study. The subjects were seated in an armchair during the whole experiment and kept their eyes closed. They were asked to keep their bodies relaxed and to avoid any hand movements to protect the EEG signals clear from the electromyogram (EMG) and EOG. The experiment comprised two experimental runs of 40 trials each (20 left and 20 right trials).

4.2 Data Analysis

Figure 4 shows the processing results EEG signal from one single trial when the subject imaging left-hand movement. In Fig. 4, the Hilbert spectrum and marginal

Fig. 3 The flow chart of the experiment

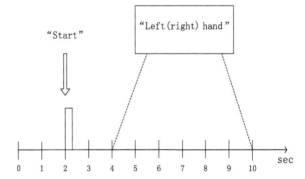

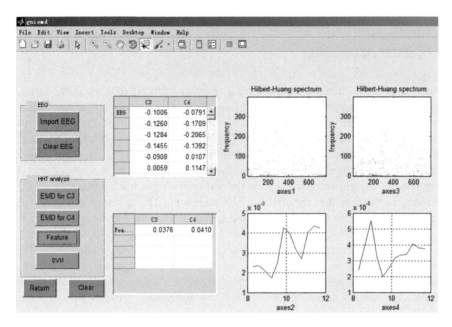

Fig. 4 EEG processing results

spectrum extracted from C3 electrode are displayed on axes1 and axes2, and C4 electrode on axes3 and axes4. As the results show, when the left-hand movement imagery starts from 4 s, the energy of the mu rhythm on the C3 electrode is enhanced while energy on the C4 electrode is inhibited. The difference value on the two electrodes is obvious and easy to extract.

For proper estimation of the classification accuracy (Müller-Gerking et al. 1999), the data set of each subject is divided into a training and testing set. The training set is used to calculate a classifier, which is used to classify the testing set. This training/testing procedure is repeated 20 times with different random partitions into training and testing sets. In the 80 trials each, the classification rate, as well as the success rate of the VR arm movement for each subject has all reached above 80 % and differs little as Table 1 shows. It has been shown that the gender difference has little effect on the recognition accuracy. The excellent recognition has also proved the effectiveness of the proposed experiment method and the processing of the EEG signals.

Table 1 Success rate of VR arm movement	Subject	Gender	Sample	Success rate (%)
	1	Male	80	87.5
	2	Female	80	82.5

5 Conclusion

This paper proposes a novel auditory evoked method and an EEG processing platform based on the characteristics of the EEG when imaging the hand movement. As the experiment proved, the proposed evoked method is feasible to realize various rehabilitation trainings of stroke patients by designing different auditory stimulation instructions. This research has the following advantages:

(1) Low rehabilitation cost. The proposed experimental method is not only suitable for the stroke patients with visual impairment, but also for the ones with normal vision. Compared with the visual command generating equipment, the generating of the auditory command is easier and with no need for complex experimental equipment. It also has the advantages of small volume and low cost.

(2) The auditory instruction is clear that can lead to high quality imaging EEG extracting only through a short training. In this experiment, the subjects can produce imaging EEG signals with obvious features after 1–2 times early training. After the preprocessing and HHT transform, the energy difference of EEG signals from C3 and C4 electrodes is clear which has obvious advantages in the precise control of the stroke rehabilitation robot.

(3) Based on the characteristics of Matlab GUI and VR sink, this paper develops an integrated platform and realizes the integration, visualization, interaction, and secondary development of the EEG processing algorithm. This platform takes full advantage of every component in Matlab GUI and VR sink, and primely displays the integrative function of the signal processing. Users can realize various necessary operations during the EEG signal processing efficiently, especially on feature extraction, user interaction, and secondary development, etc.

Acknowledgments Project supported by the National Natural Science Foundation of China (No. 51275101).

References

Chen X (2014) Automatic control system for GUI simulation design of frequency analysis. Electron Des Eng 22(22):10–15

Cui Y (2013) Research on arm rehabilitation based on motor imagery EEG. Beijing University of Technology

Gan Z (2011) Design and research of electric and control system for a hand rehabilitation robot. Harbin Institute of Technology

Huang Q, Li L, Lai S (2014) Personalized modeling of head-related transfer function based on RBF neural nerwork. J Shanghai Univ Nat Sci 20(2):157–164. doi:10.3969/j.issn.1007-2861. 2013.07.019

Johnson JA, Zatorre RJ (2006) Neural substrates for dividing and focusing attention between simultaneous auditory and visual events. Neuroimage 31(4):1673–1681. doi:10.1080/09500349514551581

Liang J, Xu G, Li M (2012) Study of auditory perception tasks based on EEG driving. Chin J Biomed Eng 31(2):217–221. doi:10.3969/j.issn.0258-8021.2012.02.009

Müller-Gerking J, Pfurtscheller G, Flyvbjerg H (1999) Designing optimal spatial filters for single-trial EEG classification in a movement task. Electroenc Clin Neurophys 110 (1999):787–798

Shen X (2011) Study on EEG feature exaction and pattern recognition based on time-frequency. Hangzhou Dianzi University

Sun H (2010) Study of brain computer interface based on left and right hand motor imagery in different visual and auditory stimulation. Hebei University of Technology

Wu Y (2007) Recognition classification and BCI experiments for mental EEG of imaging left-right hands movement. Third Military Medical University

Xu B, Si P, Song A (2011) Upper-limb rehabilitation robot based on motor imagery EEG. J Robot 33(3):307–313

Yan N (2012) The research of evoked EEG fearture by audio-visual stimulus. Shanxi University

Yuan L, Yang BH, Ma S (2010) Discrimination of movement imagery EEG based on HHT and SVM. Chin J Sci Instrum 31(3):649–654

Zhou C, Liu Z (2005) Research of single channel speed noise reduction algorithm based on independent component analysis. Laser Infrared 35(5):374–377

Part IV
Visual Recognition Application

An Efficient Detection Method for Text of Arbitrary Orientations in Natural Images

Lanfang Dong, Zhongdi Chao and Jianfu Wang

Abstract Due to the high complexity of natural scenes, text detection is always a critical yet challenging task. On the basis of existing character detection method, a novel text line detection method is proposed in this paper, which can localize text of arbitrary orientation by using related information of character regions in candidate text line. First, inspired by the Hough transform, text line detection problem is regarded as line detection problem in candidate characters set obtained by Most Stable Extremal Regions (MSERs). Second, in order to find out the relationship of adjacent candidate regions, a graph model is built based on some constraints and adjacent candidates are linked into pairs to obtain search domain. Then, to avoid repeated calculation of the same line, some strategies need to be used. Finally, as some of the potential text lines are incorrect, we use a new text line descriptor to exclude the non-text areas. Experimental results on the ICDAR 2013 competition dataset and MSRA-TD500 show that the proposed approach is favorable no matter for non-horizontal text or horizontal text.

Keywords MSERs · Graph model · Text line detection · Text line descriptor

1 Introduction

Wearable device refers to a portable device that can be directly worn on the body or integrated into the clothes or accessories. In order to realize user interaction, life entertainment, human monitoring, and other functions, this kind of equipment

L. Dong (✉) · Z. Chao · J. Wang
School of Computer Science and Technology, University of Science and Technology of China, Hefei 230027, China
e-mail: lfdong@ustc.edu.cn

Z. Chao
e-mail: chaozhd@mail.ustc.edu.cn

J. Wang
e-mail: wangjf55@mail.ustc.edu.cn

© Zhejiang University Press and Springer Science+Business Media Singapore 2017   447
C. Yang et al. (eds.), *Wearable Sensors and Robots*, Lecture Notes in Electrical Engineering 399, DOI 10.1007/978-981-10-2404-7_35

integrates various types of recognition, sensing, connection, cloud services, and other interactive technology and storage technology to replace the handheld and other devices. With the development of computer technology, it is possible to apply image processing and computer vision technology into wearable devices, such as portable intelligent navigation glasses for the blind, automatic translation and explanation tools for tourists. The key issue of these applications is extracting information from the scene accurately and efficiently. As a form of information with high expressiveness, texts in natural images get more and more attention in recent years.

Because of the complexity of background and randomicity of text color, font, orientation, and position, the recognition accuracy of OCR system is relatively low in natural image or in video image. In order to localize texts accurately, three problems need to be resolved: (1) texts in scene image may appear in any form, such as font varies, color changeable, etc. (2) texts can be embedded in any background, such as landscape, architecture, window, etc. (3) text may be similar to the background, such as handwritten characters and comics, characters and columns, pencils, trunks, etc. Therefore, how to solve these problems is the main difficulty of existing methods.

In this paper, we present an unconstrained text localization method as Fig. 1 shows. Rather than deleting all noise character regions in the phase of character detection, we use text line descriptor to exclude incorrect text lines. In the first phase, character regions are detected by MSERs. As some candidate regions only contain part of a character, a preprocessing is executed first. Then noise regions are removed by character descriptor. As it is almost impossible to discriminate text regions from non-text regions completely just by using character descriptor, post-processing needs to be considered. In most previous papers, text lines are detected by pruned exhaustive search or some heuristic rules. The exhaustive method is computational as it detects every potential line in the candidate set, and methods based on heuristic rules are more applicable for horizontal text extraction. In this paper, a quick text line extraction method is proposed, which can localize text of arbitrary orientation by detecting lines in the graph model built by adjacent relationship. Non-text areas are excluded by descriptor that integrates information (like size, center, color, etc.) of regions in candidate text lines.

Fig. 1 Block structure of the proposed method

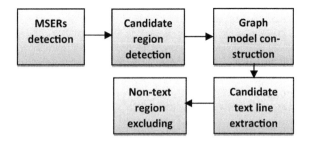

The rest of this paper is organized as follows. An overview of previous published methods is given in Sects. 2, 3 details the proposed method. Experiments and results are presented in Sect. 4 and conclusions are drawn in Sect. 5.

2 Previous Work

Present text localization approaches can be roughly divided into two categories: texture-based and component-based. The texture-based approaches assume that compared to non-text area, there are some unique textures in the text areas. Based on this assumption, Xiaodong Huang thought that text area was rougher than background (Huang 2012). In order to predict the location of text, an edge map was calculated by wavelet transform, then text rows were detected based on statistic coarseness of the edge map. Shekar B H, Smitha M L detected text by using discrete wavelet transform and gradient difference (Shekar et al. 2014). They calculated the gradient map with Laplacian template and MGD on the edge map extracted by discrete wavelet transform. Yatong Zhou distinguished text and non-text areas by discrete cosine transform. First image was divided into N * N pieces, and then based on the gradient information of each piece, the text areas were obtained. Although above approaches perform perfectly in image with regular texts, such as video frame, their detection results are not ideal in scene images because of the morphological diversity and randomness of text. Jung-Jin Lee, Kai Wang, Xiangrong Chen, etc., scanned images with sliding windows of different scales and distinguished text and non-text regions by descriptors (Mancas-Thillou and Gosselin 2006, Wen et al. 2009, Wang et al. 2011). As text areas are texture-unique, these methods are more favorable, but the amount of computation is also considerable.

In contrast to texture-based method, region-based approaches pay more attention on features of characters. Yuning Du, Shihong Lao, etc., detected dot text based on FAST points (Chen and Yuille 2004); Mingcheng Wan, Fengli Zhang, etc., thought that the distribution of corners on character edge were special and regular, which could be used to discriminate the non-text edges (Du et al. 2011). Wonder Alexandre Luz Alves and Ronaldo Fumio Hashimoto used Ultimate Attribute Opening to extract the set of characters candidates. Neumann L localized text with oriented stroke detection (Yao et al. 2012). Although various methods have been proposed, SWT (Wan et al. 2008, Epshtein et al. 2010, Huang et al. 2013) and MSERs (Neumann and Matas 2013, Iqbal et al. 2014, Neumann and Matas 2012, Matas et al. 2004) are by far the most popular methods as they are more favorable in the case of various complex scenes.

Although using MSERs to detect candidate character regions can achieve good result in majority cases, one considerable factor is that the cardinality of the MSERs set is exponential in the number of pixels in the image, so that it is time consuming for subsequent processing. In order to reduce the unnecessary regions, Neumann L designed a character descriptor to evaluate the region. He recorded the value of descriptor when the regions changed, and the region with biggest value is regarded as best rectangle for a character. In Matas' paper (Matas et al. 2004), a MSERs lattice is built by the inclusion relationship. The character region is seen as the region that satisfies some specific connecting relationship. Although these methods are effective, the computation is complex and rules in them are difficult to be obtained.

3 The Proposed Method

3.1 MSERs Detection

We assume that each character is a continuous region and has similar color, and the optimal candidate region just covers outer boundary of a character. As the MSER lattice (Fig. 2) induced by the inclusion relation shows, smaller MSERs are imbedded into bigger ones. In other words, one MSER will experience several changes (like embedded into another, or become larger or barely budged). In order to obtain the optimal MSER, growth rate of MSER must be calculated in every phase. We think that, the region with the lowest growth rate is set as the optimal MSER. The Fig. 2 shows the results of proposed approach. As smaller MSERs will be embedded into bigger ones eventually, some small and stable MSERs without containing any characters cannot be excluded (as show in Fig. 2), otherwise, the one just covering character also should be removed, which plays the same role as its successor node in MSERs lattice.

Fig. 2 Process instance of character detection method

3.2 Two-Level Character Descriptor

As the high diversity of characters and backgrounds, there are many noise regions after MSERs detection, and it is hard to exclude all of them just through one-level classifier. In this paper, a two-level character descriptor is designed.

Considering that most of irrelevant regions are of color-closing overall or rough details, and these features can be quantified easily, a coarse classifier constituted by some constraints is designed to remove part of noise regions in the first level. A rule of thumb holds that the area of character cannot exceed a certain percentage of the total area, a character is always a continuous region, and the edge points and connected domain are generally finite. Based on this, three factors are taken into account: (1) color distribution after binarization; (2) number of connected domains; (3) number of edge points. We set l as MSER, $p(i)$ denotes the proportion of pixels have value i, conn(l, i) be the number of connected domains of pixels of value i, edge(l) be the number of edge points and w and h be the width and height of MSER. Then following constraints need to be satisfied.

$$P(i) < 0.78 \ (i = 0, 1) \tag{1}$$

$$(w + h) < \text{edge}\,(l) < (w + h) * 5 \tag{2}$$

$$\sum_i \text{conn}\,(l, i) \le 7 \ \&\& \ (\text{conn}\,(l, 0) \le 3 \,||\, \text{conn}\,(i, 1)) \le 3 \tag{3}$$

Constraint (1) can be applied to exclude flat areas like whitespace between characters in Fig. 3. Constraint (2) and constraint (3) are used to eliminate MSERs with rough details. Figure 3 shows the results after the coarse classifier that integrates these constraints. We can see that majority of non-text regions are removed.

In the second level, a neural network classifier is adopted. Different from the non-text region, text regions always possess following features: (1) containing two main colors; (2) high contrast between character region and background; (3) pixels with the same gray value always located in the same connected domain; (4) smooth edge, etc. Based on these characteristics, following features are extracted:

(1) HOG feature of 8 orders;
(2) Color difference. The color difference measures roughness of the candidate region, and it is calculated as follows:

$$\sum_{i=0}^{N} (i - \text{average Gray})^2 \times p(i) \tag{4}$$

(3) Entropy of histogram
(4) Ratio of edge and the sum of width and height

Fig. 3 Result after character descriptor (from the *left* to *right* are original map, map after MSERs detection, result of coarse classifier, map after neural network classifier)

(5) Ratio of edge and area
(6) Proportion of foreground in the region
(7) Edge difference. The edge difference describes the roughness of edges in a region. As most edges of character are smooth, the values of them are smaller than regions with rough details. The calculation formula is as follows:

$$\nabla X = \sum_{x=i-1}^{i+1} x; \quad \nabla Y = \sum_{y=j-1}^{j+1} y;$$
$$\mathrm{eCom} = \sum_{i,j} \sqrt{\left(\frac{\nabla X}{\mathrm{count}} - i\right)^2 + \left(\frac{\nabla Y}{\mathrm{count}} - j\right)^2} \tag{5}$$

$\nabla X, \nabla Y$, are the sum of X-axis and Y-axis of edge points in the $3 * 3$ window with center (i, j), count is the number of edge points in the window. When the edge is straight, the edge difference tends to zero. With the increase of rough degree, the value will be bigger.

3.3 Text Line Detection

Although above method is effective to eliminate noise regions, there are still a lot of non-text regions existing. In order to solve this problem, a new text line extraction method is proposed. It not only can find out the text line quickly, but also can exclude non-text lines easily.

Characters in a text line are always neatly arranged and in a straight line (as Fig. 4). Inspired by Hough transform, the following hypothesis is put forward: candidate text line detection can be regarded as the straight line detection in the image. As all lines with different slopes need to be checked, the amount of computation is the square of number of candidate regions. It is very time consuming when the quantity of candidates is tremendous, moreover, as text lines cannot be neglected just by length like Hough transform, a lot of candidate lines will be extract. In order to decrease unnecessary computation, search space in the proposed

Fig. 4 Instances of pictures containing text line

method is narrowed by preset search paths. As regions in a text line are similar in some aspects, we can restrain search paths by constructing a graph model.

3.4 Graph Model Construction

The key of constructing graph is definite of neighboring relations. As to make the search path proceed along those regions most likely in the same line, in this module, the neighboring relation is not only determined by Euclidean distance, but also affected by size of regions. Concretely, in the graph, each node will point to no more than two nodes which are successor nodes with the shortest distance from their precursor, and have to satisfy following constraints.

$$
\begin{aligned}
\left| W(l_i) - W(l_j) \right| &< 0.5 * \min\left(W(l_i),\ W(l_j) \right) \\
\left| H(l_i) - H(l_j) \right| &< 0.5 * \min\left(H(l_i),\ H(l_j) \right) \\
\text{distance}\ (l_i, l_j) &< 2 * \min\left(\&\ \text{diag}\ (l_i),\ \&\ \text{diag}\ (l_j) \right)
\end{aligned}
\tag{6}
$$

where $L(l_1, l_2, l_3, l_4, \ldots, l_n)$ denotes the set of MSERs, n is the size of the set, $W(l_i)$ and $H(l_i)$ are the width and height of l_i. diag (l_i) is the length of diagonal of l_i.

The node with no successors will be treated as a sub graph. As two nodes are probably the nearest node to each other, that makes the retrieval unable to continue, like Fig. 5. Some measures need to be taken to avoid this.

3.5 Text Line Detection

After the construction of graph model, the retrieval of text lines only needs to be carried out along the line in the graph. Two lists (CN and LN) must be created first, CN is used to save candidate nodes need be checked and LN is used to save nodes in the search line. The search algorithm contains three steps

Fig. 5 Structures interrupting search process

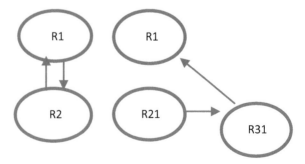

(1) Traverse node pairs comprised by adjacent nodes in the graph.

(2) Calculate the line constituted by the nodes pair, and put the nodes pair into the LN and successor nodes of them into CN.

(3) If CN is not empty, get a node in CN and checked it, else if it is in the line, put the node into LN and its successor nodes are added into CN. Repeat this step until there is no nodes in CN.

(4) The candidate text line constructed by nodes in LN is added to the candidate line list.

To avoid repeated calculation for the same line, we stipulate that if two nodes have the relation of precursor-successor in the graph and are in a straight line that has been extracted, the line constructed by them will not be considered. As to avoid adding the irrelevant regions into the line, region that will be searched must be the successor of one region in the line. Result of this text line detection method in Fig. 6 shows that most text lines were seek out. It proves that this method can extract text line effectively. However, a lot of non-text regions were also extracted. A text line classifier trained by SVM is designed to eliminate these noise text lines.

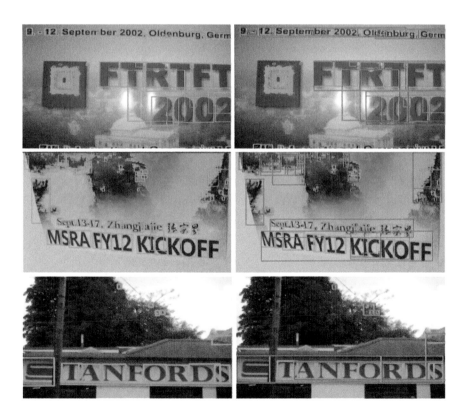

Fig. 6 Result of text line search method

Because regions in a text line are similar to each other, the comprehensive features are extracted, like

(1) Distance difference: the variance of distances between adjacent regions in the line.
(2) Height difference: the variance of height of regions in the line.
(3) Difference of ratio of foreground after binarization.
(4) Difference of average gray value of foreground. Due to color of regions in a text line are almost the same, the value will be smaller than regions have variance colors.
(5) Difference of the ratio of edge pixels to foreground pixels.
(6) Weighted color variance: the variance of $WP(l_i)$. As characters in a text line always have similar color distribution, this feature value of a real text line is lower than others theoretically, and non-text regions always have multicolor and irregular distributions. Let $P(i)$ be the proportion of pixels of value i, then $WP(l_i) = \sum_{i=0}^{255} i * P(i)$.
(7) Difference of LBP feature.
(8) Difference of wavelet energy.
(9) Hog features of text line.

4 Experimental Results

The performance of the proposed method was evaluated on ICDAR 2013 and MSRA-TD500 dataset. There are totally 509 fully annotated text images in ICDAR, of which 258 images are used for training and others for testing. There are 500 images with inclined text lines in MSRA-TD500 dataset with 300 images for training and 200 for testing. The testing procedure is split into three parts in this paper. The first part is testing the performance of character detection method; the second part is to test text line extraction method and test lines screening method is tested in the third part.

4.1 Character Regions Detection

Candidate character regions are firstly generated by MSERs, and then a character descriptor is designed as there are a lot of non-text regions. The character descriptor in this paper is a two-level classifier. In the first level, majority of noise regions are removed by constraints, and in the second level, a neural network classifier is used. Samples used to train neural network classifier in the paper are extracted in images used for training in ICDAR and MSRA-TD500. 12093 samples are produced,

Table 1 Result of character regions detection

Dataset	Samples	Recall (%)	Precision (%)
ICDAR	209	85.4	43.3
MSRA-TD500	200	83.3	50.6

including 5324 positive samples and 6769 negative samples. We test our extraction approach on ICDAR and MSRA-TD500, the results are shown in Table 1.

As most pictures in ICDAR are scene image and character regions in them are obvious, majority of characters can be extracted. By contrast, pictures in MSRA-TD500 are mostly sign images, and backgrounds are always simple. Although the recall is lower than ICDAR, the precision is higher.

4.2 Text Lines Retrieval

The text lines extraction algorithm is introduced in Sect. 3.5. In order to test this algorithm, we carried on the experiments on these two datasets above and result is shown in Table 2.

4.3 Text Lines Screening

The text line classifier used in this paper is trained by samples generated by the text line retrieval algorithm we proposed. As relevant information is utilized, we compute these nine features after extracting candidate text lines, and pick out 1400 samples (includes 600 positive samples and 800 negative samples) to train the text line classifier. Figure 7 shows the result after screening. We can see that part of text regions are extracted repeatedly. In order to avoid this, the text regions are confined to the region that has larger value generated by text line descriptor. The relative features of text lines that consist of single character are assigned as average values of corresponding values of samples.

4.4 Performance Analysis

As is known to all, the MSERs is of high efficiency, so the computational complexity of the method in this paper hinges on the complexity of character descriptor

Table 2 Result of text lines detection

Dataset	Samples	Recall (%)	Precision (%)
ICDAR	209	82.4	44.4
MSRA-TD500	200	76.3	48.3

Fig. 7 Text locating results

and text line detection. According to the description of this paper, we know that the computation amount of character descriptor is $O(n^2)$ (n is the number of pixels in an image), and complexity of text line retrieval is far lower than $O(N^2)$ (N is the number of candidate character regions), so it is much quicker than methods detecting text lines by exhaustive search or other search methods.

5 Conclusion

A new text line detection method is proposed in this paper which not only works for horizontal text, but also performs well in detecting text of arbitrary orientations in complex scenes. With constructing a graph model, a lot of unnecessary computation is omitted and the detection time is shortened significantly. In order to exclude non-text regions, a two-level character descriptor is designed and employed in character detection module, and relative features are extracted to remove noise text lines. Experimental results show that the method we prosed can extract text lines effectively no matter for horizontal or inclined texts.

References

Chen X, Yuille AL (2004) Detecting and reading text in natural scenes. In: Computer vision and pattern recognition. CVPR 2004. Conference on Proceedings of the 2004 IEEE computer society. IEEE 2004, vol 2, pp II-366-II-373. doi:10.1109/CVPR.2004.1315187

Du Y, Ai H, Lao S (2011) Dot text detection based on fast points. In: International conference on document analysis and recognition (ICDAR). IEEE 2011, pp 435–439. doi:10.1109/ICDAR.2011.94

Epshtein B, Ofek E, Wexler Y (2010) Detecting text in natural scenes with stroke width transform. In: IEEE Conference on computer vision and pattern recognition (CVPR). IEEE 2010, pp 2963–2970. doi:10.1109/CVPR.2010.5540041

Huang W, Lin Z, Yang J, Wang J (2013). Text localization in natural images using stroke feature transform and text covariance descriptors. In: IEEE international conference on computer vision (ICCV). IEEE 2013, pp 1241–1248. doi:10.1109/ICCV.2013.157

Huang X (2012) Automatic video text detection and localization based on coarseness texture. In: Fifth international conference on intelligent computation technology and automation (ICICTA). IEEE 2012, pp 398–401. doi:10.1109/ICICTA.2012.106

Iqbal K, Yin XC, Hao HW, Asghar S, Ali H (2014) Bayesian network scores based text localization in scene images. In: International joint conference on neural networks (IJCNN). IEEE 2014, pp 2218–2225. doi:10.1109/IJCNN.2014.6889731

Matas J, Chum O, Urban M, Pajdla T (2004) Robust wide-baseline stereo from maximally stable extremal regions. Image Vis Comput 22(10):761–767

Mancas-Thillou C, Gosselin B (2006) Natural scene text understanding. na, Ann Arbor

Neumann L, Matas J (2012) Real-time scene text localization and recognition. In: IEEE conference on computer vision and pattern recognition (CVPR). IEEE 2012, pp 3538–3545. doi:10.1109/CVPR.2012.6248097

Neumann L, Matas J (2013) Scene text localization and recognition with oriented stroke detection. In: IEEE international conference on computer vision (ICCV). IEEE 2013, pp 97–104. doi:10.1109/ICCV.2013.19

Shekar BH, Smitha ML, Shivakumara P (2014) Discrete wavelet transform and gradient difference based approach for text localization in videos. In: 2014 fifth international conference on signal and image processing (ICSIP). IEEE 2014, pp 280–284. doi:10.1109/ICSIP.2014.50

Wan M, Zhang F, Cheng H, Liu Q (2008) Text localization in spam image using edge features. In: International conference on communications, circuits and systems. ICCCAS 2008. IEEE 2008, pp 838–842. doi:10.1109/ICCCAS.2008.4657900

Wang K, Babenko B, Belongie S (2011) End-to-end scene text recognition. In: IEEE International Conference on Computer Vision (ICCV). IEEE 2011, pp 1457–1464. doi:10.1109/ICCV.2011. 6126402

Wen W, Huang X, Yang L, Yang Z, Zhang P (2009) An efficient method for text location and segmentation. In:. WRI world congress on software engineering. WCSE'09. IEEE 2009, vol 3, pp 3–7. doi:10.1109/WCSE.2009.292

Yao C, Bai X, Liu W, Ma Y, Tu Z (2012) Detecting texts of arbitrary orientations in natural images. In: IEEE Conference on computer vision and pattern recognition (CVPR). IEEE 2012, pp 1083–1090. doi:10.1109/CVPR.2012.6247787

Research of a Framework for Flow Objects Detection and Tracking in Video

Lanfang Dong, Jiakui Yu, Jianfu Wang and Weinan Gao

Abstract The flow objects are ubiquitous in nature, and the detection and tracking of flow objects is very important in the field of machine vision and public safety, so building a framework for the detection and tracking is more advantageous for this research. For this demand, a systematic framework is proposed. First, the foreground can be detected by GMM (gaussian mixture model) and SNP (statistical nonparametric) algorithm, and candidate regions can be determined by static features extracted in the foreground. Second, all these candidate regions should be combined and tracked. At last, dynamic features of the tracked regions should be extracted and whether it is flow objects or not should be confirmed. To solve the problem of combination of adjacent small regions and the multi-objects matching, similar regional growth algorithm and the method for tracking multiple targets are put forward. To verify the effect of the framework, a lot of experiments about smoke, fire, and rain are implemented.

Keywords Detecting and tracking flow objects · Common framework · GMM · Matching degree

1 Introduction

Flow objects mean the objects with flow characteristics, such as smoke, fog, water, fire, and so on. They are ubiquitous in nature, and the detection and tracking of flow objects is widely used in the field of machine vision, intelligent monitoring and

L. Dong · J. Yu · J. Wang · W. Gao (✉)
Vision Computing and Visualization Laboratory, University of Science
and Technology of China, Hefei 230027, China
e-mail: gwny@mail.ustc.edu.cn

L. Dong
The State Key Laboratory of Fluid Power Transmission and Control,
Zhejiang University, Hangzhou 310027, China

© Zhejiang University Press and Springer Science+Business Media Singapore 2017
C. Yang et al. (eds.), *Wearable Sensors and Robots*, Lecture Notes in Electrical
Engineering 399, DOI 10.1007/978-981-10-2404-7_36

public safety, especially in the monitoring of flood. In some areas, such as in forest, we must detect the regions for 24 h to prevent fire, it is difficult for people to do it, we can use the wearable robots, they can also be used in some dangerous areas. Since the detection and tracking of flow objects is so important, it is meaningful to build a common framework for it. However, most of present researches aim to certain flow objects, such as smoke, fog, water, fire, etc., there is not a unified framework for them to extract and assemble common information of flow objects. If we can build a common framework for detection and tracking of different flow objects, researchers and developers can avoid duplication of tedious steps, more time and energy will be saved to research and develop better features of flow objects. Moreover, we can verify the proposed features quickly.

Most of present detection methods for the fire in the video take advantage of static features, like color and translucency characteristics (Ugur et al. 2005). As the edge in the image will be blurred and corresponding high frequency energy will be declined when fire exists, various features of wavelet energy can be used (Zheng and Yang 2013; Lee and Lin 2012). Dynamic features are combined to detect and track smoke, such as the various features of smoke luminance histogram (Lai 2007), the diffusion feature (Chen and Yin 2006), and the irregularity of smoke (Chen and Yin 2006; Pagar and Shaikh 2013). The color feature (Liang and Wang 2012) and dynamic features, such as flashing feature of flame (Yuan and Zhang 2008), the rate of area change (Celik and Demirel 2007), and the roundness of flame region (Chen et al. 2004) are also used to detect fire in the video. Researchers also utilize the static feature, like color (Zhang and Li 2006) and dynamic features, like histogram feature (Garg and Nayar 2004) to detect rain. It is obvious that the detection process of fire, fog and rain are pretty much the same.

In addition, after a lot of researches, analysis and experiments, we found that the detection of flow objects can be summarized as three parts: at first, getting foreground region is necessary for the detection and tracking of various flow objects; second, candidate regions could be determined by static features extracted in the foreground; third, all these candidate regions should be combined and tracked; at last, dynamic features of the tracked regions should be extracted and whether it is flow objects or not should be confirmed, as shown in Fig. 1. From the above analysis we can know that, the main difference of detecting and tracking various flow objects is the extraction of different static and dynamic features of flow objects, so it is possible to build a common framework for the detection and tracking of various flow objects. In order to enhance the robustness and make the framework in real time, the paper utilize background subtraction and three frame difference method to detect foreground, and then an algorithm similar to region growing method is presented for merging adjacent small regions, at last, we use two-stage marking algorithm and the matching algorithm that based on matching degree to mark and track the target.

Fig. 1 The process of detecting and tracking flow objects

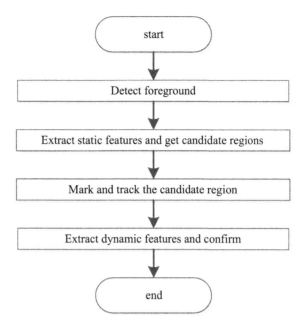

2 General Description of Framework

Establishing a unified framework by extracting common process is an abstraction of detecting and tracking flow objects in video from a higher level, it can reduce complexity and researchers need not to consider the common process any more, then more time and energy will be saved to research and develop better features for detecting and tracking certain flow object. In order to build a robust and real-time common framework, we propose a flow chat as shown in Fig. 2 for general detection and tracking.

First, a frame is read from video sequences and Gaussian Mixture Model (GMM) (Zivkovic 2004) is utilized for background modeling. In the course of modeling, statistical nonparametric (SNP) approach is used to remove the cast shadow of the detected foreground. Second, we combine the foreground above with the foreground which obtained using three frame difference method to get a complete foreground, and then several candidate regions can be located according to the static features which extracted from foreground. Third, we merge the adjacent small regions with our proposed algorithm similar to region growing method, about the combined regions, we mark them with two-stage marking algorithm (Wu 2009) and track them with the matching algorithm which is proposed in this paper based on matching degree, marking and tracking aim to extract dynamic features of flow objects from candidate and judge further. At last, in order to enhance the accuracy in judgment, we use a variety of decision and accumulated technology.

Fig. 2 Flow chart of general
detection and tracking

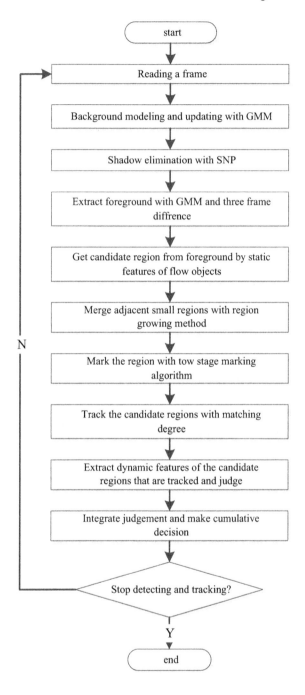

3 Detailed Description of Framework

From the previous analysis and discussion, we can conclude that the process of detecting and tracking flow objects in video can be divided into three modules mainly: the first part is to extract candidate regions, these regions are tracked in the second part and the last part is to judge these candidate regions. The following will introduce these three modules, respectively.

3.1 Extracting Candidate Regions

In order to get candidate regions, we should detect foreground first, and then extract static features of flow objects from foreground, candidate regions can be obtained after judging and merging.

3.1.1 Foreground Detection

Detecting foreground in video sequences has always been one of the important and difficult problems in video processing and analysis. Although many methods have been proposed for foreground detection, such as frame difference method, background subtraction, and optical flow method (Horn and Schunck 1981), each of these methods has advantages and disadvantages. Adaptive background subtraction method is more sensitive about sudden movement of a stationary object in scene (Toyama et al. 1999), these objects can be detected, but it will leave "empty holes" in a short period of time, which is shown in Fig. 3a. However, frame difference method can deal with this problem, but it cannot get complete moving object, which is shown in Fig. 3b. Optical flow method can detect moving object well, but it is complicated and time consuming, and the flow object has its own characteristics,

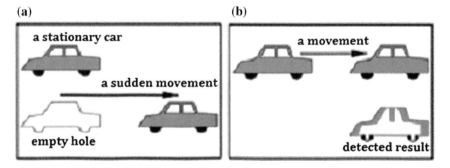

Fig. 3 Background subtraction method (**a**), frame difference method (**b**)

such as translucence. So, in consideration of real time and accuracy, this paper uses a combination of background subtraction and frame difference method.

About background subtraction, we adopt GMM for video background modeling. Since the foreground obtained by using GMM is an object containing corresponding shadow, we utilize SNP to remove the shadow. The principle is that shadow has similar chroma but lower brightness compared to background. This characteristic can be used to eliminate shadow in the foreground object. For frame difference method, we adopt dynamic three frame difference method. Dynamic three frame difference method uses the adjacent three frames $I_{t-1}(x, y)$, $I_t(x, y)$ and $I_{t+1}(x, y)$ in video sequences to detect moving object, it uses the absolute value of difference between adjacent frames comparing with preset threshold, and then whether the pixel is moving pixel will be determined according to the result. Its equation is as follows:

$$
\begin{aligned}
D_{t-1,t}(x,y) &= |I_t(x,y) - I_{t-1}(x,y)| \\
D_{t,t+1}(x,y) &= |I_{t+1}(x,y) - I_t(x,y)| \\
B_{t+1} &= \begin{cases} 1 & D_{t-1,t}(x,y), D_{t,t+1}(x,y) > T_{\text{threshold}} \\ 0 & \text{otherwise} \end{cases}
\end{aligned} \tag{1}
$$

After using a combination of GMM, SNP, and dynamic three frame difference method, the experimental results are shown in Fig. 4. Figure 4a is an original video frame, Fig. 4b is the result of foreground without using SNP, Fig. 4c is the result of foreground which detected using the method proposed in this paper, we can see clearly from the figure, the foreground of moving objects is detected well, and the shadows of the foreground objects are removed well too. About the people in red oval circle, he cannot be detected, because this person has always been stationary in consecutive video frames, GMM looks him as background, for the other people without detected in Fig. 4a is also the same reason. Figure 5a, b show an original video frame with smoke and corresponding foreground, respectively.

3.1.2 The Judgment of Static Features

Since different flow objects have various static features, we should build different static decision rules for different flow objects, and judge the foreground with these

(a)　　　　　　　　**(b)**　　　　　　　　**(c)**

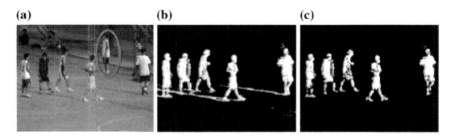

Fig. 4 An original video frame (**a**), the result of foreground with shadows (**b**), the result after eliminating shadows (**c**)

(a)   **(b)**

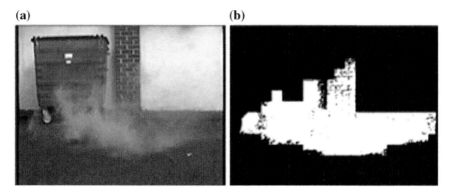

Fig. 5 An original video frame with smoke (**a**), the result of foreground (**b**)

rules to extract candidate regions. One of the most commonly used static features of flow objects is color feature.

3.1.3 Region Merging

These obtained candidate regions of flow objects are always fragmented; there are a lot of adjacent small areas, as shown in Fig. 6b. Therefore we need a way to merge those regions which should have been in the same region, there are some methods such as clustering algorithm (Liang and Wang 2012) can deal with it while its complexity is too high, that cannot meet the requirement of real time, so a method which similar to region growing algorithm is proposed in this paper, we call it SRGA.

As the name suggests, the method similar to region growing algorithm is learned from region growing algorithm (Gonzalez and Woods 2011) as shown in Fig. 7, there are two regions, one on the top left corner and the other on the bottom right corner, but in fact these two region belong to a same object, how to merge the two regions quickly For each square which labeled 1 in Fig. 7, we can execute the following actions:

(1) We locate the edge of the square as center, radius is preset size R, form a circle in image.
(2) The foreground pixel in this circle will be marked as the same label of square, and the edge pixel, that is there is one background pixel at least in the corresponding four areas of this pixel, will be added to a queue.
(3) The queue will be traversed, for each pixel in this queue, we process it with the same method of (1), (2), until the queue is empty, an object region merging is completed.
(4) Repeat the above process until all pixels are processed.
(5) Remove the region whose number of foreground pixels is less than a certain threshold.

(a)   (b)

(c)   (d)

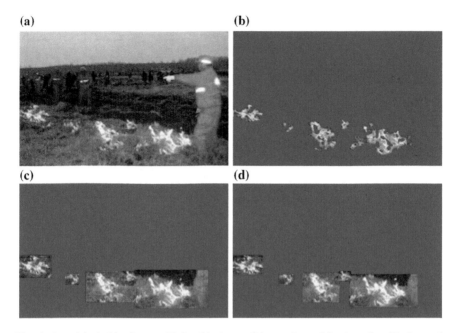

Fig. 6 An original video frame with fire (**a**), the candidate regions of fire detection (**b**), the result after merging which using DBSCAN method (**c**), the result after merging which using the method proposed in this paper (**d**)

Fig. 7 The principle of the method proposed in this paper for merging regions

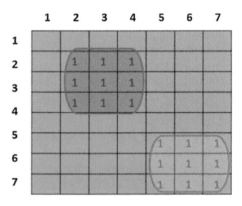

The algorithm also implements the merger action for combined regions which makes preparation for subsequent regions tracking. The experimental result is showed in Fig. 6, Fig. 6a shows an original frame in fire video, Fig. 6b shows the result after foreground detection and using static decision rules, we can see from the result image, after these two steps, there are lots of adjacent but not united regions, Fig. 6c is the result of using a clustering algorithm named DNSCAN, Fig. 6d is the result after using the SRGA method proposed in this paper, we can see clearly from Fig. 6c and d, the adjacent regions are merged well, but in Fig. 6c the bigger regions are also merged, additionally, comparing to 0.127 s the DNSCAN algorithm spends on processing a frame of size 480 * 300, our method spends only 0.0075 s for the same frame.

3.2 Candidate Regions Tracking

In order to extract the corresponding dynamic features, which would be used for judging, of flow objects from candidate regions, we must track the candidate regions. Before tracking these candidates, we should mark it first, the marking, tracking and feature extraction method will be introduce in the following paragraphs.

3.2.1 Regions Marking

In order to mark the candidate regions quickly, we adopt two-pass connected-component labeling algorithm (Wu 2009), it can implement the marking of candidate regions just by scanning these regions twice, the process as follows:

(1) Scanning phase: At this stage, provisional labels will be assigned to foreground pixels while scanning and equivalence information among provisional labels will be recorded.
(2) Analysis phase: By analyzing equivalence information, real labels are obtained in this phase.
(3) Marking phase: the second scan begins at the phase and real labels are assigned while scanning.

After these three stages, foreground pixels are already marked. Because this method can implement marking candidate regions in twice scanning and the process is simple, its speed is very high.

3.2.2 Regions Tracking

In order to get the dynamic features of flow objects, we should track the candidate regions of flow objects in video. Take the factor of real time into consideration, we cannot use these target tracking algorithm with high complexity, such as Kalman Filter (Li and Liu 2008) and Camshift algorithm (Wu and Zheng 2009). So we propose a target matching algorithm based on matching degree, it can not only satisfy the need of real time but also has high accuracy, the detail content is as follows.

In Fig. 8, the moving object B in previous frame intersects with the moving object C and D in current frame, we can see that the area of region A is greater than region E, but B and D is the best match. As there are multiple moving objects in current frame matching with the moving object in previous frame, the best match must be determined. In this paper, a decision function is proposed.

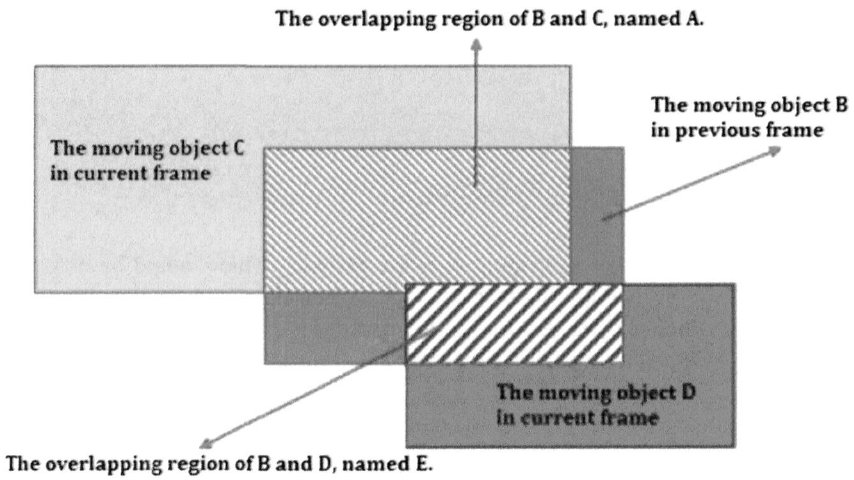

Fig. 8 The description of matching degree

$$F = \frac{\frac{X}{Y} * \frac{X}{Z}}{\frac{\max(Y,Z)}{\min(Y,Z)}} \tag{2}$$

where X is the area of overlapping region; Y is the area of original moving object; Z is the area of the moving object which for matching; F denotes matching degree.

The following will explain the meaning of each part in the formula

$$F1 = \frac{X}{Y}$$

This part is represented as $F1$, it means the more the area of overlapping region X is close to the area of original object Y, the higher of matching degree will be.

$$F2 = \frac{X}{Z}$$

This part is represented as $F2$, it means the more the area of overlapping region X is close to the area of the moving object which for matching Z, the higher of matching degree will be.

$$F3 = \frac{\max(Y,Z)}{\min(Y,Z)}$$

This part is represented as $F3$, it shows the difference between the original object and matched object, the value is greater than 1, when the relative change is larger, the value of $F3$ is bigger, and the match probability will be smaller. The best match is the one that has highest matching degree.

Because the algorithm needs far less computation than the common tracking algorithm such as Camshift, it can meet the requirement of real-time well, and from the formula and lots of experiments we can conclude that the accuracy is high.

3.2.3 Dynamic Feature Extraction

Because different flow objects have various dynamic features, we should extract the corresponding dynamic features from the candidate regions of flow object to evaluate candidate regions. The frequently used dynamic features are listed

(1) The variation of histogram.
(2) The variation of energy.
(3) The diffusion characteristic.
(4) The irregularity.
(5) The movement accumulation and the main direction.

3.3 The Judgment of Candidate Regions

Because the process of detection may be affected by other factors, for instance, the object that has similar characteristic to the object detected, we make a comprehensive judgment by extracting various dynamic features from the candidate regions first, then we will accumulate the judgment of every kind of features to improve the accuracy.

3.3.1 Comprehensive Judgment

As multiple dynamic features can be extracted from candidate regions, so we can give every verdict of the extracting dynamic features a weight, when the value of the judgment is greater than a threshold, this candidate region will be judged as the region of flow objects, the formula is as follows:

$$\text{AccumR}_i = \sum_{j=1}^{N} w_j * \text{DR}_j \quad \sum_{j=1}^{N} w_j = 1$$

$$R_i = \begin{cases} 1 & \text{if AccumR}_i > \text{Th} \\ 0 & \text{otherwise} \end{cases} \tag{3}$$

where N is the number of extracting dynamic features, w_j means the weight of the dynamic feature index j in candidate region, the sum of the weights is 1. Based on the dynamic feature index j in candidate region, DR_j means whether the candidate region is flow object or not, if it is, DR_j equals to 1. $AccumR_i$ means the probability that the candidate region index i will be judged as flow object. Th is a threshold which obtained according to experiments, generally taking 0.6. If $AccumR_i$ is greater than this Th value, that is R_i equals to 1, the candidate region will be judged as flow object. The various dynamic features of different objects have different weights, and the values of weights should be obtained through experiments, but in this paper, we use a same weight for different dynamic features for convenience, that is $w_j = \frac{1}{N}$.

3.3.2 Accumulating the Judgment

If we get the judgment just by one frame, the result will be affected by some external factors easily and the corresponding accuracy will not be too high, so we will accumulate the judgments of multiple frames and make a final decision, it can improve the accuracy and robustness of detection.

4 The Validation of the Framework

To validate the proposed framework, we test multiple video, different content videos are used to detect and track smoke, fire, and rain.

4.1 The Detection and Tracking of Smoke in Framework

4.1.1 Static Analysis

In all the static characteristics of smoke, only the color feature is analyzed. Through a large number of experiments, we found that for most burning materials, when the temperature is low, the color of smoke will change from bluish white to white; With the temperature rise until the materials burning, the color of smoke will gradually deepened and change from gray to gray-black, which means in RGB color space, the smoke color is near the diagonal line of the TGB cube. According to this character, we can get the following:

The R, G, and B components form the three mutually perpendicular axes, the origin of coordinate is O, OD is the diagonal line of the RGB cube, so $\overrightarrow{OD} = (1,1,1)$. We suppose x is an arbitrary pixel in image, the corresponding RGB value is expressed as $x = (r, g, b)$, therefore $\overrightarrow{Ox} = (r, g, b)$, θ is the angle between $\overrightarrow{OD}$ and $\overrightarrow{Ox}$, OM is the projection of $\overrightarrow{Ox}$ on the $\overrightarrow{OD}$, σ is the length of xM, p is the length of OM, as shown in Fig. 9.

According to the above analysis, θ and the brightness value are used to judge whether a pixel is smoke pixel or not, the corresponding formula is given below

Fig. 9 The analysis diagram of the static feature about smoke in RGB color space

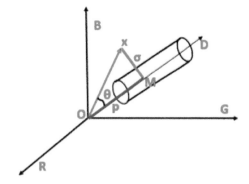

$$\cos\left(\theta\right) = \frac{\overrightarrow{Ox} \cdot \overrightarrow{OD}}{\left\|\overrightarrow{Ox}\right\| \cdot \left\|\overrightarrow{OD}\right\|}$$

$$= \frac{r+g+b}{\sqrt{r^2+g^2+b^2} \times \sqrt{1^2+1^2+1^2}} \qquad (4)$$

$$= \frac{r+g+b}{\sqrt{3\left(r^2+g^2+b^2\right)}}$$

The smaller the angle, the bigger the $\cos\left(\theta\right)$. According to a lot of experiments, θ takes $8°$ commonly, so $\cos\left(\theta\right)$ ranges from 0.99 to 1. For the color of white smoke, its corresponding brightness value I should meet the following formula:

$$I \geq \max\left(I_{\mathrm{mean}}, R_{\mathrm{T}}\right), \quad I_{\mathrm{mean}} = \frac{1}{K}\sum_{k=1}^{K} I_k \qquad (5)$$

where K is the number of pixels in the whole image, I_{mean} is the average value of luminance of the whole image, R_{T} is a threshold, based on experience, R_{T} is set to 160.

4.1.2 Dynamic Analysis

About the dynamic features of smoke, characteristic of luminance histogram, characteristic of wavelet energy, characteristic of diffusion, and irregularity are discussed.

(1) The characteristic of luminance histogram

The characteristic of luminance histogram (Lai 2007) is calculated in spatial domain and frequency domain. According to experience, when smoke exist in a picture, the gray-scale histogram of smoke region is concentrated, so we can detect the smoke region by calculating the concentration degree of the histogram, which means if luminance histogram variance of a region is greater than a threshold, the region will be considered as candidate smoke region, histogram variance is obtained by the following formula:

$$v = \frac{1}{n} \sum_{i=1}^{n} (x_i - \bar{x})^2 \tag{6}$$

where x_i is the number of gray pixel index i in histogram, $\bar{x}$ is the average of gray value in the region that might be smoke region. If the luminance histogram variance of current region is greater than the background, this region is marked as candidate smoke region.

(2) The characteristic of wavelet energy

In a video, smoke will make its coverage regions blurred. Compared with the corresponding background, the edge and detail information of the smoke regions is reduced, these information correspond to the high frequency information of image, so the high frequency energy will be reduced when the smoke exists. We can get the high frequency energy by using wavelet transform to detect smoke regions, the wavelet transform is shown in Fig. 10.

The calculation formula of wavelet energy using weights is shown below:

$$w_n(x, y) = |LH_n(x, y)|^2 + |HL_n(x, y)|^2 + 2 * |HH_n(x, y)|^2 \tag{7}$$

To improve accuracy, we divide the sub image of wavelet into smaller blocks, the size of each block is (K_1, K_2), and the energy of each block can be calculated by the following formula:

$$e(l_1, l_2) = \sum_{(x,y) \in R_i} w_n(x + l_1 K_1, y + l_2 K_2) \tag{8}$$

where $e(l_1, l_2)$ is the energy, R_i is the block, (l_1, l_2) is the location of the block, w_n is the wavelet energy shown in Eq. (7).

(3) The characteristic of diffusion

In the generation process of smoke, because of the diffusion characteristic (Chen and Yin 2006), the smoke area increases gradually, the growth rate of smoke area is defined as follows:

$$V_{area} = \frac{N_t - N_{t-\Delta t}}{n} \tag{9}$$

where V_{area} is the area growth rate between time t and $t - \Delta t$, N_t and $N_{t-\Delta t}$ is the number of pixels in smoke region, n is the number of frames in Δt time. In the diffusion process of smoke, the smoke regions are unlikely to change quickly or completely unchanged, so there are upper limit and lower limit for the growth rate, it can be expressed as $V_{min} < V_{area} < V_{max}$, the threshold V_{min} and V_{max} is set by experiments.

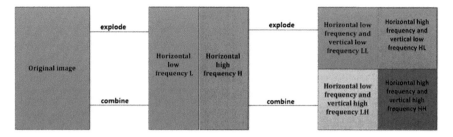

Fig. 10 Two-dimension wavelet transform

(4) The characteristic of irregularity

Affected by airflow, the shape of smoke changes constantly, so it is difficult to measure the shape of smoke accurately, we describe the shape of smoke by the irregularity (Chen and Yin 2006; Pagar and Shaikh 2013). The formula is as follows:

$$\frac{\text{The perimeter of the smoke area}}{\text{The smoke area}} \geq \text{threshold (STD)} \tag{10}$$

We get a ratio by comparing the perimeter with the smoke area, if the ratio is greater than a certain threshold, this area should be the smoke area.

4.1.3 Experimental Results

The top left picture is a current frame, the top right picture is the background of current frame getting by GMM, the bottom left picture is the foreground of current frame, the bottom right picture is the region obtained by the dynamic features of smoke, and the regions have been tracked (Fig 11).

The four images in Fig. 12 are the detection and tracking of smoke in two videos, respectively, the left two images are current frame in two videos, the right images are the smoke regions obtained by the framework, also using the static and the dynamic features of smoke, and the regions have been tracked, we can see from the figure, the result of marking and tracking is well.

4.2 The Detection and Tracking of Fire in Framework

For the detection of fire in video, we utilize the framework in the same way, except the static and the dynamic features of fire (Liang and Wang 2012; Yuan and Zhang 2008; Celik and Demirel 2007; Chen et al. 2004) are used, the result of detection and tracking as follows (Fig. 13).

Fig. 11 The detection and tracking of smoke 1

Fig. 12 The detection and tracking of smoke 2

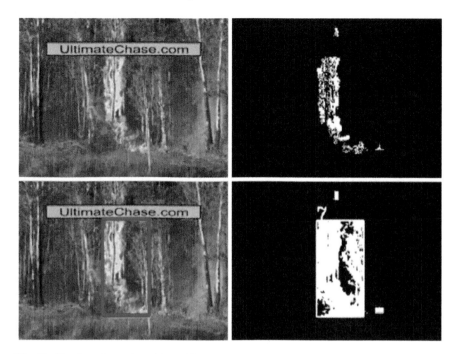

Fig. 13 The detection and tracking of fire

The top left picture is an original frame in video, the top right picture is its foreground, the bottom left picture is the result of marking, the bottom right picture is the result of marking and tracking. We can see from the figures that the result of the fire detection and tracking is well by using the framework and the features of fire.

4.3 The Detection and Tracking of Rain in Framework

About the detection of rain (Garg and Nayar 2004; Zhang and Li 2006; Zivkovic 2004), we can get the following result using this framework.

In Fig. 14, the top left picture is an original frame in video, the top right picture is the detected pixels of rain, the bottom left picture is the result showing the detected pixels in the original frame, the bottom right picture is background. We can see from the figures that the result of detecting rain is well.

4.4 Analysis

The test videos come from the database in (A.Enis; G.Gera), experiment shows that, on average, the foreground detection takes 18.5 ms, the regions marking and

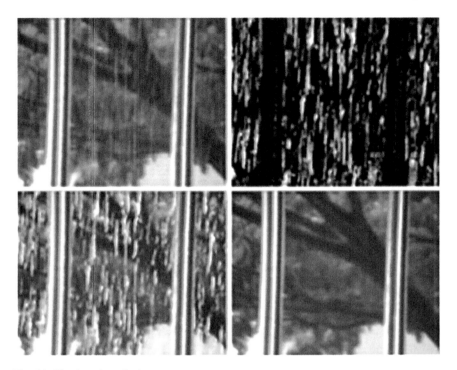

Fig. 14 The detection of rain

combination take 8.7 ms, the region tracking takes 5.4 ms. For different flow objects, the total time is different because different flow objects use different static and dynamic features. But it can meet the requirement of real time in general, and the false detection rate and missing detection rate is not high, which depends on the features we chose. From above analysis we can get that, the proposed framework can detect the smoke, fire and rain in videos well, and its computational complexity is low. On the whole, the proposed framework has nice commonality, high accuracy, and high speed.

5 Conclusions

Establishing a unified framework by extracting common process is an abstraction of detecting and tracking flow objects in video from a higher level. It can reduce complexity and researchers need not to consider the common process any more, then more time and energy will be saved to research and develop better features for detecting and tracking certain flow object.

The mainly contribution of the paper is as follows:

(1) A unified framework for the detection and tracking of flow objects in videos is established.
(2) A method for merging regions is proposed.
(3) A multiple target tracking algorithm based on match degree is proposed.
(4) Extracting moving object by combining the background detection algorithm and the three frame difference method.
(5) Integrating the SNP method into GMM to eliminate the shadows caused by foreground.
(6) Accumulating judgments by weights.
(7) Realizing the detection and tracking of smoke, fire, and rain using the proposed framework.

References

Celik T, Demirel H (2007) Fire detection in video sequences using statistical color model. J Vis Commun Image Represent, pp 176–185. doi:10.110/ICASSP.2006.1660317

Chen T-H, Wu P-H, Chiou Y-C (2004) An early fire-detection method based on image processing. In: Proceedings of IEEE international conference on image processing, Singapore: IEEE, vol 3, pp 1707–1710. doi:10.1109/ICIP.2004.1421401

Chen TH (C-H), Yin Y (2006) The smoke detection for early fire-alarming system base on video processing. In: Proceedings of the 2006 international conference on intelligent information hiding and multimedia signal processing, California: IEEE computer society, pp 427–430. doi:10.1109/IIH-MSP.2006.265033

Garg K, Nayar SK (2004) Detection and removal of rain from videos. In: Proceedings of the 2004 IEEE computer society conference, Washington: IEEE computer society, vol 1, pp 528–535. doi:10.1109/CVPR.2004.1315077

Gonzalez RC, Woods RE (2011) Digital image processing, 3rd edn. Publishing House of Electronics Industry, Beijing, pp 493–495 (in Chinese)

Horn BKP, Schunck BG (1981) Determing optical flow. Artif Intell 17:185–203. doi:10.1016/0004-3702(81)90024-2

Lai CL (2007) A real time video processing based surveillance system for early fire and flood detection. In: Proceedings of the IEEE instrumentation and measurement technology, Warsaw: IEEE, pp 1–6. doi:10.1109/IMTC.2007.379190

Lee CY, Lin CT (2012) Smoke detection using spatial and temporal analyses. Int J Innov Comput Inf Control 8(7):4749–4770

Li Z, Liu G (2008) Target tracking based on mean-shift and Kalman filter. J Proj Rocket Missiles Guidance 28(1):71–74

Liang J, Wang H (2012) The fire detection of image based on fuzzy clustering. Comput Eng, pp 196–198

Pagar PB, Shaikh AN (2013) Real time based fire and smoke detection without sensor by image processing. Int J Adv Electr Electron Eng 6(2)

Toyama K, Krumm J, Brumitt B, Meyers B (1999) Wallflower: principles and practice of background maintenance. In: Proceedings of international conference on computer vision, Kerkyra, vol 1, pp 255–261. doi:10.1109/ICCV.1999.791228

Ugur TB, Dedeoglu Y, Cetin AE (2005) Wavelet based real-time smoke detection in video. In: Proceedings of 13th European signal process conference, Antalya, Turkey, 2:105–117

Wu K (2009) Optimizing two-pass connected-component labeling algorithms. Pattern Anal Appl 12(2):117–135. doi:10.1007/s10044-008-0109-y

Wu H, Zheng X (2009) Improved and efficient object tracking algorithm based on Camshift. Comput Eng Appl 45(27):178–180

Yuan F, Zhang Y (2008) The smoke detection of video based on cumulant and main direction. J Image Graph Chin pp 808–813

Zhang X, Li H (2006) Rain removal in video by combining temporal and chromatic properties. In: 2006 IEEE International Conference on Proceedings of multimedia and expo, Toronto, pp 461–464

Zheng X, Yang S (2013) A method of smoke detection based on Various Features combination. Comput Sci Appl 3(5):239–243

Zivkovic Z (2004) Improved adaptive Gaussian mixture model for background subtraction. In: Proceedings of 17th international conference pattern recognition, UK, vol 2, pp 28–41. doi:10.1109/ICPR.2004.1333992

A Kinect-Based Motion Capture Method for Assessment of Lower Extremity Exoskeleton

Min-hang Zhu, Can-jun Yang, Wei Yang and Qian Bi

Abstract Rehabilitation exoskeleton provides a new method for therapy of stroke patients, but it needs a portable and effective method to analyze the validity of exoskeleton rehabilitation training. This paper proposes a motion capture system based on Kinect to measure the variation of joint angles of a patient's lower extremity during rehabilitation training. Comparing the measured angles with the input rehabilitation trajectory, human–machine coupling property of exoskeleton can be achieved to analyze its validity. The system used image sequence motion detection algorithm based on markers and arranged eight marker bars on the surface of lower extremity. Then it got bars' spatial direction vectors by clustering algorithm (DBSCAN) and least square method. With lower extremity 5-bar model built, the system finally gained the joint angles. Meanwhile an experiment was designed to verify the precision of Kinect motion capture system. Results showed that the maximal static error was 2.76° and correlation coefficient of dynamic track was 0.9917. This proved that the Kinect motion capture system was feasible and reliable to provide parameter foundation to assess the validity of exoskeleton.

Keywords Rehabilitation exoskeleton · Human–machine coupling · Motion capture · DBSCAN clustering

1 Introduction

Stroke is one of the acute diseases caused by cerebral blood circulation, which is of high lethality and disability rate. More importantly, its incidence is growing gradually every year and destructs human health and family happiness. Lower

M. Zhu · C. Yang (✉) · W. Yang · Q. Bi
State Key Laboratory of Fluid Power Transmission and Control, Zhejiang University,
Hangzhou 310027, China
e-mail: ycj@zju.edu.cn

M. Zhu
e-mail: zoomingh@zju.edu.cn

© Zhejiang University Press and Springer Science+Business Media Singapore 2017
C. Yang et al. (eds.), *Wearable Sensors and Robots*, Lecture Notes in Electrical
Engineering 399, DOI 10.1007/978-981-10-2404-7_37

extremity hemiplegics caused by stroke need intensive rehabilitation training in the early period of the disease. However, manual training of therapists always cannot meet the need of patients. Lower extremity rehabilitation exoskeleton, combining the theory of robot and rehabilitation medicine, can give stroke patients intensive and pointed training and improve the situation of lacking therapists to some extent.

One of design goals of rehabilitation exoskeleton is realizing the training mode of medicine science precisely, which means the exoskeleton should drive the lower extremity training on the given trajectory. One of the most successful exoskeletons is Lokomat (Lunenburger et al. 2005) which can impose the patient on the treadmill and suspensory system by parallel quadrangular link mechanism, and haul the patient walking along the training gait using bandages. Other existing rehabilitation exoskeletons are designed in the same way. The bandage connection brings the human–machine error, which means there are some deviation between the given gait angles and actual joint angles. A big deviation will induce the reduction of exoskeleton rehabilitation ability. The measurement of this deviation will not only help estimate the exoskeleton property of human–machine coupling, but help improve the rehabilitation effect by inputting the deviation as a feedback signal into the control model.

The measurement of deviation needs to capture the rotating angles of patient's joints. And motion capture method is mainly divided into two parts: contact type and noncontact type. Yan et al. (2014) designed a set of mechanical exoskeleton to collect posture information of upper extremity by installing a high precise encoder on the exoskeleton joint. Kobashi et al. (2009) arranged IMU on the lower extremity to detect joint angles in different conditions. Vlasic et al. (2007) introduced an electromagnetic motion capture system, in which stable distributed electromagnetic field was built in the space to locate the sensors linked to the human body to estimate human posture. Koyama et al. (2011) designed one type of optical angular sensor which could be fixed to the two bars of joint to get the angle. Zhuojun et al. (2013) built the mapping mode between sEMG and joint angles by fuzzy neural network model to survey angles. Sun et al. (2012) designed a measurement platform which needed several ultrasonic sensors arranged in front of the human body and transformed the distances from sensors to body into angle information.

These measuring methods above require people to wear corresponding sensors, which largely interferes the nature of motion and is not suitable to be deployed on patients. As the technology of image processing is developing, some new methods based on image sequences are widely being researched. Thang et al. (2011) proposed an estimating method of joint angles based on 3D human model, in which people did not need to wear anything and the joint angles could be estimated by the binocular vision system. However, the resolution of this method was not very high (10 mm) and could not meet the need of getting the deviation of human–machine coupling. Kim et al. (2011), Windolf et al. (2008) adopted motion capture method based on markers, which were arranged on the surface of human body and could be tracked by multi-camera system. The motion information could be computed by the geometrical relation of the markers. The resolution was very high, up to the level of um, but apparently the measuring range was limited. Recent years, Kinect, as a 3D camera, is used widely to obtain the 3D information of the environment, based on which the coordinates and angles of joints can be computed with some machine vision algorithms. Nevertheless, the location of joints Kinect gives out is not very stable, which significantly affects the conversion of joint angles.

To reduce the weight patient should wear as much as possible and achieve measuring precision as high as possible, this paper adopted the video motion capture system based on markers, collected the 3D environment data by Kinect, realized a system to measure the joint angles of lower extremity based on 5-bar model. Then a precision experiment was designed to estimate the static error and dynamic error of the system. And finally the system measured the joint angles of a patient in the rehabilitation process which could give some parameter suggestions to subsequent design and control of lower extremity exoskeleton.

2 5-Bar Model of Lower Extremity

2.1 D-H Bar Model

In the 5-bar model of lower extremity (Fig. 1), there are three DOFs in hip joint and one in knee joint, 8 DOFs in total in the legs (Pons 2008). The base coordinate frame $X_0 Y_0 Z_0$ is cohered to the center of the pelvis between the hips. The hip joint is corresponded with 3 frames $X_1 Y_1 Z_1$, $X_2 Y_2 Z_2$, $X_3 Y_3 Z_3$, representing circumduction, adduction-abduction and flexion-extension, respectively. The origins of the three frames are positioned at the center of hip joint. There is one frame $X_4 Y_4 Z_4$ located in

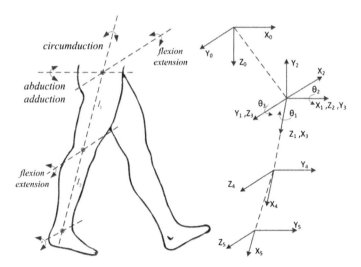

Fig. 1 D-H notation for a human leg

Table 1 D-H parameters for leg segments

Number	β_i	θ_i	d_i	a_i	α_i
0 (0–1)	0°	0°	d	a	0°
1 (1–2)	$-50° - 40°$	$\beta_1 + 90°$	0	0	$-90°$
2 (2–3)	$-20° - 45°$	$\beta_2 + 90°$	0	0	90°
3 (3–4)	$-30° - 120°$	β_3	0	l_1	0°
4 (4–5)	$0° - 150°$	$\beta_4 + 90°$	0	l_2	0°

β_1 Hip circumduction; β_2 Hip abduction/adduction
β_3 Hip flexion/extension; β_4 Knee flexion/extension

the knee and one effector coordinate frame $X_5 Y_5 Z_5$ in the end of shank. Table 1 shows D-H parameters of leg segments.

2.2 Calculating Method of Joint Angles

The method included hip angles calculating and knee angle calculating. The first one was measured by kinematic inverse solution and the second one was computed by angular separation of the thigh and shank.

To figure out angles of hip joint, the effector coordinate frame E was deployed on the thigh. Transform Matrixes of adjacent DOFs could be achieved by substituting D-H parameters of hip into Matrix A:

$$A_1 = \begin{bmatrix} 1 & 0 & 0 & a \\ 0 & 1 & 0 & 0 \\ 0 & 0 & 1 & d \\ 0 & 0 & 0 & 1 \end{bmatrix}. \tag{1}$$

$$A_2 = \begin{bmatrix} -\sin \beta_1 & 0 & -\cos \beta_1 & 0 \\ \cos \beta_1 & 0 & -\sin \beta_1 & 0 \\ 0 & -1 & 0 & 0 \\ 0 & 0 & 0 & 1 \end{bmatrix}. \tag{2}$$

$$A_3 = \begin{bmatrix} -\sin \beta_2 & 0 & \cos \beta_2 & 0 \\ \cos \beta_2 & 0 & \sin \beta_2 & 0 \\ 0 & 1 & 0 & 0 \\ 0 & 0 & 0 & 1 \end{bmatrix}. \tag{3}$$

$$A_4 = \begin{bmatrix} \cos \beta_3 & -\sin \beta_3 & 0 & l_1 \cdot \cos \beta_3 \\ \sin \beta_3 & \cos \beta_3 & 0 & l_1 \cdot \sin \beta_3 \\ 0 & 0 & 1 & 0 \\ 0 & 0 & 0 & 1 \end{bmatrix}. \tag{4}$$

By the machine vision algorithm, the transform matrixes from Kinect frame to base frame and effector frame could be figured out.

$$
{}^C T_0 = \begin{bmatrix} bn_x & bo_x & ba_x & bp_x \\ bn_y & bo_y & ba_y & bp_y \\ bn_z & bo_z & ba_z & bp_z \\ 0 & 0 & 0 & 1 \end{bmatrix}. \tag{5}
$$

$$
{}^C T_E = \begin{bmatrix} en_x & eo_x & ea_x & ep_x \\ en_y & eo_y & ea_y & ep_y \\ en_z & eo_z & ea_z & ep_z \\ 0 & 0 & 0 & 1 \end{bmatrix}. \tag{6}
$$

Then, the location stance matrix effector frame relative to base frame could be calculated by ${}^C T_0^{-1} \cdot {}^C T_E$. The robot kinematic equation was built to compute angles of 3DOFs in hip joint.

$$
A_1 \cdot A_2 \cdot A_3 \cdot A_4 = {}^C T_0^{-1} \cdot {}^C T_E = \begin{bmatrix} n_x & o_x & a_x & p_x \\ n_y & o_y & a_y & p_y \\ n_z & o_z & a_z & p_z \\ 0 & 0 & 0 & 1 \end{bmatrix}. \tag{7}
$$

The angles were shown as below

$$
\begin{cases} \beta_1 = - \arctan(a_x/a_z) \\ \beta_2 = - \arcsin(a_z) \\ \beta_3 = - \arctan(o_z/n_z) \end{cases}. \tag{8}
$$

where

$$
\begin{aligned}
a_x &= ea_x \cdot bn_x + ea_y \cdot bn_y + ea_z \cdot bn_z. \\
a_z &= ea_x \cdot ba_x + ea_y \cdot ba_y + ea_z \cdot ba_z. \\
o_z &= eo_x \cdot ba_x + eo_y \cdot ba_y + eo_z \cdot ba_z. \\
n_z &= en_x \cdot ba_x + en_y \cdot ba_y + en_z \cdot ba_z.
\end{aligned}
$$

To measure the angles of knee β_4, given $\mathbf{T}$ and $\mathbf{C}$ as the direction vectors of thigh and shank respectively, we can infer

$$
\beta_4 = \arccos \frac{\mathbf{T} \cdot \mathbf{C}}{|\mathbf{T}||\mathbf{C}|}. \tag{9}
$$

3   Motion Detecting Method Based on Kinect

The measuring system chose Kinect as the device to perceive the environment. Kinect could provide IR image and depth image of the environment and the two images can be matched after calibration. Then the depth image gave 3D coordinate of every pixel in IR image while IR image included the information of marker bars linked to human body. The glistening maker bars were split by the density-based spatial clustering algorithm of applications with noise (DBSCAN), and then the spatial direction vectors of these bars could be fitted out using least square method. Combining calculating method introduced in 1.2, the motion parameters can be detected with the proposed measuring system.

3.1   Arrangement of Glistening Marker Bars

According to the coordinate frame building of D-H bar model, marker bars were arranged as Fig. 2. The base frame consisted bar 0 and bar 1, whose origin was located in the pelvis center. Considering the convenience of deploying, we combined the two bars into one T bar. Similarly, one T bar was put on the thigh to build the effector frame E to calculate the angles of hip joint. To be simple, the shank was just arranged with one bar.

Fig. 2 Reflective marker bar arrangement

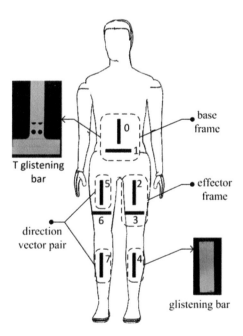

3.2   Algorithm of Motion Capture

Generally, there was not a little salt and pepper noise in the IR image extracted from Kinect. Although the median filter has a good effect in filtering this noise, some of the noise was still reserved to influence the clustering of glistening bars. Additionally, using filter could easily make the glistening marker points sparser as well, which led to broken bars in the IR image and wrong clustering. Due to the inevitable noise, the often used K-means algorithm could not cluster accurately and effectively. Considering the factors above, the density-based clustering algorithm was adopted to reserve enough points to cluster as well as filtering image noise effectively.

DBSCAN (Ester et al. 1996) is a typical clustering algorithm based on density. DBSCAN classifies areas of high density as clusters and those of low density as noise. The main strong point is that the algorithm does not need inputting the number of clusters prior to clustering. In addition, DBSCAN performs well in a spatial database with noise and can identify clusters of any shape.

Figure 3 illustrates the clustering process. The aim of DBSCAN is to find out all density-connected points in the database. The implement of the algorithm can be described as followings: (1) pick up one core point which is not included into any cluster, then grope out all points density-reachable from the core point and cluster them as the same cluster of this core point. (2) repeat (1) until all core points are clustered.

DBSCAN needs two parameters: the radius of Eps-neighborhood ε and the minimal point number Minpts in the Eps-neighborhood. These two parameters are used to define the core point which can directly determine the effect of clustering. Too small ε could lead to divide the one cluster which is not very concrete into several while too large ε will make the noise into the cluster. Similarly, excessively large Minpts contributed to mistake too many points as noise. Accordingly, excessively small Minpts could easily regard the noise as a new cluster. Considering all above, different application should choose different parameters to get the optimal effect.

Fig. 3 DBSCAN clustering algorithm

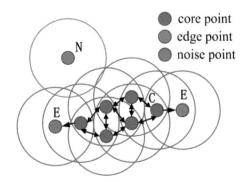

Fig. 4 DBSCAN clustering
rendering

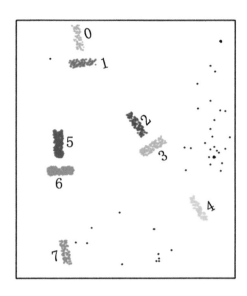

Figure 4 gives the effect of DBSCAN in our experiment with ε 10 and Minpts 20. In IR image, every glistening bar consisted of discrete but dense points meanwhile the noise was sparse salt and pepper points. As is shown in Fig. 4, the bars were separated into eight parts obviously and the noise was filtered effectively.

In the IR image, every point in the bar had a 3D coordinate derived from the corresponding depth image. Afterwards, the spatial direction vector of each bar could be fitted by the least square method. One T bar included two vertical direction vectors which could form a rectangular coordinate frame in space by adding the third vector using cross multiplication. After such transformation, the base frame and effector frame described previously could be built.

4 Introduction of Lower Extremity Rehabilitation Exoskeleton

To construct the kinematic and dynamic models of the system conveniently, isomorphic design was taken to hatch up the exoskeleton. As required by isomorphism, the structure of the exoskeleton and distribution of DOFs were designed as same or same proportion as human body's. In Fig. 5, different from previous rehabilitation system, treadmill and suspension system, our exoskeleton utilized a supporting stick to help patients get rehabilitation training in the imitation of walking. Based on the analyses foundation of biomechanics and human–machine coupling, an exoskeleton for rehabilitation was redesigned optimally. Disk motors were used in hip and knee joints decelerated with harmonic speed changers such that the exoskeleton owned the compact structure and enough torque. Additionally,

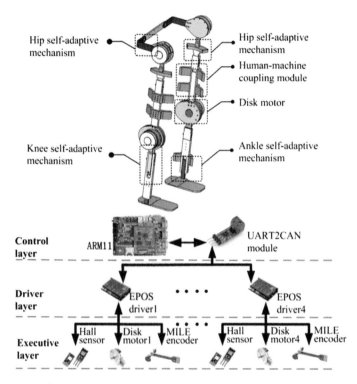

Fig. 5 ZJU rehabilitation exoskeleton of lower extremity

to realize flexible bionic joint design, the axes of the three DOFs of hip joint were configured to intersect at the hip center. Furthermore, cam path mechanism was considered to adapt to the glide of the knee center. Bandages were adopted to attach the patient with the exoskeleton to implement rehabilitation.

The exoskeleton was controlled by ARM microprocessor with embedded Linux. Train data were stored in SD which was convenient to be changed and read by the control module. The controller with motor drivers EPOS constituted the CAN to control the lower extremity exoskeleton cooperatively.

5 System Experiment and Analyses

5.1 Precision Verification Experiment

Figure 6 shows the precision verification platform which included Maxon disk motor EC90 with MILE encoder of 0.1125°. The axis of disk motor was connected with a round panel on which there was a glistening bar. The other bar was deployed beneath the round panel vertically. The motor was driven in some regulations to

Fig. 6 Precision verification
platform

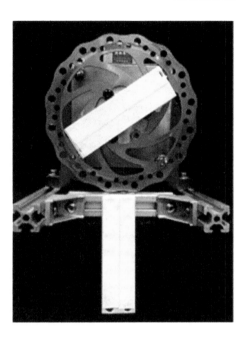

make the angle between the two bar changed correspondingly. Kinect could figure
out the changing separation angle in real time according to 0 for verifying the static
and dynamic precision of motion capture algorithm.

Static precision experiment was illustrated in Fig. 7. The disk motor was driven
to make the separation angle between the panel bar and the fixed bar fluttering from
0 degree to 80 degree, rotating 20° every 3 s. Kinect was put at the position 1.5 m
away from the platform, gauging the changing angle. The result showed that the
maximal static error is 2.76° and the standard deviation is 1.51.

The dynamic precision experiment was demonstrated in Fig. 8. Frequency
sweep signal was used to drive the motor, meaning the angle would vary from low
frequency to high. The maximal angular velocity in the normal gait is 60 degree per
second, so the frequency used in the experiment was from 0.2 to 1 Hz. Similarly,
Kinect was also deployed at the position 1.5 m before the platform. The sampling
period measured was 60–70 ms and the relevant coefficient between measured
value and given signal was 0.9917. The parameters proved that the measuring

Fig. 7 Static precision test

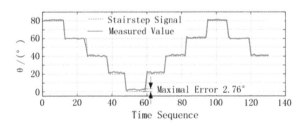

Fig. 8 Dynamic precision test

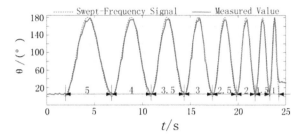

system had high tracking property and could detect the changing angles of joints in the rehabilitation.

5.2 Human–Machine Coupling Experiment

To assess the coupling character of our exoskeleton, an experiment was designed as Fig. 9. This study was approved by the Institutional Review Board of Zhejiang University. Informed written consent was obtained from all subjects. According to the normal gait, the exoskeleton was driven to haul the patient with the bandages, and finally the patient walked slowly assisted by two supporting sticks. Kinect was also put at the position 1.5 m before the patient. To maximize the possibility to get the IR image including the 8 bars, Kinect was configured with the optic angle oblique upward. Then Kinect helped arrange the bars whose positions were corrected slightly until all angles measured were closed to 0°. Figure 10 describes the measured angles of hip and knee and their tendency were in accordance with the normal gait.

Biomechanics research reveals that power consumed on the vertical plane is more than on the frontal plane and horizontal plane (Racine 2003). Thereby the

Fig. 9 Lower extremity exoskeleton human–machine coupling experiment

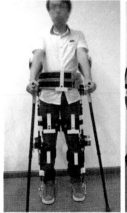

Fig. 10 Joint angles in rehabilitation training

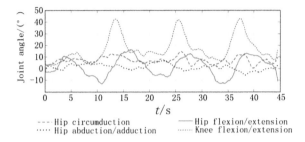

Fig. 11 Results of human–machine coupling

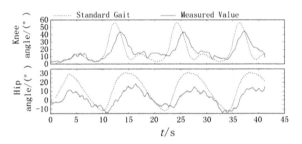

tracking character of flexion-extension angles of hip and knee joint were chosen to evaluate the human–machine coupling of our exoskeleton. It is demonstrated in Fig. 11 that in terms of knee joint, there was an average error of 12° between actual knee joint angle and the angle input to the knee joint motor, about 1 s in delay and the relevant coefficient was 0.7278. Correspondingly, as for hip joint, the average error was about 15°, and it also delayed for about 1 s. The relevant coefficient was 0.8981, a little more than the one of hip joint. The error represented the coupling character of the exoskeleton. From the parameters above, it could be inferred that the exoskeleton still had some defects in human–machine coupling.

There were several reasons for the unsatisfactory performance in coupling. The joint motors were working under load, which the lower extremity gave the exoskeleton. Besides, the bandages inflicting the patient to the exoskeleton were flexible. Hence, it could not be guaranteed that the patient moved in complete accordance with what the exoskeleton wanted. To improve the coupling performance, suitable control algorithm and reasonable structure were required.

6 Conclusions

This paper proposes a motion capture system based on Kinect to evaluate the validity of the exoskeleton in the rehabilitation process. The system arranged the eight glistening bars on the patient body, and then fitted the spatial vector of each bar by the DBSCAN and the least square method. The angles of hip and knee joints were measured by the solution of D-H model of lower extremity. To assess the precision of the measuring system, the verification platform was designed. The result showed that the system had a relatively low static error and high dynamic tracking performance. In the rehabilitation experiment, the angles of lower extremity of the patient wearing the exoskeleton were captured. By comparison with the standard gait angles, the paper analyzed the human–machine coupling performance of exoskeleton.

In the subsequent design of exoskeleton, for one thing, it is positive to increase the stiffness of human–machine connection module on the premise of comforting the patient. For another, some control methods should also be considered to improve the coupling performance, such as feedforward control, impedance control, and visual servo control.

Acknowledgments Project supported by the Science Fund for Creative Research Groups of National Natural Science Foundation of China (No. 51221004).

References

Ester M, Kriegel HP, Sander J et al (1996) A density-based algorithm for discovering clusters in large spatial databases with noise. KDD 96(34):226–231

Kim JY, Kim YS (2011) Development of motion capture system using dual video cameras for the gait design of a biped robot. Int J Humanoid Rob 8(02):275–299. doi:10.1142/S0219843611002502

Kobashi S, Kawano K, Tsumori Y et al (2009) Wearable joint kinematic monitoring system using inertial and magnetic sensors. RIISS'09. IEEE workshop on robotic intelligence in informationally structured space. IEEE, pp 25–29. doi:10.1109/RIISS.2009.4937902

Koyama Y, Nishiyama M, Watanabe K (2011) Wearable motion capturing with the flexing and turning based on a hetero-core fiber optic stretching sensor. In: 21st international conference on optical fibre sensors (OFS21). International Society for Optics and Photonics, pp 77534B–77534B-4. doi:10.1117/12.885068

Lunenburger L, Colombo G, Riener R et al (2005) Clinical assessments performed during robotic rehabilitation by the gait training robot Lokomat. In: 9th international conference on rehabilitation robotics, 2005. ICORR 2005. IEEE, pp 345–348. doi:10.1109/ICORR2005.1501116

Pons JL (2008) Wearable robots: biomechatronic exoskeletons. Souther Gate. Wiley, Chichester

Racine JL (2003) Control of a lower extremity exoskeleton for human performance amplification. University of California, Berkeley, California, pp 23–24

Sun B, Liu X, Shen J et al (2012) Joint angle measurements based on omni-directional lower limb rehabilitation platform. In: 2012 international symposium on micro-nanomechatronics and human science (MHS), IEEE, pp 337–341. doi:10.1109/MHS.2012.6492433

Thang ND, Kim TS, Lee YK et al (2011) Estimation of 3-D human body posture via co-registration of 3-D human model and sequential stereo information. Appl Intell 35(2):163–177. doi:10.1007/s10489-009-0209-4

Vlasic D, Adelsberger R, Vannucci G et al (2007) Practical motion capture in everyday surroundings. ACM Trans Graph (TOG) ACM 26(3):35. doi:10.1145/1276377.1276421

Windolf M, Götzen N, Morlock M (2008) Systematic accuracy and precision analysis of video motion capturing systems—exemplified on the Vicon-460 system. J Biomech 41(12):2776–2780. doi:10.1016/j.jbiomech.2008.06.024

Yan H, Yang C-J, Chen J (2014) Optimal design on shoulder joint of upper limb exoskeleton robot for motor rehabilitation and system application. J Zhejiang Univ (Engineering Science) 6:017 (in Chinese). doi:10.3785/j.issn.1008-973X.2014.06.01

Zhuojun X, Yantao T, Li Z et al (2013) Feasibility study on extraction finger joint angle information from sEMG signal. In: 8th IEEE conference on industrial electronics and applications (ICIEA), IEEE, pp 120–125. doi:10.1109/ICIEA.2013.6566351

Table Tennis Service Umpiring System Based on Video Identification—for Height and Angle of the Throwing Ball

Yun-feng Ji, Chao-li Wang, Zhi-hao Shi, Jie Ren and Ling Zhu

Abstract Table tennis is a very rapid movement in competitive sports, and its referee needs to make a judgment based on observation and experience within a few seconds, which is a very challenge task. Due to the subjective judgment of the referee's uncertainty and imprecision, it will lead to some false judgment, resulting in a significant impact on the results of the competition. In the current competitions, the most controversial penalty is the height and angle of the throwing ball's problem. To solve this controversial problem, this paper designs a complementary video-based recognition of table tennis service umpiring system that can identify and track the location of the table tennis in the video and determine the exact moment when the ball leaves the hand. As a new umpiring system, this paper first introduces into the machine learning methods to identify in table tennis, and discusses a new method of predicting the region of interest to reduce the recognition error rate and improve the recognition efficiency. This system will eventually get through the data of height and angle of the throwing ball, which not only provides a scientific reference for the referees, but also provides a supplementary basis for the athletes to question the referee.

Keywords Table tennis · Umpiring system · Region of interest · Machine learning · Recognition

Y. Ji · Z. Shi · J. Ren (✉) · L. Zhu
Physical Education and Sports Training, Shanghai University of Sport,
Shanghai 200093, China
e-mail: renjiemail2008@163.com

Y. Ji
e-mail: jyf123456789@126.com

C. Wang
Control Science and Engineering, University of Shanghai for Science
and Technology, Shanghai 200093, China
e-mail: clclwang@126.com

© Zhejiang University Press and Springer Science+Business Media Singapore 2017
C. Yang et al. (eds.), *Wearable Sensors and Robots*, Lecture Notes in Electrical
Engineering 399, DOI 10.1007/978-981-10-2404-7_38

1   Introduction

As a competitive sport, the technology development of table tennis has great relevance to the rules, and the technical competitions must meet the requirements of the rules on the legality of technology. To determine whether compliance with the rules, table tennis is the same with most competitive sports as tennis, football, it depends on the subjective judgment of the referee. Referee mostly subjective judgment Based on experience, referee's subjective judgment always has the uncertainty and imprecision, which will more or less lead to some false results. A typical example is in the 2012 London Olympic Games women's table tennis single final match that Ding Ning's serve was fined violation, which has become one of the key factors that led to her losing.

Compared with the table tennis referee's imperfect system, a number of other sports have some auxiliary referee's tools, such as the tool in tennis matches named Hawkeye and in football matches named gate line technology. Hawkeye uses multiple high-resolution cameras simultaneously from different angles on the game screen to continuous shooting. Then the image processing techniques is exploited to identify and locate the ball by using 3D simulation of tennis imaging, venue, and virtual reconstruction. Hawkeye eventually can reconstruct the trajectory of the tennis movement, and the use of "Eagle Eye" technology always takes no more than 10 s with the correct rate of 99.9 %.

Now the dispute of table tennis is focused on the serve penalty. Among the rules, the main controversial questions include two. One is the lowest value at allowing height for throwing balls. The other is whether the trajectory of throwing balls is vertical. The two specifications are the basis of Ding Ning's serve illegal:

> The players have to toss a ping-pang ball almost vertical upwards with a hand, without imparting spin to the ball, and the ball must rise at least 16 cm after leaving hand. Before hitting it, the ball can't touch anything.

The importance of serve stage by using sensors in table tennis has needless suspicious, which has become a large number of major research directions. Xu Shaofa has put forward that serve is a very important technology, which occupies an extremely important position in the game (Xu 2007). Tang Jianjun explores the relationship between technological development and the rules limits from the perspective of sports technology social development, and he thought that the core problem was legality and fairness (Tang 2003). Zhang Jiancheng discussed the different levels of women table tennis players in the rotating cognition of serve. Meanwhile, for the problem in the 2012 London (Zhang 2013) Olympic Games women's table tennis single final match that Ding Ning's serve was fined violation, Chen Qiuxi used a two-dimensional Cartesian coordinate system to study her serve drop angle and tossing height of the ball. And he concluded: that the drop height is higher than 40 cm, and the angle is about 80° (Chen and Deng 2013).

Obviously, the importance and legitimacy of table tennis have been recognized by the majority scholars, but the table tennis referee's penalty is still based on the experience of the referees. Up to now, there has not been a fairly complete and scientific system to judge the matches correctly. How to a design scientific and accurate table tennis referee system based on computer technology has become one of much pressing problems in table tennis sports. Dr. Patrick KC Wong used a large number of match photographs to design an identification method, but the method proposed has weak anti-interference and high error rates (Wong, Patrick K.C). In order to design scientific and accurate table tennis referee system, this paper will need to overcome the following difficulties:

(a) How to identify the table tennis accurately and rapidly in the video?
(b) How to track the trajectory of table tennis accurately?
(c) How to determine the time that the ball leaves the hand?
(d) How to reduce the interference, improve accuracy, reduce the amount of calculation and shorten the identification time?

Therefore, this paper mainly designs a table tennis referee system according to the above four difficult problems based on video identified, and its main purpose is to calculate the drop height and drop angle of players' serve.

2 The Identification Based on Image

2.1 Experimental Device

In this study, the photograph mode of high-speed camera is 200 fps (640 * 360). That is, the video capture 200 pictures per second with a resolution of 640 * 360. The position of camera is beside the referee with the distance of 3 m to the referee and the height of 1.5 m to the floor. This study was approved by the Shanghai University of Sport Human Subjects Research Review Committee. Consent was obtained to publish the images used for this study.

2.2 Set ROI

According to athletes serve video from camera, we intercept an image, shown in Fig. 1.

It is known that the video includes many multiframe continuous images, and this paper extracts each frame images from video to process first. For this article experiments, if the pictures are processed on the whole, there will be computationally intensive, long processing time, error interference and other problems. To solve these problems, we set up a ROI region, named the region of interest, as shown in the red box below. We extract the red box to be dealt with separately and the other part ignored, which can solve these problems to some extent. Among

Fig. 1 Set ROI

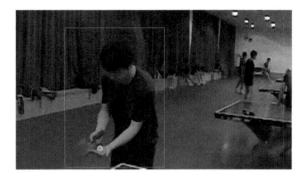

them, the ROI selection is related to the position of camera, with the purpose that includes the possible trajectory of ball.

2.3   Serve Process

According to serve video, the serve process is divided into the following three steps, shown in Fig. 2.

According to Fig. 2, three pictures represent the three steps of the serve:

Step 1:   The moment that table tennis ball is placed on the hand naturally, then open the hand and remain stationary;

Step 2:   The moment that table tennis ball leaves hand, which will be the starting point of the detection height of the ball;

Step 3:   The moment that table tennis ball entirely leaves hand, which will be the ending point of the detection height of the ball when the speed of the ball is 0.

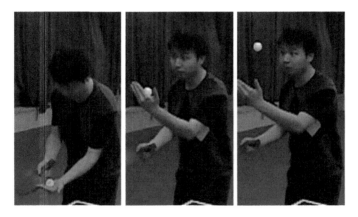

Fig. 2 Serve process

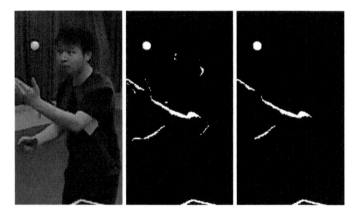

Fig. 3 Image pre-processing

2.4 Identification of Ball in Picture

The main purpose of this paper is to trace the moving table tennis ball, get its trajectory, and then calculate the angle and the height. So the primary purpose of this paper is to accurately identify the table tennis ball in the video. First, this paper extract a frame images from video to process, shown in Fig. 3.

According to Fig. 3, in order to identify the ball, we do the image pre-processing firstly. The image pre-processing steps are as follows:

Step 1: The binary of image processing that the threshold is set to 100 RGB values.

Step 2: The expansion and etching of image processing, which can remove some irrelevant small white pixel block.

After image preprocessing, we will begin to identify pictures the ball. This article tries to determine whether the white pixel block is table tennis ball or not by six characteristics in Table 1.

This paper sets the threshold through the above six characteristics. If the detected values of the object are within the threshold we set, we will assume that the object is the table tennis ball. Among six characteristics, we need Squareness to ensure whether the ball in hand or leave hand. When the ball is at the hand of the players, due to the palm will naturally slightly curved inward, there will be a small part of the bottom half of the ball block. And that will lead to the result that $L_X > L_Y$, so let $T > 0$. According to the threshold, we begin to test the preprocessed picture, and the results are as follows in Fig. 4.

As shown in Fig. 4, we calculate the six characteristics of each white pixel block. Shown on the right, there are yellow fonts displayed in the pixel block on the right. Depending on the threshold we set above, we filter out the pixel block we

Table 1 Six characteristics of recognition of table tennis ball

Characteristic	Description
Roundness	A metric is employed to estimate the roundness of an object: $M = 4\pi A/C^2$ The metric will be equal to 1 if the object is a circle because A will be πr^2 and C will be $2\pi r$
Perimeter (C)	The length of the object's boundary
Area (A)	Number of pixels the object possess
Maximum width (L_X)	The distance between the leftmost and rightmost pixels of the object
Maximum height (L_Y)	The distance between the top and bottom pixels of the object
Squareness	A metric is employed to estimate the Squareness of an object: $T = |L_X - L_Y|/L_X$ The metric will be equal to 0 if L_X is closer to L_Y

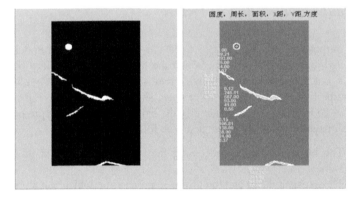

Fig. 4 Results of recognition of table tennis ball

need, and mark a black circle in the pixel block. Then the pixel block marked is the table tennis ball that we identify.

So far, we have completed the task of the identification of table tennis balls in a still picture.

3 The Identification Based on Video

In the video, often due to the light, shadow, background or other interferences, it will led the identification of the correct rate and speed to decline. So we propose a new method for identifying table tennis balls in the video. According to the position of table tennis ball in the previous frame image, we can predict the position of table tennis ball in the next frame. This paper innovatively integrates face recognition, machine learning and other aspects of the use into the identification so that the recognition accuracy could be much improved.

3.1 Recognition of the Ball on the Hand

3.1.1 Training Samples

In this experiment, we collect 200 different pictures that table tennis players are ready to serve as positive samples for machine learning and training. Also we collect 200 different arbitrarily selected images of non-serving preparations as negative sample for machine learning and trainings. Then, we collect other 20 different positive samples and 20 different negative samples for testing.

As shown in Fig. 5, in positive samples for training, the palm of the hand need to open naturally to hold table tennis ball so that each picture has a similar feature. This paper uses LGP algorithm and adaboost algorithm for machine learning and training (Li and Wang 2014).

(a) LGP algorithm

Firstly we introduce the definition of the gradient. Given a grayscale image matrix, the submatrix of which (named matrix A) is the 3×3. We use i_n to represent the center of the pixel values, $n = 0, \ldots, 7$. There is a provision that the numbers in the clockwise direction starting from the upper left corner are 0–7. The gradient value of each point around the center is defined as follows:

$$g_n = |i_n - i_c| \quad n = 0, \ldots, 7 \tag{1}$$

The definition of average gradient of 3×3 submatrix is:

$$\bar{g} = \frac{1}{8} \sum_{n=0}^{7} g_n \tag{2}$$

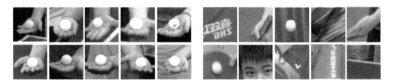

Fig. 5 Part of the training positive samples and negative samples

For any point (x_c, y_c) of the image, its LGP value is expressed as follows:

$$\text{LGP}\,(x_c, y_c) = \sum_{n=0}^{7} s(g_n - \bar{g})2^n \tag{3}$$

The function s is defined as follows:

$$s(x) = \begin{cases} 0, & x < 0 \\ 1, & others \end{cases} \tag{4}$$

LGP operation is to transform the original matrix A to the substitution matrix A with the converted value according to Eq. (3). The specific procedure is shown in Fig. 6.

(b) Adaboost Algorithm

This article takes adaboost algorithm as the main algorithm of training samples (Li and Wang 2014). As a way to achieve the purpose of this article, it is not the focus. The training and testing process are as follows:

(a) Image pre-processing: Firstly convert the input image to grayscale image. Then put the resulting grayscale image in accordance with a successively smaller proportion using image pyramid technique. Here, the proportion is

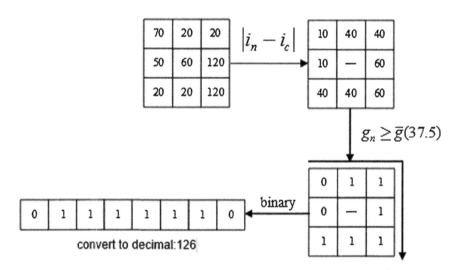

Fig. 6 The specific procedure of LGP

successively reduced in accordance with the ratio of 0.8. After 5 times, length and width of the image are only 1/3 of the original and the area is of only 1/9 of the original image. Thus the image can be scaled to very small, which will reduce the processing time.

(b) Feature extraction: Convert the pyramids of the resulting image into the respective LGP images, and then replace the gray value of each pixel of the image with LGP values. After conversion, the edge will be removed.

(c) Category detected: This article uses a specific adaboost level classifier as learning algorithm references to detect the category(Li and Wang 2014).

(d) Final Detection Area: Due to the presence of position detection error and the incomplete training samples, the picture into the overclassification after detection usually occurs many overlapping areas, and the wrong areas. In order to further determine the most appropriate area, this article uses over-lapping area consolidation method (SOMM) to determine the final detection area.

3.1.2 Test Result

After taking 200 positive sample images and 200 negative sample images into training, this article uses an image that the player is ready to serve to testing. The test image is not in the training samples, and the test result is shown in Fig. 7.

As showing in Fig. 7, the second picture shows the effect that has been tested after adaboost level classifier training. Due to the presence of positive detection error and the incomplete training samples, there are a lot of overlapping areas and detect errors areas. We need to use overlapping area consolidation method (SOMM) to determine the final area. The third picture is the effect that through SOMM test. As shown, it can accurately identify the table tennis ball in the hands of preparatory actions, and it also gets the coordinates of the initial position of table tennis ball.

Fig. 7 Test result

3.2 Set Predict ROI

After solving the problem of identifying the preparatory actions, we need to begin
to identify and track the ball in the video. This article proposes a new method to
predict the ROI through previous test results to forecast the position of the ball in
the next frame. So we can set up a ROI area in advance. In the next frame, we just
identify the image in the ROI area we set up, which will greatly reduce recognition
time and improve recognition accuracy, The specific algorithm is as follows:

(a) Positive and negative samples will be prepared in advance. First, we detect
    each frame of the video, and then use SOMM method to determine the final
    inspection area.
(b) Detect the image within the region with image recognition method, and then
    calculate the six major characteristics of the object. Compared to the estab-
    lished threshold, if the six characteristics are within the threshold range, we
    believe that the object is what the desired ball is. Otherwise, the identification
    object is not the ball.
(c) If in the three consecutive frames we can determine that the target object is the
    desired ball, we believe that the detection is right at this time, and we will no
    longer execute (a). We can set up a ROI area in advance according to the
    previous two frames. In the next frame, we just identify the image in the ROI
    area we set up. As shown in Fig. 8, the blue boxes represent the predict ROI
    that we set up. The predict method is to find out the location of the center of
    the balls in the two former image, that is, (X_{n-2}, Y_{n-2}) and (X_{n-1}, Y_{n-1}). We
    define $M = Y_{n-1} - Y_{n-2}$. Then we take $(X_{n-1}, Y_{n-1} + M)$ as the location of the
    center of the ball in the next frame. Since the center axis $(0, 0)$ is in the upper
    left of the picture, so if $M > 0$, the ball is dropped at this time; if $M < 0$, the
    ball is rising at this time; if $M = 0$, indicating that at this time the ball still in
    the longitudinal direction. In addition, in order to ensure it not happens that the
    speed of the ball is too fast that will be beyond the set range ROI, we need to
    ensure that the side length of ROI region is longer more than $2 * M$ to the

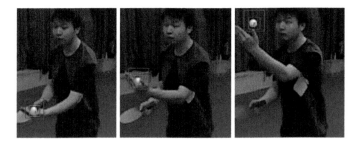

Fig. 8 Set predict ROI

perimeter of the ball. We can note that the center of the blue box is significantly higher than in the center of the red box in Fig. 8, which means the moment that the ball is on the rise.

3.3 Determine the Moment that the Ball Away from the Hand

According to the above method, we can trace the trajectory of the table tennis ball in the video, then we are going to achieve our purpose of determining whether the Serve conforms to the rules, which is that calculates the height and angle of table tennis ball in the serve. According to the rules, the calculation requires to determine the moment that the ball away from the hand completely as a start moment, and the moment that the ball reaches the highest point as the end moment. During the time, the height of serve is the subtracted value of maximum ordinate and minimum ordinate and the angle of serve is the trajectory slope. In order to determine the start moment and the end moment accurately, we put the identification of the ball in some key frames in the video to place in the following figure, as shown in Fig. 9.

As showing in Fig. 9, these images show that that the ball detected from the moment on the hand to the moment away from the hand. The seven images of the first row show table tennis ball in the preparation period. The seven images of the second row show the ball detected from the moment ready to leave the hand to the moment away from the hand completely. After observing the test results, we find that when the ball is in the hand, the detected ball is a substantially ellipse. At this time, since the influence of the words "Double Happiness" on the ball, the detected ball has some defects, which leads to the test results that has some differences with the ball in the air. Then we need to revise the original threshold value T to the new threshold value T_1, whose range is wider than the original threshold value T. In the second row of seven pictures, it can be observed that at the moment the ball away from the hand, due to some shelter of the hand, the six characteristics of the ball

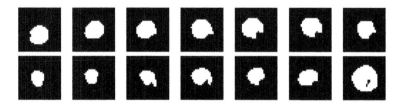

Fig. 9 The identification of the ball in some key frames in the video, the numbers of frames in sequence are as follows: 271 276 281 286 291 296 301 304 307 310 313 316 319 322

Fig. 10 Set the box's color according to threshold value

occur to change. In this case we set the threshold be adjusted accordingly, continue to relax the threshold range to T_2. The last picture represents the image that the ball fully in the air, and at this time the threshold value is set to T.

As it mentioned above, the article needs to determine the starting point of serve namely the moment when the ball leaves hands. According to Fig. 8, when the ball will leave hands, the test results must have some changes. In theory, the program can detect the complete ball in the moment when the ball leaves hands. So the article can set threshold value to determine that moment. In that moment, every feature of test subjects will change from range of threshold T_2 to range of threshold T. But according to a lot of experiments, the program cannot test the complete ball in the moment when table tennis leaves hands because of unconscious hands overlap, so the article loosen the threshold value and determine the leaving moment is that when every feature of test subjects change from range of threshold T_2 to range of threshold T_1.In order to distinguish the state of ball, the article marks the tested ball with box in video. When the features of test subject is in range of threshold T_1 or T, the color of box will be red, and when the features of test subject is in range of threshold T_2, the color of box will be green which is shown in Fig. 10.

4   Experimental Results

According to the test results in video, we can display the detected table tennis center in the same coordinate system in every frame image which is shown in Fig. 11.

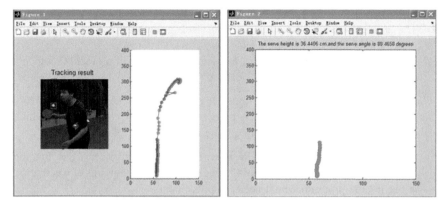

Fig. 11 The experimental results

According to the color of the box, the article attaches the same color dots in a line which is displayed in the same coordinate system. The results show that after several continuous green dots displayed, many red dots will display for a period of time linked as a line. It is known that this red line is trajectory of table tennis in the air after leaving hands.

Figure 11 shows the trajectory of table tennis singly. According to the table tennis' pixel coordinate in video, the height of serve is the subtracted value of maximum ordinate and minimum ordinate and the angle of serve is the trajectory slope. At last, according to the ratio of video length and reality length, the real height and angle of serve can be got. In Fig. 11, the real serve height is 36.4406 cm and the real serve angle is 89.4658° which is a good service according to the table tennis rules.

5 Conclusions and Recommendations

Aiming at the height and angle of throwing balls, the article designed the supporting table tennis service judge system based on video identification, which could judge the legitimacy of serving a ball. The system will be the scientific reference for the referees to penalty rightly and the supporting basis for athletes to query referees.

As a new type of table tennis service judge system, the innovation points of the article are as follows:

(a) The article carried out the legal issues of serving a ball based on video identification for table tennis players for the first time. Besides, the system could calculate the height and angel of throwing a ball.
(b) The article applied the machine learning knowledge to the table tennis recognition for the first time.

(c) The article provided a new way to identify table tennis.
(d) The article used a new method to reduce recognition error rate and improve recognition efficiency through setting up predict interest area.
(e) The article solved the table tennis ball leaving the hand of no racket moments problems in the process of serve balls by experiments.

As a new coming into research project, the designed referee system had many deficiencies due to the short research time and insufficient consideration which needs further improvements. The main shortcomings are as follows:

(a) The calculating time is long. The intercept serve video time is 8 s, but the handled video time is nearly 2 min. It maybe means the amount of calculation is big for the program which needs improvements.
(b) The experimental environment is single. The main experimental subject is the author himself who is as the serve player. The program is not promoted to formal games and only used in experimental environment.
(c) The selected positive and negative sample size is small in the process of utilizing the knowledge of machine learning which may be cause large error during identification process.
(d) In the article, various factors such as the light and the environment are considered insufficient. There are still some recognition errors in the experiment although the way of setting up predicts interest area may reduce the error rate in a certain extent. So the article needs further research.
(e) The setting threshold value is not fixed which could be used in every video in the process of recognition. But in order to improve recognition rate, enlarging the range of threshold value may cause some other recognition errors which needs a large number of experiments to improve it.

Based on these problems mentioned above, the article will continue to improve the table tennis service judge system through a large number of experiments so that the program could be applied in actual game as soon as possible and contribute to the development of table tennis!

Acknowledgements Project supported partly by the National Natural Science Foundation (61374040), Science and Technology Commission of Shanghai Municipality (15490503100), Chenguang Program (13CG55).

References

Chen Q, Deng D (2013) Irregularity risk of service in table tennis: taking debatable service of table tennis women's singles final in 2012 London Olympic Games for example. J Shenyang Sport Univ 32(6):101–103
Li L, Wang C (2014) Face detection based on local gradient patterns and square overlap merge method. Comput Simul 31(5):279–283

Tang J (2003) Development of serve technique of table tennis and technique limit in rules—visual angle of society in technique development of competitive sport. J Beijing Sport Univ 26 (3):423–424, 431

Xu S (2007) Serve and return of serve. People's Sports Publisher, Beijing, pp 1–7

Zhang J, Shi Z (2013) The feature of cognitive processing during serving-rotation judgment of table tennis players. China Sport Sci 33(1):42–51

Visual Servo-Based Control of Mobile Robots for Opening Doors

Xiao-mei Ma, Chao-li Wang and Lei Cao

Abstract Fingerprint lock cannot work effectively in the case of finger cut or desquamating. Besides, the structure of electronic locks is different from most wooden doors' and security doors', and it is also too time-consuming and laborious to change the existing structures. This paper presents a method of opening the door by a mobile robot at home. The robot receives wireless commands and calls the door fixed camera to capture a photo of the visitor. After that, it makes a visual servo based path planning after the verification of the visitor autonomously, then moves from a certain place to the interior door and opens the door. The mobile robot consists of a fixed base 4 degrees of freedom manipulator and a camera. This article focuses on the extraction of work space of the robot, path planning, face recognition, lock positioning, and robotic motion control. Using visual C++ development environment and image processing technology, a controller of the mobile robot to open the door is proposed, the effectiveness of the proposed method is also demonstrated through experiments.

Keywords Visual servo · Path planning · Image segmentation · Face recognition · Inverse kinematics

X. Ma (✉) · C. Wang (✉)
Control Science and Engineering, University of Shanghai for Science
and Technology, Shanghai 200093 China
e-mail: April_2016@163.com

C. Wang
e-mail: clclwang@126.com

L. Cao
Electrical and Electronic Experiment Center, Henan University of Science
and Technology, Luoyang 471023, China

© Zhejiang University Press and Springer Science+Business Media Singapore 2017   511
C. Yang et al. (eds.), *Wearable Sensors and Robots*, Lecture Notes in Electrical
Engineering 399, DOI 10.1007/978-981-10-2404-7_39

1 Introduction

With the development of computers, sensors and artificial intelligence technology, the performance of intelligent robots continues to improve. Robots become more and more important in contemporary society.

People often return home or office without keys, it is time-consuming to get keys or find a locksmith, placing a spare key in the door shelter is also unsafe. How to open the door conveniently is the key problem. Fingerprint lock is convenient. However, it cannot work effectively in the case of finger cut or desquamating. Besides, the structure of electronic locks is different from most wooden doors' and security doors', it is too laborious to change the existing structures. Considering that robots play a more and more important role in contemporary society, robots will do much more work to help humans to enjoy life. This paper presents a method of opening doors by a mobile robot at home. The opening service provided by robots is not only convenient, but also easy for owners to enjoy the owners' pleasure. Therefore, it has a very good practical significance.

LBP is a simple and effective texture feature extraction (Ojala et al. 1996, 2002) operator, it is applied to background modeling (Heikkilä and Pietikäinen 2006), facial analysis (Zhao and Pietikäinen 2007) and face recognition (Ahonen et al. 2006) successfully for its notable feature, which is insensitive to light and owns a certain anti-rotation ability. This article uses LBP to conduct face recognition.

Currently, the main navigation methods include magnetic navigation, GPS navigation, signpost navigation, visual navigation (Blanc et al. 2005), etc. Compared to other navigation methods, visual navigation has the advantage of a wide range of signal detection, complete access to information, without secondary pollution, etc. This article separates the camera and robot to obtain the global ground image so as to extract the floor area in order to achieve obstacle avoidance and path planning.

In addition, the article also discusses the positioning of the lock knob, the inverse kinematics of how to control the manipulator to conduct the open action with the known positon of the manipulator's end. The experiment platform consists of a MT-AR robot and a MT-ARM manipulator developed by Shanghai ingenious automation technology Co., Ltd., a PC machine, three cameras and a wireless router. Finally, a summary and outlook is conducted.

2 Face Recognition

With the development of face recognition technology, face recognition system in commercial use has become increasingly widespread (Zhao et al. 2003; Phillips et al. 2003). This article also uses face recognition to identify the visitor.

LBP is an effective texture description operator, it measures and extracts image's local texture information, and it is gray scale invariant. The first step is to convert images to grayscale images, then calculate the histogram. Similarity function is used to calculate the similarity between two histograms. In this paper, chi-square statistics is used, which is defined as

$$\chi^2(S,M) = \sum_i \frac{(S_i - M_i)^2}{S_i + M_i}, \quad i = 0, 1, \ldots, L-1, L \tag{1}$$

where L represents the number of the discrete gray level, S_i and M_i represent the number of the pixel which gray value equals i in the database image and the detected image, respectively.

The global image's histogram calculated above does not contain the spatial information, which can be obtained by calculating the spatial enhancement histogram. The image is spatially divided into equal regions represented by $R_1, R_2, \ldots, R_T$. Calculate each histogram and assign weights according to the importance in face recognition, such as eyes are very important (Zhao et al. 2003). Chi-square statistics with weights is defined as

$$\chi^2(S,M) = \sum_{i,j} \omega_j \frac{(S_{i,j} - M_{i,j})^2}{S_{i,j} + M_{i,j}}, \quad i = 0, 1, \ldots, L-1, L;$$
$$j = 1, 2, \ldots, T; \; \omega_j \in (0, 1) \tag{2}$$

where ω_j is the weight of R_j.

3 Navigation and Path Planning

In this paper, the camera is mounted on the ceiling to obtain a global ground image. Image processing technology is used to extract area where robot could walk and position the robot. Assumed that the location of the door is fixed, a planner of the robot will generate a path automatically with the known start point and target point, and the proposed controller makes the robot go to the destination.

3.1 Floor Area Extraction

Mounting camera on the ceiling can effectively avoid obstacles blocking the camera's field of view. Considering the noise and geometric distortion due to light, camera lens and other reasons, the image need to be smoothed, filtered, and geometric distortion correction (Li and Li 2007; Zhu et al. 2014) before segmentation.

Image segmentation algorithm is based on YCbCr color space feature. Chrominance signals are viewed as the characteristic value. The method of selecting the normal, lighter, and darker areas of floor and each obstacle as samples is adopted in order to improve the accuracy. Segment the image into blocks first, view the adjacent 3×3 pixels as a unit and calculate the average value as the unit pixel value. The average of all the pixels is calculated as a color characteristic value when extracting floor and obstacle samples' character. Providing that the color characteristic value sets of the floor and obstacles, respectively, are $M\{m_1, m_2, m_3, \ldots\}$ and $N = \{n_1, n_2, n_3, \ldots\}$. Euclidean distance between a unit area of the original image and the elements of the sets M, N is calculated. If the minimum distance occurs on a floor sample, the unit area belongs to floor area, vice versa belongs to obstacle areas. A binary image is obtained after the segmentation, black areas are obstacle areas and white are the areas where robot could walk.

Figure 1 illustrates an example of segmentation. Figure 1a is the original image, the rectangles mark the obstacle samples and the circles mark the floor samples. Figure 1b shows the binary image after segmentation based on color character. It is easy to find the disturbances caused by the region near the edge of the obstacle

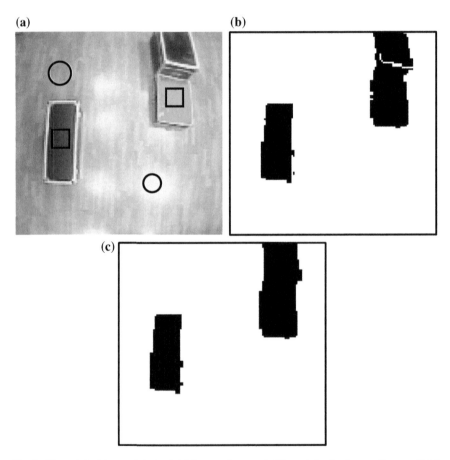

Fig. 1 The original image (**a**), the initial extraction result (**b**), and the final extraction result (**c**)

whose color is similar to floor, and the contours of obstacles are not smooth. Thus, morphological processing is necessary (Cui 2002). As shown in Fig. 1c, a better segmentation result is obtained after morphological processing.

Many experiments show that the method is robust and relatively stable, and the accuracy is high.

3.2 Path Planning

In this paper, the A star algorithm (Hart et al. 1968) is used for path planning. The algorithm is a heuristic search method. The evaluation function is the core of the algorithm, which is defined as

$$f(n) = g(n) + h(n), \tag{3}$$

where $g(n)$ represents the actual cost of moving from the initial node S_0 to a certain node N, the cost of a node movement in the horizontal or vertical is 10, a diagonal direction is 14. $h(n)$ represents the evaluated cost of moving from the node N to the target node, the method of Manhattan is used to evaluate.

The algorithm is described as follows:

Set up two tables named OPEN and CLOZE.

Step 1 Put initial node S_0 into the table OPEN.

Step 2 Check the 8-neighborhood nodes around
        S_0 and add the node which can reach or can go up to into the table OPEN
        while ignoring obstacle nodes, and set S_0 as father of these nodes.

Step 3 Remove node S_0 from the table OPEN and add it to the table CLOZE.

Step 4 Elect the node with smallest $f(n)$ in table OPEN, remove it from the table
        OPEN and add it to the table CLOZE.

Step 5 Check all the neighborhood nodes around the node selected in Step 4.
        Ignoring obstacle nodes, add the nodes to the table OPEN which are not in
        and set the currently selected node as father of them. If already in the table
        OPEN, check whether arrives the node via currently selected node has a
        smaller value or not, if not, do nothing, and if so, put the current node as
        father of the node. Then recalculate the value and $g(n)$ of the node $h(n)$.

Step 6 Check the 8-neighborhood nodes around currently selected node, repeat
        Step 2 until the target node is added to the table OPEN, then the path has
        been found. Or lookup fails and the table OPEN is empty, which means
        there is no path.

Step 7 Save the path.

Tracing the position of the robot is precondition. A marker is installed on robot in order to obtain robot's initial position by detecting the marker. The marker is shown in Fig. 2. Set the green rectangle pointing as the positive direction of the

Fig. 2 The marker of the
robot

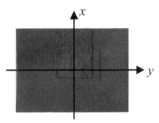

Fig. 3 An example of path
planning

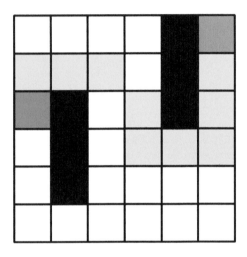

robot movement. When conducting spatial modeling, the robot is viewed as a particle; the robot's work space is divided into same grids with the robot actual size as a reference. Expand the obstacles appropriately to ensure that the robot must be able to walk along its borders. The excess pixels in the left and the bottom of the image are removed when the size of the image cannot be divided into boxes exactly. A path planning diagram is shown in Fig. 3. The black boxes represent obstacles, red represents robot, green represents door, and yellow represents the shortest path.

Detect the angle between the positive direction of the robot and the planned path to control robot to move to the destination.

4 Control Algorithm of the Manipulator

4.1 Manipulator's Kinematics

This paper adopts the industrial manipulator with four degrees of freedom, and the end is designed as a gripper. D-H method is used to establish the coordinate frames (Huo 2005), as shown in Fig. 4.

Fig. 4 The manipulator with
D-H link coordinate frames

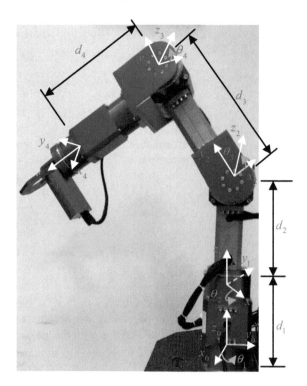

Table 1 D-H parameter

i	a_i	α_i	d_i	θ_i
1	0	0	d_1 (constant)	θ_1
2	0	90°	d_2 (constant)	θ_2
3	0	0	d_3 (constant)	θ_3
4	0	90°	d_4 (constant)	θ_4

Table 1 displays the D-H link parameter. Coordinate frame $i-1$ can be transformed into coordinate frame i through the following consecutive relative movement.

Step 1   Translate d_i from x_{i-1} to x_i along z_{i-1}.

Step 2   Rotate θ_i angle from x_{i-1} to x_i about z_{i-1}, $\theta_i \in (-\pi, \pi]$.

Step 3   Translate a_i from z_{i-1} to z_i along x_i.

Step 4   Rotate angle α_i from z_{i-1} to z_i about x_i, $a_i \in (-\pi, \pi]$.

The homogeneous transformation matrix of the continuous relative transformation is defined as

$$A_i^{i-1} = \text{Trans}_z(d_i)\text{Rot}_z(\theta_i)\text{Trans}_x(a_i)\text{Rot}_x(\alpha_i)$$

$$= \begin{bmatrix} c_i & -c\alpha_i s_i & s\alpha_i s_i & a_i c_i \\ s_i & c\alpha_i c_i & -s\alpha_i c_i & a_i s_i \\ 0 & s\alpha_i & c\alpha_i & d_i \\ 0 & 0 & 0 & 1 \end{bmatrix} \tag{4}$$

where $s_i \triangleq \sin\theta_i$, $c_i \triangleq \cos\theta_i$, $s\alpha_i \triangleq \sin\alpha_i$, $c\alpha_i \triangleq \cos\alpha_i$.

Substituting D-H parameter in Table 1 into Eq. (4) leads to

$$A_4^0 = A_1^0 A_2^1 A_3^2 A_4^3$$

$$= \begin{bmatrix} c_{12}c_{34} & s_{12} & c_{12}s_{34} & d_3 s_{12} + d_4 s_{12} \\ s_{12}c_{34} & -c_{12} & s_{12}s_{34} & -d_3 c_{12} - d_4 c_{12} \\ s_{34} & 0 & -c_{34} & d_1 + d_2 \\ 0 & 0 & 0 & 1 \end{bmatrix} \tag{5}$$

where $c_{ij} \triangleq \cos(\theta_i + \theta_j)$, $s_{ij} \triangleq \sin(\theta_i + \theta_j)$.

It needs to control the gripper of the manipulator to move to the known position and keep it at the same horizontal line with the lock knob in order to grab the fixed height lock knob. Namely, A_4^0 is known, we should solve each joint variable $q = [q_1, q_2, q_3, q_4]^T$ according to the joint conversion relationship.

This paper chooses the recursive inverse transformation technique (Huo 2005) to solve the inverse kinematics. Both sides of the corresponding matrix elements are equal in Eqs. (5)–(8). Select the constant or single variable elements and list corresponding equations. Using bivariate arctangent function $a \tan 2$ determine the angle. Multiple solutions may occur in the processing of solving, which is to be judged according to the structure characteristics of the manipulator and selected the ultimate possibilities.

$$(A_1^0)^{-1} A_4^0 = A_2^1 A_3^2 A_4^3 \tag{6}$$

$$(A_2^1)^{-1}(A_1^0)^{-1} A_4^0 = A_3^2 A_4^3 \tag{7}$$

$$(A_3^2)^{-1}(A_2^1)^{-1}(A_1^0)^{-1} A_4^0 = A_4^3. \tag{8}$$

4.2 Target Recognition and Capture

Opening the door is realized by making robot rotate the lock knob and pull the door. The expected lock knobs are those with grooves so as to be fixed by the gripper. Considering that the small lock knob and the interference caused by environment, a red square marker is recognized instead of the lock knob. The centroids of the red square and the lock knob are designed at the same vertical line as well as the camera and the base of the manipulator. Thus, when the camera and the centroid of the red square are at the same vertical line, the base of the manipulator is also at the same vertical line with the lock knob.

As it is shown in Fig. 5a, the centroid of the red square being at the central vertical line of the window ensures that the camera and the centroid of the red square are at the same vertical line. Figure 5b shows extraction of the edges of the square.

The process is divided into three steps: lock knob alignment, approaching, and grab (Wu 2010).

(1) Alignment: When robot arrives at the destination, camera usually misaligns the marker. At this point, the robot is required to move to the marker so as to the base of the manipulator is at the same vertical line with the lock knob.

(2) Approaching: The robot approaches to the lock knob after the lock knob is aligned. As it shown in Fig. 6, d_1, d_2 represents two distance between the camera and the marker, a represents side length of the marker, a_1, a_2 represents the length of marker's projection in the imaging plane, f represents the focal length of the camera.

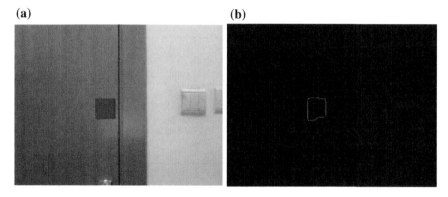

Fig. 5 The *red square* marker (**a**), and the edges of the square (**b**)

Fig. 6 The pinhole imaging principle

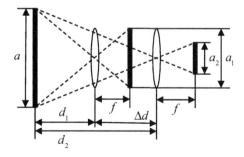

Doing geometric analysis according to Fig. 6 leads to

$$d_1 = \frac{\Delta d \cdot a_2}{a_1 - a_2} \tag{9}$$

Equation (9) shows that the projection of side length of the marker in camera imaging plane can reflect distance between the marker and the camera. Approaching completes when the side length of the marker is equal to the expected value. Set the length of the marker in camera imaging plane equal to the desired value when the distance between the target and end of the manipulator is 10 cm. The error of directly controlling the robot to go forward 10 cm is very small even without visual feedback. It ensures that the lock knob is in the center position of the gripper exactly.

(3) Grab: Close the gripper as long as the lock knob is in central position of the gripper. Rotate counterclockwise after the lock knob is seized, then release the lock knob. Pull the door after positioning the door handle through simple translation transformation. Control the robot to retreat aside after completing the opening action. Voice prompts "come in, please." Then the visitor could come in.

Fig. 7 The procedure of the experiment

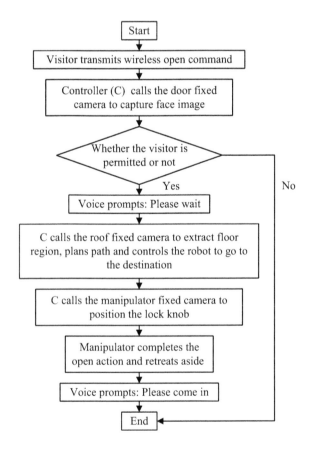

5 The Procedure of the Experiment System

The proposed experimental procedure is shown in Fig. 7.

Figure 8 shows the movie snapshots of the process of navigation. The annotated arrow indicates the planned path. The grab of the lock knob is shown in Fig. 9.

Fig. 8 The movie snapshots of the process of navigation

Fig. 9 The grab of the lock knob

6 Conclusion and Future Work

Taking into account the situation that people often return home or office without keys, the proposed method of opening the door by a mobile robot indoors is convenient and meaningful. All the visitor need to do is to press the button on the door to send wireless command. Mobile robot receives the wireless command sent by visitor indoors and conducts visual servo-based path planning after the verification of visitors autonomously, then moves from a certain place to the interior door and opens the door.

It should be noted that the accuracy of face recognition is not 100 %, how to further identify the visitors after they enter the room is still a problem. Besides, due to the influence of light, the stability of the system is still needed to explore and improve.

Acknowledgments Project supported partly by the National Natural Science Foundation (61374040), Scientific Innovation program (13ZZ115), Hujiang Foundation of China (C14002), Graduate Innovation program of Shanghai (54-13-302-102).

References

Ahonen T, Hadid A, Pietikäinen M (2006) Face description with local binary patterns: application to face recognition. IEEE Trans Pattern Anal Mach Intell 28(12):2037–2041. doi:10.1109/TPAMI.2006.244

Blanc G, Mezouar Y, Martinet P (2005) Indoor navigation of a wheeled mobile robot along visual routes. In: Proceedings of the 2005 IEEE international conference on robotics and automation, pp 3354–3359. doi:10.1109/ROBOT.2005.1570628

Cui Y (2002) Image processing and analysis: the method and application of mathematical morphology. Science Press, Beijing (in Chinese)

Hart PE, Nilsson NJ, Raphael B (1968) A formal basis for the heuristic determination of minimum cost paths. IEEE Trans Syst Sci Cybern 4(2):100–107. doi:10.1109/TSSC.1968.300136

Heikkilä M, Pietikäinen M (2006) A texture-based method for modeling the background and detecting moving ojects. IEEE Trans Pattern Anal Mach Intell 28(4):657–662. doi:10.1109/TPAMI.2006.68

Huo W (2005) Robots dynamics and control. Higher Education Press, Beijing (in Chinese)

Li JS, Li XH (2007) Digital image processing. Tsinghua University Press, Beijing (in Chinese)

Ojala T, Pietikäinen M, Harwood D (1996) A comparative study of texture measures with classification based on feature distributions. Pattern Recogn 29(1):51–59. doi:10.1016/0031-3203(95)00067-4

Ojala T, Pietikäinen M, Mäenpää T (2002) Multiresolution gray-scale and rotation invariant texture classification with local binary patterns. IEEE Trans Pattern Anal Mach Intell 24(7):971–987. doi:10.1109/TPAMI.2002.1017623

Philips PJ, Grother P, Micheals RJ et al (2003) Face recognition vendor test 2002 results. IEEE Int Workshop Anal Model Faces Gestures 44. doi:10.1109/AMFG.2003.1240822

Wu G (2010) The research of service robot vision-based object grasping and software development. A dissertation submitted to Southeast University for the academic degree of master of engineering (in Chinese)

Zhao G, Pietikäinen M (2007) Dynamic texture recognition using local binary patterns with an application to facial expressions. IEEE Trans Pattern Anal Mach Intell 29(6):915–928. doi:10.1109/TPAMI.2007.1110

Zhao W, Chellappa R, Rosenfeld A, Phillips PJ (2003) Face recognition: a literature survey. ACM Comput Surv 35(4):399–458. doi:10.1145/954339.954342

Zhu XC, Liu F, Hu D (2014) Digital image processing and image communication. Beijing University of Posts and Telecommunications Press, Beijing (in Chinese)

Control of Two-Wheel Self-balancing Robots Based on Gesture Recognition

Jie-han Liu, Lei Cao and Chao-li Wang

Abstract Currently, wearable devices have been paid a great attention in lots of practical fields. This paper discusses the design of two-wheel self-balancing robots based on gesture recognition. The objective is to control the robot's movement by using gestures. Static gesture recognition without wearing data gloves is exploited to control the motion of the robot. The gesture recognition involves the calculation the distance between the user gesture and template gesture through Hu geometric features. The remote terminal is the Two-wheel Self-balancing Robots. These are unstable and a class of special wheeled mobile robots which are multivariable, nonlinear, strong coupling dynamic system, but it has the characteristics of small, simple, flexible movement. This paper proposes a controller design of Two-wheel Self-balancing Robots, and the simulation results show the effectiveness of the proposed method, which is verified also on the DSP platform at the same time.

Keywords Static gesture recognition · Hu geometric feature moment · Two-wheeled self-balancing robot · Multi-variable · Nonlinear · DSP

1 Introduction of the Current Situation

In 1993, B. Thomas et al. proposed a method through the data glove or paste special colors on hand to identify the gestures by using the data glove or paste special colors on hand. With the development of computer technology, the research of

J. Liu (✉) · C. Wang (✉)
Control Science and Engineering, University of Shanghai for Science and Technology,
Shanghai 200093, China
e-mail: 13564374538@126.com

C. Wang
e-mail: clclwang@126.com

L. Cao
Electrical and Electronic Experiment Center,
Henan University of Science and Technology, Luoyang 471023, China
e-mail: Caoleis@163.com

© Zhejiang University Press and Springer Science+Business Media Singapore 2017   525
C. Yang et al. (eds.), *Wearable Sensors and Robots*, Lecture Notes in Electrical
Engineering 399, DOI 10.1007/978-981-10-2404-7_40

gesture recognition has largely focused on the recognition of bare hands by means of a camera.

About the two-wheel self-balancing robots, in 1985, professor Kazuo Yamafuji of University of Electro-Communications proposed an idea about one kind of standing machines, and did the application for the patent of "Parallel two-wheeled robot". With the development of the Two-wheel Self-balancing robots, in United States, Japan, Switzerland, France, have been investigating the research in this area. In 2001, American inventor Dean Kamen invented a new two-wheel vehicle "Segway." Currently, a lot of progress has been made for the Two-wheel Self-balancing robots in China University of Technology, National Taiwan University, Beijing Institute of Technology, Henan University of Science and Technology.

In 2004, the company WowWee in the United States launched a robot "MIP," which is controlled by phone. And the company Shanghai New Century Robot launched a robot "i-ROBOT," which is also controlled by phone. And in this paper, the static gesture is used to control the movement of the robot without using phone, which is more flexible and it could better integrate the experience of human beings.

About working in outdoor scenario, communication distance of this system can reach several kilometers by equipping with a powerful Bluetooth signal transmitter. And on other hand, due to use advanced control algorithm, the robot will be able to walk on all kinds of ground. So, this integration still works well in outdoor scenario.

2   Gesture Recognition

2.1   Gesture Recognition Overview

2.1.1   Input Media and Method

Currently, gesture recognition includes two main ways, the first one is based on data glove, and the other is based on computer vision (Bretzner et al. 2002).

Gesture recognition method comprising the following: template matching, neural networks, dynamic time warping, hidden Markov model. This paper studies static gesture recognition (Su et al. 1996).

2.1.2   Steps of Static Gesture Recognition

First, the user's hand image is collected, and the mean shift image segmentation algorithm is used to filter out noise after RGB color space being transformed into HSV color space. The threshold is determinated in order to obtain a binary image. The binary image requires to be corroded and expanded, which could remove noise and make the border more clearer and smoother. Then after the gestures profile is searched and the false contour is removed, the convex envelope is obtained. Finally, the outline of an image as Hu moment feature vectors is used to achieve gesture recognition by calculating distance between the user gesture and template gesture (Fig. 1).

2.2 Gesture Segmentation

2.2.1 HSV Color Extraction

RGB color space is required especially to be transformed into HSV color space (Zhang et al. 2010) after mean shift image segmentation algorithm is used to filter out noise.

About the cause of the transformation, it is that in computer vision, RGB is not stable, and the color subjects susceptible to strong light, low light, shadow, etc. But for HSV color space, the color is more stable, and it can well reflect the nature of color.

Figs. 2 and 3 show the RGB space and HSV color space.

Some typical equations are followed as:

$$V = \max(R, G, B), \tag{1}$$

$$S = \begin{cases} V - \min(R, G, B)/V, & \text{if } V \neq 0, \\ 0, & \text{otherwise,} \end{cases} \tag{2}$$

$$H = \begin{cases} 60(G - B)/T, & \text{if } V = R, \\ 120 + 60(B - R)/T, & \text{if } V = G, \\ 240 + 60(R - G)/T, & \text{if } V = B, \\ T = V - \min(R, G, B). \end{cases} \tag{3}$$

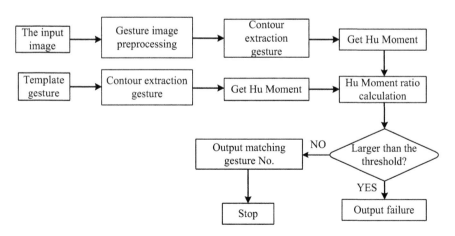

Fig. 1 Gesture recognition flowchart

Fig. 2 RGB color space

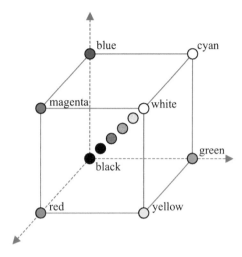

Fig. 3 HSV color space

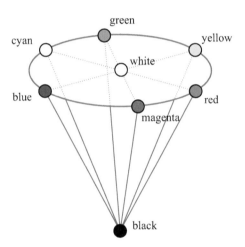

It should be noted that in the HSV color space, variable range is: $0 \leq V \leq 1$, $0 \leq S \leq 1$, $0 \leq H \leq 360$, if $H < 0$, and $H = H + 360$.

Figure 4a and b are the results about transforming RGB color space into HSV color space.

2.2.2 Corrode and Expand

After image binarization, corrode technology and expand technology are used to process the image. The first is used to eliminate the boundary points of the object, and the other means that all the background point of contact with the object will merge to the object, and increase the target and fill in the empty goal.

(a) (b)

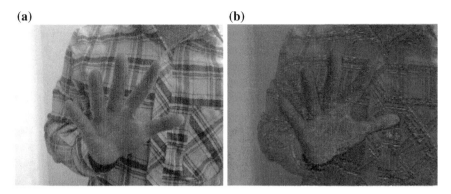

Fig. 4 The original image (**a**), the HSV image (**b**)

2.2.3 Find Profile

The various ways of contour extraction could be implemented in the code of Open Source Computer Vision Library. The paper analyzes one way of the various ways about how to find profile, and stipulate that traversal the image from top to bottom and left to right.

As shown in Fig. 5a, A is a first outer contour point, and has not been marked with a new tag number. Then start from point A, tracking the entire outer contour point of A, and back to A point, and all points on the path should be marked A's tag.

As shown in Fig. 5b, if you have already marked the outer contour point A^*, start from A^* to right, the point at the right should be marked a tag of A^* until black pixels meet.

As shown in Fig. 5c, if you have already marked point B, and B is a point of inner contour (in the vertical direction which under B, it is black pixel and B is not a point of the outer contour), start from B, track the inner contour, all points on the path should be marked B's tag until black pixels meet.

As shown in Fig. 5d, if you have already marked the inner contour point B^*, start from B^* to right, the point at the right should be marked a tag of B^* until black pixels meet.

Figure 6 and b are the results about drawing the contours of the hand.

2.3 Gesture Recognition

2.3.1 Hu Geometric Feature Moment

In 1962, Hu proposed geometric moments (Hu 1962), and proved that seven moments will not change in the case of image translation, rotation, and proportional change.

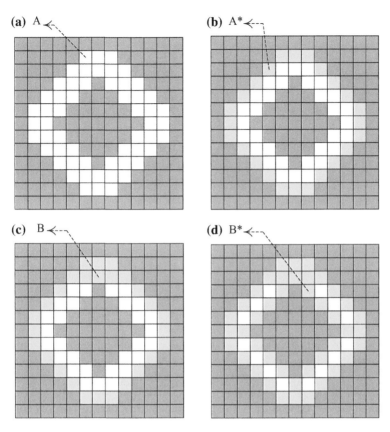

Fig. 5 First outer contour point (**a**), outer contour points marked (**b**), first inner contour point (**c**), inner contour points marked (**d**)

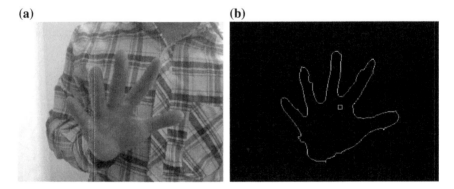

Fig. 6 The original image (**a**), the contour of hand (**b**)

For binary image $f(x, y)$, $p+q$ order geometric moment, it is defined as the following typical equations:

$$u_{pq} = \sum_x \sum_y x^p y^q f(x, y), \tag{4}$$

1-order moment (u_{01}, u_{10}), which is used to determine the centroid $(\bar{x}, \bar{y})$ of the image, the relationship between (u_{01}, u_{10}) and $(\bar{x}, \bar{y})$ is as follows:

$$\begin{cases} \bar{x} = u_{10}/u_{00}, \\ \bar{y} = u_{01}/u_{00}, \end{cases} \tag{5}$$

If you move $\bar{x}$ and $\bar{y}$ to the coordinate origin, then get unchanged central moment for image translation:

$$u_{pq} = \sum_x \sum_y (x - \bar{x})^p (y - \bar{y})^q f(x, y), \tag{6}$$

(x, y) represents a spatial position coordinates of the image pixels, $(\bar{x}, \bar{y})$ is the centroid of image.

Taking into account the actual effects and computational complexity, here option four moments as feature vectors, they are combined into a feature space $(\varphi_1, \varphi_2, \varphi_3, \varphi_4)$:

$$\varphi_1 = \eta_{20} + \eta_{02}, \tag{7}$$

$$\varphi_2 = (\eta_{20} - \eta_{02})^2 + 4\eta_{11}^2, \tag{8}$$

$$\varphi_3 = (\eta_{30} - 3\eta_{12})^2 + (\eta_{03} - 3\eta_{21})^2, \tag{9}$$

$$\varphi_4 = (\eta_{30} + \eta_{12})^2 + (\eta_{03} + \eta_{21})^2, \tag{10}$$

In order to maintain scale invariance of the geometric moments, normalized geometric moments:

$$\eta_{pq} = M_{pq}/u_{00}^{1/2(p+q+2)}. \tag{11}$$

2.3.2 Template Matching

Distance between the user gesture and template gesture of d_g:

$$d_g = \sum_{i=1}^{4} \alpha_i |\varphi_i - m_i|. \tag{12}$$

Among them, m_i is Hu moments characteristic component of the four users' gestures, φ_i is Hu moments characteristic component of template gesture, α_i is the weight of each characteristic component, in order to adjust the inconsistent number of each characteristic component, in practice, value of α_i should be 10^4, 10^{10}, 10^{15}, 10^{16}.

2.3.3 Recognition Results

The gesture recognition is implemented by using opencv library which runs on vs2010 platform.

Figure 7a–e defined the gesture control command of Two-wheel Self-balancing Robots.

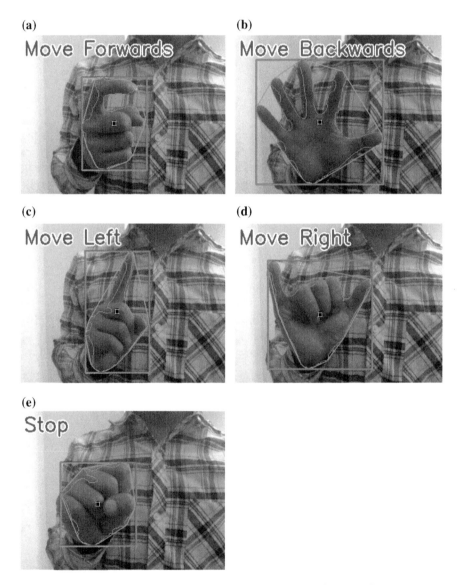

Fig. 7 Forward gesture (**a**), back gesture (**b**), turn left gesture (**c**), turn left gesture (**d**), stop gesture (**e**)

3 Control of Two-Wheel Self-balancing Robots

3.1 System of Two-Wheel Self-balancing Robots

3.1.1 The Physical Mode of Two-Wheel Self-balancing Robots

Figure 6 is a model of Two-wheel Self-balancing robots (Fig. 8).

Physical parameters of Two-wheel Self-balancing robots are the following:

(w_m, u_m, v_m)	The axis midpoint coordinates of the two wheels,
(w_r, u_r, v_r)	The coordinates of right wheel,
(w_l, u_l, v_l)	The coordinates of right wheel,
(w_b, u_b, v_b)	The coordinate of robot's center,
ε	Body pitch angle,
$\phi_{l,r}$	Wheel angle (l, r indicates left and right),
η	Body angle,
J_β	Wheel inertia moment,
M	Body weight,
L	Distance of the center of mass from the wheel axle,
J_α	Body pitch inertia moment,
$\phi_{m_{l,r}}$	DC motor angle,

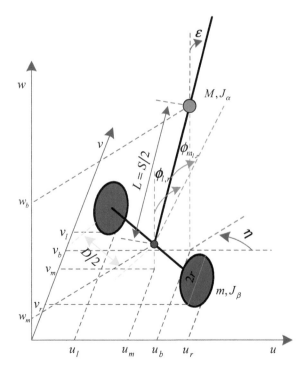

Fig. 8 The model of two-wheel self-balancing robots

In accordance with these parameters, the equations of motion of the robot could be established by Lagrangian theory, and state equations are formed after proporate choice of status variables.

3.1.2 Control System Block Diagram

Figure 7 is a system block diagram. The outputs are wheel angle, tilt angle, wheel angular velocity, and tilt angular velocity of Two-wheel Self-balancing robots (Fig. 9).

3.1.3 Output Results

Figs. 10 and 11 show the curve of tilt angle (deg) and tilt angular velocity (deg/s) of robot body.

And it can be seen from the picture that the two variables quickly reach equilibrium state with small overshoot, this shows that the system has good robustness.

3.2 Platform of Two-Wheel Self-balancing Robots

3.2.1 Angle Conversion

Angle sensor is adxl335, the conversion relationship is as follows:

$$acceler = acce\_Ad - 4096 * 1.623/3 - 3$$

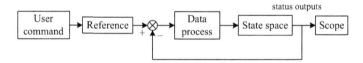

Fig. 9 Control system block diagram

Fig. 10 Deg of robot body

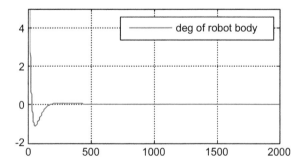

Fig. 11 Deg/s of robot body

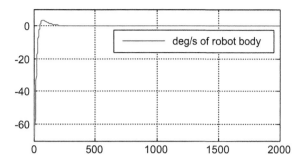

The data acceler is x-axis digital increment, and the x-axis voltage of adxl335 is 1.623 v when x-axis is horizontal, and digital value -3 as a compensation.

The next step is to obtain the sine value, on the x-axis direction as follows:

$$0.32 \sin(\text{theta}) = \text{acceler} * 3/4096$$

So sine value is as follows:

$$\sin(\text{theta}) = \text{acceler} * 0.0022888$$

Get the angle (in radians):

$$\text{theta} = a\sin(\text{acceler} * 0.0022888).$$

3.2.2 Angle Velocity Conversion

The angular velocity sensor is ewts82. Here ewts82 and adxl335 are used complementarily each other. The angular velocity could be obtained by using the digital increment of ewts82 after voltage divider circuits. The angular velocity digital increment is as follows:

$$\text{gyro} = \text{AdcAin1} - (4096 * 1.23)/3$$

The angular velocity value (in radians) is as follows:

$$\text{gyro}* = 0.05859375 * (3.1415926/180).$$

3.2.3 Kalman Filter

Getting the current actual angle by Kalman filter (Kwon et al. 2006).

3.2.4   Position and Speed Acquisition

Getting the position and speed by incremental optical encoder (2342 encoder).

3.2.5   Motor and Drive Plate

Using faulhaber2342L012 coreless gear motor, use lmd18200 driver board with optocoupler isolation which isolates interference of motor power, so control signal will output accurate PWM ware.

3.2.6   Hardware Platform

Using the Texas advanced DSP2000 type controller, which is built in enhanced PWM, CAP, SPI, CAN, ADC module, etc.

3.2.7   Bluetooth

The baud rate of Bluetooth host is set to 9600 bit/s and the Bluetooth is set to be slave module. And the communications between computer and Bluetooth host by using serial module, the communications between DSP and Bluetooth slave by using com port.

As is shown in the Fig. 12, it is a Bluetooth module which includes serial module.

3.2.8   Robot Platform

About robot platform, it includes many peripheral devices, like shaping circuit, driver board, motor, encoder, Bluetooth slave module, angle sensor, power module, battery.

Fig. 12 Bluetooth module

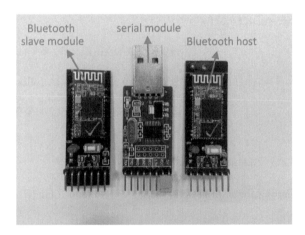

Fig. 13 Platform of
two-wheel self-balancing
robots

As is shown in the Fig. 13, it is a platform of two-wheel self-balancing robots.

4 System Debugging

Initialize the self-balancing robot gain parameters, and then repeatedly adjust the four parameters so that robots achieve stable control effect. If it is sensitive to changes in a number of angles, adjust parameter of the angle and angular velocity. If it is sensitive to movement speed, in addition to changing incremental as well as adjust the position and velocity gain.

Using Bluetooth communication between gesture recognition client and two-wheel self-balancing robots, get the best system performance from repeated experiments. Finally, gesture recognition rate is relatively high, balance performance of two-wheel self-balancing robots also very good, this achieves the desired effect.

Acknowledgment This paper is supported partly by The National Natural Science Foundation (61374040), Scientific Innovation program (13ZZ115), Hujiang Foundation of China (C14002), Graduate Innovation program of Shanghai (54-13-302-102).

References

Bretzner L, Laptev I, Lindeberg T (2002) Hand gesture recognition using multi-scale colour features, hierarchical models and particle filtering. In: Fifth IEEE international conference on automatic face and gesture recognition. Proceedings, pp 423–428. IEEE. doi:10.1109/AFGR. 2002.1004190

Hu MK (1962) Visual pattern recognition by moment invariants. IRE Trans Inf Theory 8(2):179–187. doi:10.1109/TIT.1962.1057692

Kwon S, Yang KW, Park S (2006) An effective Kalman filter localization method for mobile robots. In: 2006 IEEE/RSJ international conference on intelligent robots and systems, IEEE, pp 1524–1529. doi:10.1109/IROS.2006.281982

Su MC, Jean WF, Chang HT (1996) A static hand gesture recognition system using a composite neural network. In: Proceedings of the fifth ieee international conference on fuzzy systems, vol 2. IEEE, pp 786–792. doi:10.1109/FUZZY.1996.552280

Zhang GQ, Li ZM, Li XX et al (2010) Research on color image segmentation in HSV space. Comput Eng Appl 46(26):179–181. doi:10.3778/j.issn.1002-8331.2010.26.055

An Extended Kalman Filter-Based Robot Pose Estimation Approach with Vision and Odometry

Xue-bo Zhang, Cong-yuan Wang, Yong-chun Fang and Ke-xin Xing

Abstract Visual cameras and encoders are usually equipped on mobile robotic systems. In this paper, we present a robust extended Kalman filter-based pose estimation approach by fusing the information from both the onboard camera and encoders. Different from existing works, the system state is chosen in a new simplified way, including the robot pose and the depth of feature points. Moreover, a new observation model is formulated and the corresponding Jacobian matrix is derived. A robust feature association approach with an outlier removing mechanism is proposed. Experimental results are provided to demonstrate the effectiveness of the proposed approach.

Keywords Mobile robots · Pose estimation · Vision and odometry · Robust feature association

1 Introduction

For mobile robots, visual cameras and encoders gradually become indispensable onboard sensors, which are used for robot localization (Chen 2012; Zhang et al. 2014), visual servoing (Zhang et al. 2011, 2015), visual tracking (He et al. 2013), and so on. Odometry information is generally used for wheel motor control and dead reckoning, however, the accumulation error for robot localization is inevitable

X. Zhang (✉) · C. Wang · Y. Fang
Institute of Robotics and Automatic Information System (IRAIS), Nankai University, Tianjin 300071, China
e-mail: zhangxuebo@nankai.edu.cn

X. Zhang · C. Wang · Y. Fang
Tianjin Key Laboratory of Intelligent Robotics (TJKLIR), Nankai University, Tianjin 300071, China

K. Xing
College of Information Engineering, Zhejiang University of Technology, Hangzhou 310023, China

© Zhejiang University Press and Springer Science+Business Media Singapore 2017
C. Yang et al. (eds.), *Wearable Sensors and Robots*, Lecture Notes in Electrical Engineering 399, DOI 10.1007/978-981-10-2404-7_41

and it gradually increases as the robot moves. As a typical kind of exteroceptive sensors, visual cameras present several extraordinary features including rich information, low wight, small size, and acceptable power consumption. In addition, when observing the same scene, there is no accumulated pose error using the vision sensor. Hence, it is clear that encoders and visual cameras are mutually complementary sensors, especially for repetitive tasks within a small range of the workspace. Accordingly, how to fuse the information of encoders and vision, has attracted many researchers in the fields of robotics and computer vision.

In the literatures, many results have been reported for visual odometry (VO) (Naroditsky et al. 2012; Scaramuzza and Fraundorfer 2011), visual simultaneous localization and mapping (VSLAM) (Lategahn et al. 2011; Lui and Drummond 2015; McDonald et al. 2013), vision-aided inertial navigation (VAIN) (Hesch et al. 2014; Martinelli 2012; Panahandeh and Jansson 2014), Structure from Motion (SfM) (Spica et al. 2014). In these works, Kalman filter is usually utilized as an efficient tool for visual and inertial information fusion, however, most works employ full three-dimensional (3D) world coordinates of feature points to constitute the system state, which renders high-computational complexity and memory cost, especially, when the number of features is large.

In this paper, we propose a new robust extended Kalman filter-based information fusion approach for pose estimation of mobile robots with vision and odometry. The system state and the observation model are defined in a new and simplified way, which could decrease the complexity of the algorithm and alleviate the computational memory requirement for the pose estimation task. Experimental results are provided to show the effectiveness of the proposed approach.

The remainder of the paper is organized as follows. The overall framework of the proposed approach is described in Sect. 2. Section 3 describes the system model and Sect. 4 presents the robust EKF-based pose estimation approach. Experimental results are provided in Sect. 5. Section 6 gives some concluding remarks.

2 Overall Framework

The scheme of the proposed robust EKF-based pose estimation method is illustrated in Fig. 1. We start the EKF process with a number of zero feature point, and the initial robot pose is assumed to be the origin. Encoder data is used to predict the robot pose with its kinematic model. Concurrently, feature detection is performed with onboard camera to generate observation, pose prediction, and visible feature points are utilized to calculate observation prediction with the designed observation model. Then, robust feature association is executed between the prediction and observation. We use innovation between matching features to update the robot pose.

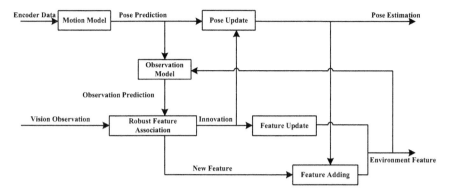

Fig. 1 Overall scheme

3 System Model

3.1 System State

One goal of this paper is to estimate the system state including the pose of mobile robot and the 3D information of feature points, by fusing the sensor data of odometer and vision. The robot pose is defined as follows:

$$X_r = \begin{bmatrix} x & y & \theta \end{bmatrix}^\mathrm{T}, \tag{1}$$

where x and y denote the robot position with respect to the reference coordinate frame, θ denotes the robot orientation respect to the x axis of the reference coordinate frame. As shown in Fig. 2, $\mathcal{F}^*$ denotes the reference robot frame, $\mathcal{F}$ denotes the current robot frame, the orientation direction as shown in Fig. 2 is positive.

Fig. 2 Frame relationship and feature depths

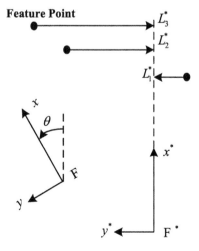

Since two-dimensional (2D) information of the feature points could be obtained from the images, this paper only use the depth information of feature points as a part of the system state:

$$L = \begin{bmatrix} L_1^* L_2^* \cdots L_N^* \end{bmatrix}^{\mathrm{T}}, \tag{2}$$

where N denotes the number of effective feature points. As shown in Fig. 2, L_i^* denotes the depth information of feature points respect to reference coordinate frame $\mathcal{F}^*$, which is the coordinate along the camera optical axis.

Remark 1 The depth of feature points, other than three-dimensional (3D) coordinates of features, are defined as the second part of the system state. This new way of state representation decreases the computational complexity and memory requirement for the pose estimation algorithm.

3.2   Plant Model

The plant model of the proposed robust EKF contains two parts: the robot kinematics model and the feature process model.

The robot kinematics model is modeled with conventional discrete time process as

$$X_r(k+1) = f[k, X(k), u(k), V(k)] \tag{3}$$

where $u(k)$ is the control input from step k to step $k+1$, $V(k)$ is the noise vector which obeys the following distribution:

$$V(k) \sim N(0, Q(k)), \tag{4}$$

where $Q(k)$ is the covariance matrix for the process noise.

The control signal is obtained using the odometer measurements as follows:

$$u(k) = \begin{bmatrix} \Delta x_k & \Delta y_k & \Delta \theta_k \end{bmatrix}^{\mathrm{T}}, \tag{5}$$

where $\begin{bmatrix} \Delta x_k & \Delta y_k \end{bmatrix}^{\mathrm{T}}$, $\Delta \theta_k$ are the translation and rotation changes of robot coordinates between step k and step $k+1$, respectively.

With odometer measurements, the kinematics model for a mobile robot can be deduced as:

$$X_r(k+1) = \begin{bmatrix} x(k) + \Delta x_k \cos \theta(k) - \Delta y_k \sin \theta(k) \\ y(k) + \Delta x_k \sin \theta(k) + \Delta y_k \cos \theta(k) \\ \theta(k) + \Delta \theta_k \end{bmatrix} \tag{6}$$

In the feature-based robust EKF, the depth of the feature points is invariant, and hence, the feature process model can be represented as,

$$L_i^*(k+1) = L_i^*(k), \tag{7}$$

where $i = 1, 2, \ldots, N$ is the feature point index.

3.3 New Observation Model

The camera is mounted on the robot, and the observation includes the robot orientation and the depth of the feature points, which is extracted using two-view geometry approach (Zhang et al. 2014) and encoder data. The newly proposed observation model is of the following form:

$$Z(k) = h(X_r(k), L(k)) + w(k), \tag{8}$$

where $L(k) = \begin{bmatrix} L_1^* & L_2^* & \cdots & L_{N_k}^* \end{bmatrix}^{\mathrm{T}}$ is the depth of feature points, $X_r(k)$ is the robot pose at step k, $Z(k)$ is the observation, and $w(k)$ is the white measurement noise with covariance $Ew(k)w(k)^{\mathrm{T}} = R$.

In this work, the observation component $z_i(k)$ of the feature point with depth $L_i^*(k)$ at step k is defined as,

$$z_i(k) = \left[\frac{x(k)}{L_i^*(k)}, \frac{y(k)}{L_i^*(k)} \right]^{\mathrm{T}} \tag{9}$$

Therefore, the observation model is defined as:

$$\begin{aligned} Z(k) &= \left[\theta(k) z_1(k)^{\mathrm{T}} z_2(k)^{\mathrm{T}} \cdots z_{N_k}(k)^{\mathrm{T}} \right]^{\mathrm{T}} \\ &= \left[\theta(k)s \frac{x(k)}{L_1^*(k)} \frac{y(k)}{L_1^*(k)} \frac{x(k)}{L_2^*(k)} \frac{y(k)}{L_2^*(k)} \cdots \frac{x(k)}{L_{N_k}^*(k)} \frac{y(k)}{L_{N_k}^*(k)} \right]^{\mathrm{T}}, \end{aligned} \tag{10}$$

where $\begin{bmatrix} x(k) & y(k) \end{bmatrix}^{\mathrm{T}}$, $\theta(k)$ are the translation and rotation of the mobile robot at step k.

4   Robust EKF-Based Pose Estimation

In this paper, SIFT algorithm (Lowe 2004) is used for feature extraction and matching, which could be implemented with graphics processing unit (GPU) to achieve real-time performance. The task of the proposed feature-based EKF method is to estimate the system state and the covariance matrix:

$$X = \begin{bmatrix} X_r^T & L_1^* & L_2^* & \cdots & L_N^* \end{bmatrix}^T \tag{11}$$

$$P = \begin{bmatrix} P_{RR} & P_{RL} \\ P_{RL}^T & P_{LL} \end{bmatrix}, \tag{12}$$

where P_{RR} is the robot pose covariance, P_{RL} is the covariance between the robot pose and feature depth, and P_{LL} is the feature depth covariance.

4.1   State Initialization

At the beginning, we assume that the initial robot location is the origin of the reference coordinate frame, and no observation is available for feature point detection. So the initial EKF state only contains the robot pose information as

$$X(k = 0) = \begin{bmatrix} X_r^T(k = 0) \end{bmatrix}^T = \begin{bmatrix} 0 & 0 & 0 \end{bmatrix}^T \tag{13}$$

and the initial covariance matrix is

$$\mathbf{P}(k = 0) = [\mathbf{P}_{RR}(k = 0)] = \begin{bmatrix} 10 & 0 & 0 \\ 0 & 10 & 0 \\ 0 & 0 & 1 \end{bmatrix} \tag{14}$$

4.2   State Prediction

Based on the kinematics model (6), we obtain the robot pose prediction. In addition, the locations of static feature points with respect to the reference coordinate frame are invariant. Hence, the state prediction is conducted as

$$\hat{X}(k+1|k) = \begin{bmatrix} \hat{x}(k|k) + \Delta x_k \cos \hat{\theta}(k|k) - \Delta y_k \sin\hat{\theta}(k|k) \\ \hat{y}(k|k) + \Delta x_k \sin \hat{\theta}(k|k) + \Delta y_k \cos \hat{\theta}(k|k) \\ \hat{\theta}(k|k) + \Delta \theta_k \\ L_1^*(k|k) \\ \vdots \\ L_N^*(k|k) \end{bmatrix} \tag{15}$$

and the covariance of the predicted state is

$$P(k+1|k) = f_x(k)P(k|k)f_x^{\mathrm{T}}(k) + G(k)Q(k)G^{\mathrm{T}}(k) \tag{16}$$

Where $f_x(k)$ and $G(k)$ are the Jacobians of the kinematics model respect to $X_r(k)$ and $u(k)$, respectively, defined as

$$f_x(k) = \begin{bmatrix} \nabla \mathbf{F}_r(k) & 0 & \cdots & 0 \\ 0 & \mathbf{I} & & \vdots \\ \vdots & & \ddots & \\ 0 & & \cdots & \mathbf{I} \end{bmatrix} \tag{17}$$

$$G(k) = [\nabla \mathbf{F}_u(k) \quad 0 \quad \cdots \quad 0]^{\mathrm{T}} \tag{18}$$

$Q(k)$ is the covariance matrix for the process noise.

4.3 Visual Observation Prediction

Based on the observation model (10), the observation prediction is

$$\hat{Z}(k+1|k) = \left[\hat{\theta}(k+1|k) \, \frac{\hat{x}(k+1|k)}{\hat{L}_1^*(k+1|k)} \, \frac{\hat{y}(k+1|k)}{\hat{L}_1^*(k+1|k)} \cdots \frac{\hat{x}(k+1|k)}{\hat{L}_{N_k}^*(k+1|k)} \, \frac{\hat{y}(k+1|k)}{\hat{L}_{N_k}^*(k+1|k)} \right]^{\mathrm{T}} \tag{19}$$

and the covariance of the predicted observation is

$$S(k+1) = h_x(k+1)P(k+1|k)h_x^{\mathrm{T}}(k+1) + R(k+1), \tag{20}$$

where $h_x(k+1)$ is the Jacobian of the observation model, defined as

$$h_x(k+1) = [\nabla \mathbf{H}_r(k) \quad \nabla \mathbf{H}_l(k)]. \tag{21}$$

with

$$\nabla \mathbf{H}_r(k) = \frac{\partial \mathbf{Z}}{\partial \mathbf{X}_r} |\mathbf{X}_r(k+1|k),$$

$$\nabla \mathbf{H}_l(k) = \frac{\partial \mathbf{Z}}{\partial \mathbf{L}} |\mathbf{L}(k+1|k).$$

$R(k+1)$ is the covariance of the measurement noise.

4.4   Outlier Removing Mechanism

Though the performance of the SIFT algorithm is good, some outliers still appear in practical implementations, which may result in undesirable effects for the EKF-based pose estimation.

Based on the state prediction $\hat{x}(k|k+1), \hat{y}(k|k+1), \hat{\theta}(k|k+1), L_i^*(k|k+1)$, we can compute the scaled translation vector as $\left[\frac{\hat{x}(k+1|k)}{L_i^*(k+1|k)} \ \frac{\hat{y}(k+1|k)}{L_i^*(k+1|k)} \right]^{\mathrm{T}}$. This translation can be expressed in the current camera coordinate system as $\left[\frac{\hat{t}_x(k+1|k)}{L_i^*(k+1|k)}, \frac{\hat{t}_z(k+1|k)}{L_i^*(k+1|k)} \right]^{\mathrm{T}}$. Thus, the estimated image coordinate of the matched feature could be computed through reprojection as follows:

$$\bar{x}_i = \frac{x_i^* \cos \hat{\theta}(k|k+1) + \sin \hat{\theta}(k|k+1) + \frac{\hat{t}_x(k+1|k)}{L_i^*(k+1|k)}}{-x_i^* \sin \hat{\theta}(k|k+1) + \cos \hat{\theta}(k|k+1) + \frac{\hat{t}_z(k+1|k)}{L_i^*(k+1|k)}}$$

$$\bar{y}_i = \frac{y_i^*}{-x_i^* \sin \hat{\theta}(k|k+1) + \cos \hat{\theta}(k|k+1) + \frac{\hat{t}_z(k+1|k)}{L_i^*(k+1|k)}}$$

Given a prescribed threshold ε, the matched feature point is classified as an outlier if the following condition is satisfied:

$$\sqrt{(x_i - \bar{x}_i)^2 + (y_i - \bar{y}_i)^2} > \varepsilon$$

4.5   Feature Association

At step k, we assume that the system state includes N_x points' depth information, and at step $k+1$, we assume that the new vision observation includes N_z points' depth information. Then using the indices of these points, we can judge whether it is an observation of a new feature or a revisited one.

As shown in Fig. 3, the $L_i^*(k)$ in system state $X(k)$ is associated with the $L_j^*(k+1)$ in the observation vector $Z(k+1)$ by the points' indices.

In Fig. 3, the red features are associated ones. We assume that the number of the associated feature points is N_c, then for these associated points, we extract the corresponding system state prediction and observation prediction as stated in Sects. 4.2 and 4.3. So far, the dimensions of these variables becomes $\hat{X}(k+1|k) \in \mathbb{R}^{(N_c+3)\times 1}$, $f_x(k) \in \mathbb{R}^{(N_c+3)\times(N_c+3)}$, $\hat{Z}(k+1|k) \in \mathbb{R}^{(2N_c+1)\times 1}$, $h_x(k+1) \in \mathbb{R}^{(2N_c+1)\times(N_c+3)}$, and $G(k) \in \mathbb{R}^{(N_c+3)\times 3}$.

4.6 System State Update for Associated Features

Based on the mutual covariance between the system prediction error and the observation prediction error, we obtain the gain of the filter as follows:

$$K(k+1) = P(k+1|k)h_x^{\mathrm{T}}(k+1)S^{-1}(k+1) \tag{22}$$

Then the state update equation is

$$\hat{X}(k+1|k+1) = \hat{X}(k+1|k) + K(k+1)v(k+1) \tag{23}$$

where $v(k+1)$ is the innovation between the visual observation prediction and the measurement, which is calculated as

$$v(k+1) = Z(k+1) - \hat{Z}(k+1|k) \tag{24}$$

and the corresponding covariance is updated as

$$P(k+1|k+1) = P(k+1|k) - K(k+1)S(k+1)K^{\mathrm{T}}(k+1) \tag{25}$$

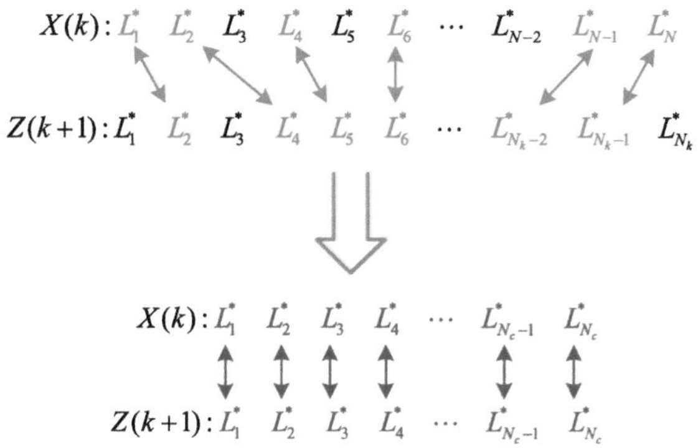

Fig. 3 Feature association process

4.7 Feature Adding

For a new feature point with index j, based on the observation equation, we obtain:

$$Z(k+1)_{2*j} = \frac{\hat{x}(k+1|k+1)}{\hat{L}_j^*(k+1|k+1)} \tag{26}$$

$$Z(k+1)_{2*j+1} = \frac{\hat{y}(k+1|k+1)}{\hat{L}_j^*(k+1|k+1)} \tag{27}$$

From (26) and (27), we compute the depth of the feature point j along the optical axis direction as follows:

$$\hat{L}_j^*(k+1|k+1) = \sqrt{\frac{(\hat{x}(k+1|k+1))^2 + (\hat{y}(k+1|k+1))^2}{(Z(k+1)_{2*j})^2 + (Z(k+1)_{2*j+1})^2}} \tag{28}$$

Then, $L_j^*(k+1)$, $\hat{L}_j^*(k+1|k+1)$ will be updated into the system state according to the indices of the features as the initial value of the new added feature.

4.8 System State Extension

The N_c feature points associated in Sect. 4.5 can be updated in Sect. 4.6; the $N_x - N_c$ unassociated feature points in system state $X(k)$ will keep the original value of the same; the $N_z - N_c$ unassociated feature points in observation vector can be computed in Sect. 4.7.

By stacking together these three kinds of feature points according to the order of their indices, the new system state is formed as follows:

$$X = \begin{bmatrix} X_r^T & L_1^* & L_2^* & \cdots & L_{N_x+N_z-N_c}^* \end{bmatrix}^T \tag{29}$$

and the covariance of the predicted state is

$$P = \begin{bmatrix} P_{RR} & P_{RL_1^*} & \cdots & P_{RL_{N_x+N_z-N_c}^*} \\ P_{L_1^*R} & P_{L_1^*L_1^*} & \cdots & P_{L_1^*L_{N_x+N_z-N_c}^*} \\ \vdots & \vdots & \ddots & \vdots \\ P_{L_{N_x+N_z-N_c}^*R} & P_{L_{N_x+N_z-N_c}^*L_1^*} & \cdots & P_{L_{N_x+N_z-N_c}^*L_{N_x+N_z-N_c}^*} \end{bmatrix}. \tag{30}$$

Fig. 4 Ideal reference trajectory for open-loop control

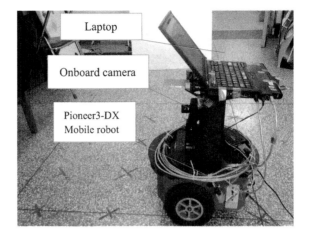

5 Experiment

To verify the proposed robust EKF-based pose estimation approach, it is implemented on Pioneer-3DX robot equipped with an onboard camera, as shown in Fig. 4. Figure 4 shows an ideal reference trajectory, which is set for the mobile robot to generate open-loop control velocities. Such open-loop control is conducted back and forth for ten times, which results in that the encoder accumulation error is large (Fig. 5).

Figure 5 demonstrates comparative localization results, wherein the blue line is the ground truth obtained by the ceiling camera, the green line is the experiment result obtained by the proposed robust EKF-based approach, and the red line is the dead reckoning results using encoder data. It is shown that the proposed approach avoids the pose drift, and it is much better than the traditional encoder-based dead reckoning result.

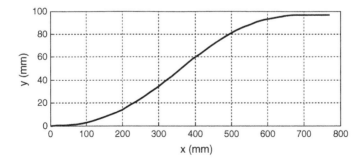

Fig. 5 Ideal reference trajectory for open-loop control

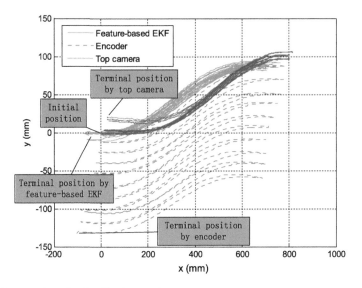

Fig. 6 Comparative robot localization results

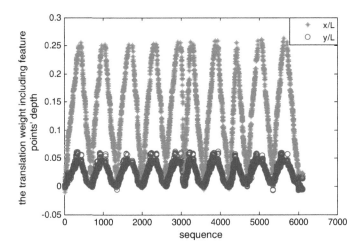

Fig. 7 Estimated scaled translation

The estimated scaled translation of the mobile robot is shown in Fig. 7, which exhibits periodical behavior as the real motion. Figure 8 shows the evolution process for the depth information of the features, which is gradually stable in the presence of image and motion noises.

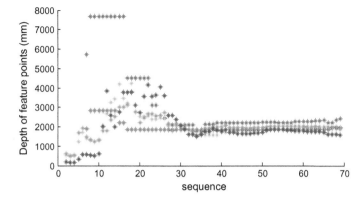

Fig. 8 Evolution of the depth information

6 Conclusion

In this paper, we propose a robust EKF-based pose estimation approach for mobile robots, wherein monocular vision and odometry information are fused to achieve better localization results. The system state model and observation model are newly proposed to decrease the computational complexity and memory requirement. An outlier removing mechanism is proposed to enhance the robustness of the proposed approach. Experimental results are provided to validate the proposed approach (Figs. 6, 7 and 8).

Acknowledgments Project supported in part by the National Natural Science Foundation of China (No. 61203333 and No. 61202203), in part by Specialized Research Fund for the Doctoral Program of Higher Education of China (No. 20120031120040).

References

Chen S (2012) Kalman filter for robot vision: a survey. IEEE Trans Ind Electron 59(11):4409–4420

He W, Fang Y, Zhang X (2013) Prediction-based interception control strategy design with a specified approach angle constraint for wheeled service robots. In: Proceedings of 2013 IEEE/RSJ international conference on intelligent robots and systems (IROS). Tokyo, Japan, pp 2725–2730

Hesch JA, Bowman DG, Kottasand SL, Roumeliotis SI (2014) Consistency analysis and improvement of vision-aided inertial navigation. IEEE Trans Rob 30(1):158–176

Lategahn H, Geiger A, Kitt B (2011) Visual SLAM for autonomous ground vehicles. In: Proceedings of IEEE international conference on robotics and automation. Shanghai, China, pp 1732–1737

Lowe DG (2004) Distinctive image features from scale-invariant keypoints. Int J Comput Vision 60(2):91–110

Lui V, Drummond T (2015) Image based optimization without global consistency for constant time monocular visual SLAM. In: Proceedings of 2015 IEEE international conference on robotics and automation (ICRA). Seattle, Washington, pp 5799–5806

Martinelli A (2012) Vision and IMU data fusion: closed-form solutions for attitude, speed, absolute scale, and bias determination. IEEE Trans Rob 28(1):44–60

McDonald J, Kaess M, Cadena C, Neira J, Leonard JJ (2013) Real-time 6-DOF multi-session visual SLAM over large-scale environments. Robot Auton Syst 61(10):1144–1158

Naroditsky O, Zhou XS, Gallier J, Roumeliotis SI, Daniilidis K (2012) Two efficient solutions for visual odometry using directional correspondence. IEEE Trans Pattern Anal Mach Intell 34 (4):818–824

Panahandeh G, Jansson M (2014) Vision-aided inertial navigation based on ground plane feature detection. IEEE/ASME Trans Mechatron 19(4):1206–1215

Scaramuzza D, Fraundorfer F (2011) Visual odometry, part I: the first 30 years and fundamentals. IEEE Robot Autom Mag 18(4):80–92

Spica R, Giordano PR, Chaumette F (2014) Active structure from motion: application to point, sphere, and cylinder. IEEE Trans Rob 30(6):1499–1513

Zhang X, Fang Y, Liu X (2011) Motion-estimation-based visual servoing of nonholonomic mobile robots. IEEE Trans Rob 27(6):1167–1175

Zhang X, Fang Y, Sun N (2015) Visual servoing of mobile robots for posture stabilization: from theory to experiments. Int J Robust Nonlinear Control 25(1):1–15

Zhang X, Wang C, Fang Y, Xing K (2014) Planar motion estimation from three-dimensional scenes. In: Proceedings of the 2014 IROS workshop on visual control of mobile robots, pp 21–26

Visual Servoing of a New Designed Inspection Robot for Autonomous Transmission Line Grasping

Tao He, He-sheng Wang, Wei-dong Chen and Wei-jie Wang

Abstract This paper presents a new visual servoing using two lines features and this method is adopted for a novel dual-arm power transmission line inspection robot to grasp transmission line autonomously. The new designed inspection robot has 5 DOFs ofeach arm, and the arms for the wheel-arm platform will cross obstacle on the transmission lines alternately. With a new mechanism, the robot can actively control the position of its center of mass, thus enhancing its stability and reducing the load on motors when changing its configuration to overcome obstacles. Then an adaptive visual servoing using line features of two power transmission lines for the robot in an uncalibrated eye-in-hand setup is developed, guaranteeing the off-line arm can grasp the line automatically after overcoming obstacles. A new Lyapunov method is used to prove the asymptotic convergence of the errors to zero. The paper ends with the presentation of several simulation results.

Keywords Inspection robot · Power transmission line inspection · Autonomous line grasping · Visual servoing

T. He · H. Wang (✉) · W. Chen · W. Wang
Department of Automation, Shanghai Jiao Tong University,
Shanghai 200240, China
e-mail: wanghesheng@sjtu.edu.cn

T. He
e-mail: sjthotel@sjtu.edu.cn

W. Chen
e-mail: wdchen@sjtu.edu.cn

W. Wang
e-mail: weijwang@sjtu.edu.cn

T. He · H. Wang · W. Chen · W. Wang
Key Laboratory of System Control and Information Processing,
Ministry of Education of China, Shanghai 200240, China

1 Introduction

The power transmission lines are one of the most important strategic assets, transmitting electric energy to the urban centers and industries. If there is any problem in the transmission line networks, millions of people will be affected and considerable economic losses will be caused to the nation. In order to avoid this, it is vital important to maintain the safety and stability of the power transmission lines. The inspection of power transmission lines is usually difficult and dangerous, especially in mountain areas, because the lines are always overhead with dozens of meters above the ground. As a result, robotic systems are introduced in this filed and the interest has increased remarkably over the past few years.

In the recent 20 years, several research teams around the world have created some power transmission line inspection robots. LineScout proposed by Serge et al. (Montambault and Pouliot 2006; Pouliot and Montambault 2009) in Hydro-Québec research institute has completed actual field testing on live lines, but it cannot inspect lines with large turning angle and it is tele-operated only. Paulo et al. (Tamura and Kimura 2010; Fonseca et al. 2012) proposed a simple and dexterous robot named Expliner, but it is also tele-operated and unstable when overcoming obstacles. Wang and Wu et al. (Wang et al. 2014) proposed a dual arm robot which can control the position of its center of mass and can grasp line autonomously using visual servo, but its structure is too complex and it needs three steps to finish the motion of grasping with additional sensors.

The grasping control of the off-line arm is the key to autonomous obstacle-crossing. Robots represented in Montambault and Pouliot (2006), Pouliot and Montambault (2009), Tamura and Kimura (2010), Fonseca et al. (2012) are all tele-operated, and the motion of grasping the line takes too much time which may not succeed just through one attempt. Robots represented in Han and Lee (2008), Chen et al. (2006) grasp lines based on proximity transducers, electromagnetic sensors. But the partial signals are liable disturbed by external interference, thus affecting the stability and reliability of the system.

Recently, several researchers (Slotine and Li 1987; Wang et al. 2007; Zhang et al. 2007; Wang and Sun 2011) have attempted to solve this problem by using visual servoing. Wang and Wu et al. (Wang et al. 2014) proposed a dual arm robot which can grasp line autonomously using visual servo, but it needs three steps to finish the motion of grasping with additional sensors and it cannot deal with problems when the power line has large turning angle behind the obstacle. Wang and Liu et al. (Liu et al. 2005, 2006; Wang et al. 2008a, b) make their effort on image-based visual servoing of robots with an uncalibrated eye-in-hand camera using line features. Performance of this system is verified through experiments. But this method is based on dynamics.

This paper presents an adaptive visual servoing using two lines features, which is useful for autonomous transmission line grasping of a new type of robot for the power transmission line inspection. And the first stage in the development of the new platform guarantees that the robot can change its configuration to overcome obstacles on the transmission power lines. The remainder of this paper is organized as follows. Section 2 covers the mechanical design of the new robot, as well as the steps for overcoming obstacles on the line. Section 3 is devoted to a new adaptive controller for dynamic image-based visual servoing of the power transmission line inspection robot using line features. In Sect. 4 we report on the simulations results to validate the performance of the proposed control method. Section 5 concludes by discussing future developments.

In this paper, there are a few points that should be noted about the notations, as follows:

- Matrices are represented bold capital letters (e.g., M).
- Vectors are written as bold lower case letters (e.g., $u1$).
- A matrix, or vector accompanied with a bracket (t) denotes that its value varies with time.
- The scalar product of a by b is denoted by direct product $a^T b$ and the cross-product of a by b is written as $a \times b$.

2 Design of the New Inspection Robot

2.1 Structure of the Robot

This new power transmission line inspection robot is designed as a wheel-arm vehicle with two antisymmetrical arms to hold the power transmission line, as shown in Fig. 1.

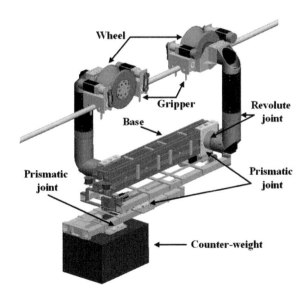

Fig. 1 Overall view of the robot

Both of the arms have five DOFs to fulfill rotating, lifting and lowering, moving on the line with a wheel as well as locking to the line with two pairs of safety grippers mounted on both sides of the wheel. Both arms can slide relative to the platform's base synchronously driven by one motor. To keep the safety and stability of the robot when overcoming the obstacles, the position of its center of mass should be controlled actively. A mass center adjustment mechanism is connected to the base, which has two prismatic joints in parallel. The counter-weight, including the battery and electronics cabinet, changes its position along the center adjustment mechanism that will also reduce the load on motors when overcoming the obstacles.

Compared with existing power transmission line inspection robots, this robot is simple and light. Flexibility and stability are taken into account to deal with the special work on power transmission lines.

2.2 *Extraordinary Gestures for the Suspension Clamps*

Under normal condition without any obstacles or with those under the lines, the new power transmission line inspection robot would move on the lines steadily through rotating the wheels. But this way is not applicable for obstacles above the lines, such as the suspension clamps.

The structure and degree of the robot would be enough for overcoming obstacles above the lines with extraordinary gestures. The robot would handle this situation by arms overcoming the obstacle alternatively. Each arm would be lifted once, rotated outside, lowered and then brought back down in contact with the line after overcoming the suspension clamp. During the process, the other arm would clamp the line to maintain stability. Based on prior information of the power transmission lines and the obstacles, we would have the rough position of the line after the obstacle, thus making it possible to generate a suitably path to the appropriate position of the off-line arm. The off-line arm would execute the path to avoid the obstacle on the line in open-loop.

Half of the sequence of the gestures is showed in Fig. 2. During Fig. 2a, Robot stops next to the suspension clamp and lock the arm I (right) to the line. Then arm II (left) is lifted by rotating the lower revolute joint of each arm in Fig. 2b. After lifting, arm II (left) is rotated outside by rotating the upper revolute joint of arm II (left) in Fig. 2c. During Fig. 2d, both arms slide relative to the platform's base synchronously and arm II (right) is lifted after overcoming the obstacle by rotating the lower revolute joint of each arm. Then arm II (right) is rotated inside by rotating the upper revolute joint of arm II (right) in Fig. 2e. After that, arm II (right) is brought back down in contact with the line by rotating the lower revolute joint of each arm in Fig. 2f. When moving and overcoming obstacles on the line, the counter-weight is controlled actively to balance the position of the center of the mass, such enhancing the robot's stability and reducing the load on motors.

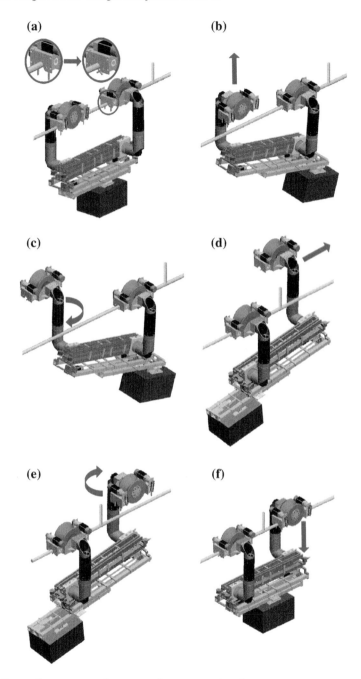

Fig. 2 Extraordinary gestures for overcoming a suspension clamp

3   Visual Servoing with Line Features

When the robot is overcoming obstacles that need extraordinary gestures in Fig. 2, arm I (right in Fig. 2a) is locked to the line and arm II (left in Fig. 2a) overcomes the obstacle first. In order to ensure the off-line arm can be brought down in contact with the line reliably, the wheel of the off-line arm shall be aligned with the line in the process.

Method provided by Wang and Liu et al. (Liu et al. 2005, 2006, 2013; Wang et al. 2007, 2008a, b) does not apply to the robot proposed in this paper, because the position of counter-weight is changed dynamically and decoupled from the visual servoing. The load on motors will be reduced by reason of the actively control of the counter-weight, so the nonlinear forces in robot dynamics can be neglected reasonably in the sequence of visual servoing. Moreover, the prismatic joint which controls the slide of two arms relative to the platform's base synchronously will not move during the sequence of grasping the line, so the robot can be simplified as a 4-DOFs manipulator without any prismatic joints, as shown in Fig. 3.

Visual servoing in this paper differs from previous works in the following aspects. First, this paper copes with visual servoing with kinematics took into account only, but the method proposed by Wang and Liu et al. (Liu et al. 2005, 2006; Wang et al. 2008a, b) addressed problems using dynamics. Second, this paper adopts a depth-independent interaction matrix (Liu et al. 2006) mapping the image errors onto the joint space and it does not depend on the depths of the line, which make it possible to grasp the line automatically when the power line has large turning angle behind the obstacle. But the model proposed by Wang and Wu et al. (Wang et al. 2014) assumes the power lines in front of the obstacle and behind the obstacle are collinear, thus it can compute the depth of the line.

Fig. 3 Module for the robot

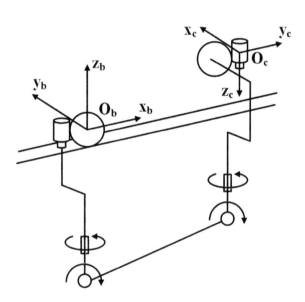

3.1   Eye-in-Hand-Vision Model

The eye-in-hand cameras are mounted on the top of both arms next to outside pair of grippers, as shown in Fig. 3. In addition, the cameras are pin-hole cameras and mounted downward to take the image to avoid the detrimental effects of sunlight, dust, and water drops. The coordinate system $\{O_b\}$ therein is for reference, the origin of the system is the center of the wheel of arm I (left in Fig. 3). The coordinate system $\{O_c\}$ is the camera coordinate frame.

In this paper, we use the representation of image lines proposed in Andreff et al. (2002). Figure 4 presents the projections of two power transmission lines. $\{O_c\}$ is the camera coordinate frame and $\{O_i\}$ is the retina coordinate frame. O_c is a point shifted by 1 from O_i along the negative z of camera coordinate frame. Let coordinate system $\{O_g\}$ be parallel to the camera coordinate frame $\{O_c\}$. Two 3-D power lines, L1 and L2, are projected onto the image plane, while $l1$ and $l2$ are their corresponding projections on the image plane. In this part, we will use only L1 and l1 for the formula derivation. P1 and P1$'$ are two arbitrary points on L1, while p1 and p1$'$ are their corresponding projections on the image plane.

$u1$ is the unit vector of the line $L1$ and $^iu1$ is its projection on the image plane which is the unit vector of $l1$. Vector $n1$ is a unit normal of the plane which is determined by vector $^iu1$ and point O_g. Refer to (Wang et al. 2008a), $n1$ can be obtained from (2) to (3). M is the perspective projection matrix and it depends on the intrinsic and extrinsic parameters only (Chaumette and Hutchinson 2006, 2007), which does not need to know necessarily. M can be represented as $M = (\Omega \quad \chi)$, where Ω is the left 3×3 submatrix of M and χ is the 4th column vector of M.

Fig. 4 Projections of two power transmission lines

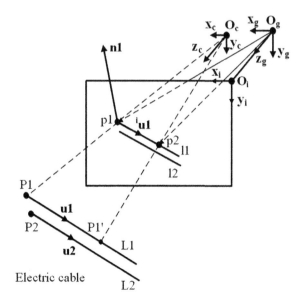

$T_{\mathrm{e}}^{-1}(t)$ is the homogeneous transform matrix from the base frame to the center of the wheel of arm II (right in Fig. 3) in Fig. 4, while $R(t)$ and $\xi(t)$ are rotation and translation matrix.

$$T_{\mathrm{e}}^{-1}(t) = \begin{pmatrix} R(t) & \xi(t) \\ 0 & 1 \end{pmatrix} \tag{1}$$

$$b(t) = \left\{ \Omega\left(R(t)^b X_{\mathrm{p}1} + \xi(t) + \chi\right) \right\} \times \Omega R(t)^b u \tag{2}$$

$$n(t) = \frac{b(t)}{||b(t)||} \tag{3}$$

Refer to Wang et al. (2008a), we have the following Eqs. (12) and (13), where $F(t)$ is written in (14).

In (12) and (13), $J(q)$ is the Jacobian matrix of q, which is the 4×1 angle vector of the revolute joints of the two arms. $L(n(\mathrm{t}), q(\mathrm{t}), \theta_l)$ is the depth-independent interaction matrix and its dimension is 3×4, while θ_l represents the unknown parameters which are the products of the components of M and lines features. In fact, the dimension of θ_l is 108×1 when one line features are took into account (refer to Wang et al. 2008a for detailed proof), or 216×1 when two lines features are included.

3.2 Controller Design

Different from the controller mentioned in Wang et al. (2008a), we take kinematics and features of two lines into account. The features used here is the unit normal vector $n(t)$. If the camera is mounted rigidly on the arm top, the target on the image plane is a constant vector n_d. The image error is defined as follows:

$$\Delta n(t) = n(t) - n_d \tag{4}$$

The controller employed will force $\Delta n(t)$ to zero with the estimation of unknown camera parameters and the 3-D position of the power line. In this part, we use features of two power lines to ensure the unique solution of the off-line arm in joint space, shown as follows:

$$\dot{q}(t) = -\sum_{i=1}^{2}\left(\hat{L}_i^T(t) + \frac{1}{2}\hat{c}_i^T(t)\Delta n_i^T(t)\right)B_i\Delta n_i(t)$$

$$= -\sum_{i=1}^{2}\left(L_i^T(t) + \frac{1}{2}c_i^T(t)\Delta n_i^T(t)\right)B_i\Delta n_i(t)$$

$$+ \sum_{i=1}^{2}\left(L_i^T(t) + \frac{1}{2}c_i^T(t)\Delta n_i^T(t)\right)B_i\Delta n_i(t)$$

$$- \sum_{i=1}^{2}\left(\hat{L}_i^T(t) + \frac{1}{2}\hat{c}_i^T(t)\Delta n_i^T(t)\right)B_i\Delta n_i(t) \qquad (5)$$

$$= -\sum_{i=1}^{2}\left(L_i^T(t) + \frac{1}{2}c_i^T(t)\Delta n_i^T(t)\right)B_i\Delta n_i(t)$$

$$+ \sum_{i=1}^{2}Y_{li}(n_i(t),q(t))\Delta\theta_{li}(t)$$

Where the subscript notation i denotes the power line number, $\Delta\theta_{li}(t) = \hat{\theta}(t) - \theta_l(t)$ is the estimation error and $Y_{li}(n_i(t),q(t))$ is a regressor matrix which does not depend on the unknown parameters of camera and the power line.

The first part of the controller is used to force the image errors of the two lines to zero, while the second part is used to force the velocity of the estimation of unknown parameters to zero synchronously.

Different from previous work, we use features of two lines to control the robot, and we estimate the unknown parameters of the lines, respectively. The algorithm to estimate the unknown parameters is the same as that proposed in Wang et al. (2008b). The error vector is defined as

$$e_{li}(t_j) = \hat{b}_i(t_j) - \left(\hat{b}_i^T(t_j)n_i(t_j)\right)n_i(t_j)$$

$$= \hat{b}_i(t_j) - b_i(t_j) + (b_i^T(t_j)n_i(t_j))n_i(t_j)$$

$$- \left(\hat{b}_i^T(t_j)n_i(t_j)\right)n_i(t_j) \qquad (6)$$

$$= W_{li}(n_i(t_j),q(t_j))\Delta\hat{\theta}_{li}(t_j)$$

The adaptive algorithm to update the estimation of the unknown parameters is defined as (15).

B_i, Γ and K_3 are positive-definite gain matrices.

3.3 Stability Analysis

To prove the asymptotic stability of the proposed controller and adaptive algorithm, we introduce the following positive function (16).

Multiplying the $\dot{q}^T(t)$ from left to (5), we obtain the Eq. (17).

Multiplying the $\Delta\theta_{li}^T(t)$ from left to (15), we obtain

$$\Delta\theta_{li}^{\mathrm{T}}(t)\boldsymbol{\Gamma}\Delta\dot{\theta}_{li}(t) = -\Delta\theta_{li}^{T}(t)\boldsymbol{Y}_{li}(\boldsymbol{n}_i(t), \boldsymbol{q}(t))\dot{\boldsymbol{q}}(t)$$
$$- \sum_{j=1}^{k} \boldsymbol{e}_{li}^T(t_j)\boldsymbol{K}_3\boldsymbol{e}_{li}(t_j) \tag{7}$$

Differentiating (16) and combining the Eqs. (17) and (7) result in (18).

Obviously, we can obtain from (18): $\dot{V}(t) \leq 0$. From (16), we know $V(t)$ is a positive function, so $V(t)$ has a finite limit as $t \to \infty$. Hence, $\Delta\boldsymbol{n}_i(t)$, $\Delta\dot{\theta}_{li}(t)$, $\dot{\boldsymbol{q}}(t)$ and $\boldsymbol{e}_{li}(t_j)$ are bounded. From (15), we known $\frac{\mathrm{d}}{\mathrm{d}t}\hat{\theta}_{li}(t)$ is bounded, which means $\dot{\boldsymbol{e}}_{li}(t_j)$ and $\ddot{\boldsymbol{q}}(t)$ are also bounded. The derivative of $\dot{V}(t)$ is as follows

$$\ddot{V}(t) = -2\ddot{\boldsymbol{q}}^T(t)\dot{\boldsymbol{q}} - 2\sum_{i=1}^{2}\sum_{j=1}^{k} \dot{\boldsymbol{e}}_{li}^T(t_j)\boldsymbol{K}_3\boldsymbol{e}_{li}(t_j) \tag{8}$$

So $\ddot{V}(t)$ is also bounded. From Barbalat's Lemma, we can conclude

$$\lim_{t\to\infty} \dot{\boldsymbol{q}}(t) = 0 \tag{9}$$

$$\lim_{t\to\infty} \boldsymbol{e}_{li}(t_j) = 0 \tag{10}$$

$$\lim_{t\to\infty} \dot{\hat{\theta}}_{li}(t) = 0 \tag{11}$$

In order to prove the convergence of the image error, we rewrite the controller in (5) and obtain (19), (20) and (21) by considering the invariant set of the system when $V(t) = 0$. Under the control of the new controller and the adaptive rule for parameters estimation, the joints velocities are convergent to zero. The dimension of the Jacobian matrix $\boldsymbol{J}(\boldsymbol{q})$ is 6×4 and it is reasonable to assume the rank of $J(q)$ is 4. Hence the first part of Eq. (19), $(\boldsymbol{J}(\boldsymbol{q}) \quad \boldsymbol{J}(\boldsymbol{q}))$, has a rank of 4. When a sufficient number of images are used to estimate the unknown parameters, the matrices in (20) and (21) both have a rank of 2. Thus the second part of Eq. (19), $\begin{pmatrix} A & 0 \\ 0 & B \end{pmatrix}$, has a rank of 4. Since $\boldsymbol{n}_1(t)$ and $\boldsymbol{n}_2(t)$ are both unit vector, the third part of Eq. (19), $\begin{pmatrix} B_1\Delta\boldsymbol{n}_1(t) \\ B_2\Delta\boldsymbol{n}_2(t) \end{pmatrix}$, has four autonomous variables. Combining Eq. (19), we can conclude that $\lim_{t\to\infty} \Delta\boldsymbol{n}_i(t) = 0$.

$$c(t) = n(t)F(t) \begin{pmatrix} R(t) & 0 \\ 0 & R(t) \end{pmatrix} J(q) \tag{12}$$

$$L(n(t), q(t), \theta_l) = (I_{3\times3} - n(t)n^T(t))n^{-1}(t)c(t) \tag{13}$$

$$F(t) = \left(sk\{\Omega R(t)^b u\}\Omega. \begin{array}{l} -sk\{\Omega R(t)^b u\}\Omega sk\{R(t)^b X_{p1} + \xi(t)\} \\ + sk\{\Omega(R(t)^b X_{p1} + \xi(t)) + \chi\}\Omega sk\{R(t)^b u\} \end{array} \right) \tag{14}$$

$$\frac{d}{dt}\hat{\theta}_{li}(t) = -\frac{1}{\Gamma}\left\{ Y_{li}(n_i(t), q(t))\dot{q}(t) + \sum_{j=1}^{k} W_{li}^T(n_i(t_j), q(t_j))K_3 e_{li}(t_j) \right\} \tag{15}$$

$$V(t) = \frac{1}{2}\left\{ \sum_{i=1}^{2} \|b_i(t)\|\Delta n_i^T(t)B\Delta n_i(t) + \sum_{i=1}^{2} \Delta\hat{\theta}_{li}^T(t)\Gamma\Delta\hat{\theta}_{li}(t) \right\} \tag{16}$$

$$\dot{q}^T(t)\dot{q}(t) = -\dot{q}^T(t)\sum_{i=1}^{2}\left(L_i^T(t) + \frac{1}{2}c_i^T(t)\Delta n_i^T(t) \right)B\Delta n_i(t)$$
$$+ \dot{q}^T(t)\sum_{i=1}^{2} Y_{li}(n_i(t), q(t))\Delta\theta_{li}(t) \tag{17}$$

$$\dot{V}(t) = \sum_{i=1}^{2}\Delta\theta_{li}^T(t)\Gamma\Delta\dot{\theta}_{li}(t) + \sum_{i=1}^{2}\frac{1}{2}\Delta n_i^T(t)B\Delta n_i(t)\frac{d}{dt}\|b_i(t)\| + \sum_{i=1}^{2}\|b_i(t)\|\Delta n_i^T(t)B\Delta\dot{n}_i(t)$$
$$= -\dot{q}^T(t)\dot{q}(t) - \sum_{i=1}^{2}\sum_{j=1}^{k} e_{li}^T(t_j)K_3 e_{li}(t_j) \tag{18}$$

$$\dot{q}(t) = -\sum_{i=1}^{2}\left(\hat{L}_i^T(t) + \frac{1}{2}\hat{c}_i^T(t)\Delta n_i^T(t) \right)B_i\Delta n_i(t)$$
$$= -\left(J^T(q) \quad J^T(q) \right)\begin{pmatrix} A & 0 \\ 0 & B \end{pmatrix}\begin{pmatrix} B_1\Delta n_1(t) \\ B_2\Delta n_2(t) \end{pmatrix} \tag{19}$$
$$= 0$$

$$A = \begin{pmatrix} R^T(t) & 0 \\ 0 & R^T(t) \end{pmatrix}\hat{F}_1^T(t)\left(I_{3\times3} - \frac{1}{2}(n_1(t) + n_{1d}(t))n_1^T(t) \right) \tag{20}$$

$$B = \begin{pmatrix} R^T(t) & 0 \\ 0 & R^T(t) \end{pmatrix}\hat{F}_2^T(t)\left(I_{3\times3} - \frac{1}{2}(n_2(t) + n_{2d}(t))n_2^T(t) \right) \tag{21}$$

4 Simulations

To verify the performance of the proposed controller for the new designed inspection robot, simulated test was carried out with MATLAB. The mechanism model of the robot is established with robotics toolbox for MATLAB, as shown in Fig. 3. The length of the arm form the center of wheel to the center of the joint for lifting and lowering is 0.7 m. The length of the link between joints for lifting and lowering of the two arms is 1.0 m. The pose of 3-D power line L1 is $\left(^b u1, ^b X_{p1}\right)$, $^b u1 = \begin{bmatrix} 1 & 0 & 0 \end{bmatrix}^T$ and $^b X_{p1} = \begin{bmatrix} 1 & 0.2 & 2 \end{bmatrix}^T$, while the pose of L2 is $\left(^b u2, ^b X_{p2}\right)$, $^b u2 = \begin{bmatrix} 1 & 0 & 0 \end{bmatrix}^T$ and $^b X_{p2} = \begin{bmatrix} 1 & -0.2 & 2 \end{bmatrix}^T$. The desired gesture of the robot is $q = \begin{bmatrix} 0 & 0 & 0 & 0 \end{bmatrix}$. Assuming that the pin-hole camera is mounted on arm II (right in Fig. 3) top, the calibration values of the intrinsic parameters of the camera are au = 650, av = 650, u0 = 250 and v0 = 250. The extrinsic parameters of the camera form the camera frame the center of wheel of arm II (right in Fig. 3) is

$$T = \begin{bmatrix} 1 & 0 & 0 & 0 \\ 0 & 1 & 0 & 0 \\ 0 & 0 & 1 & 0 \\ 0 & 0 & 0 & 1 \end{bmatrix}$$

The initial estimation of the poses of two 3-D power lines respect to the base frame are $^b u1 = \begin{bmatrix} 0.985 & 0 & 0.174 \end{bmatrix}^T$, $^b X_{p1} = \begin{bmatrix} 1 & 0 & 2 \end{bmatrix}^T$, $^b u2 = \begin{bmatrix} 0.985 & 0 & 0.174 \end{bmatrix}^T$ and $^b X_{p2} = \begin{bmatrix} 1 & 0 & 2 \end{bmatrix}^T$. The control gains are $B = 2$, $\Gamma = 0.00001$ and $K_3 = 100000$. The initial gesture of the robot is $q = \begin{bmatrix} -0.5236 & 0 & 0 & 0 \end{bmatrix}$.

Fig. 5 Trajectories of feature lines on the image plane

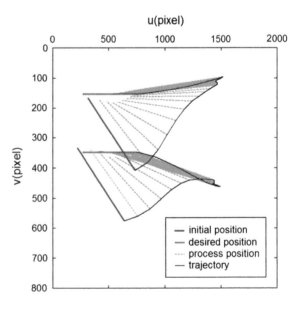

Fig. 6 3-D trajectory of the camera

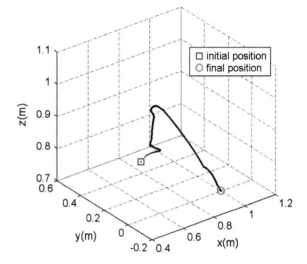

Fig. 7 Profile of the joints positions

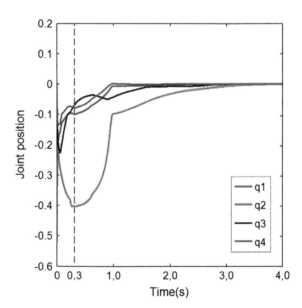

The initial estimation of the intrinsic parameters of the camera are au = 600, av = 600, u0 = 200 and v0 = 200. The initial estimation of the extrinsic parameters of the camera form the camera frame the center of wheel of arm II (right in Fig. 3) is

$$T = \begin{bmatrix} 0.996 & -0.087 & 0 & 0 \\ 0.087 & 0.996 & 0 & 0 \\ 0 & 0 & 1 & 0 \\ 0 & 0 & 0 & 1 \end{bmatrix}$$

The simulation results are demonstrated from Figs. 5, 6, 7, 8, 9, 10 and 11. The initial and desired positions of the feature lines are shown in Fig. 5. The 3-D trajectory of the camera is illustrated in Figs. 6 and 7 plots the profile of the joint position. Figure 8 shows the positions of the arm. Figure 8a is the initial position of the arm, Fig. 8b is the position at 0.3 s of the arm and Fig. 8c is the final arm of the simulation. Figures 9 and 10 plot the profiles of the normal vector errors of the feature lines $l1$ and $l2$, respectively. The profiles of Part of the estimated parameters are demonstrated in Fig. 11. The simulation results showed the expected asymptotic convergence of the errors to zero.

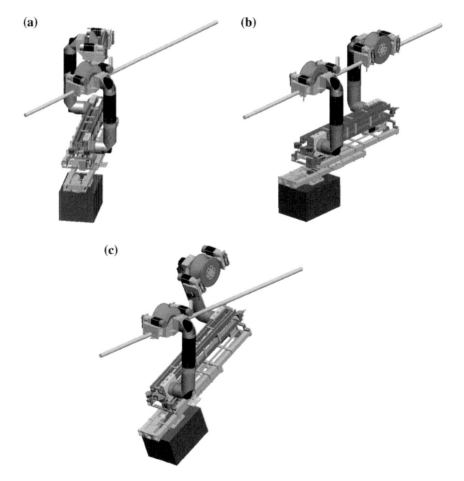

Fig. 8 Positions of the arm

Fig. 9 Normal vector errors of the feature line *l*1

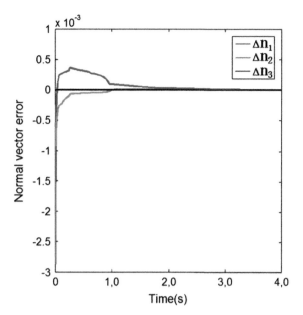

Fig. 10 Normal vector errors of the feature line *l*2

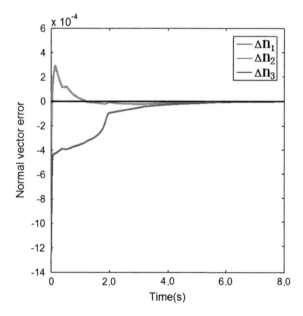

Fig. 11 Profile of the
estimated parameters

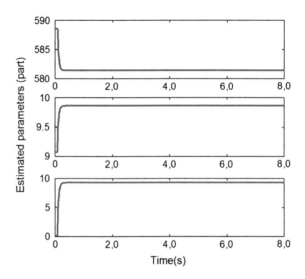

5   Conclusions

This paper presents a design of a new dual-arm power transmission lines inspection robot. Due to the novel mechanism, the robot can change its configuration to overcome obstacles on the transmission power lines. An adaptive visual servoing using line features of two power transmission lines for the robot in an uncalibrated eye-in-hand setup is proposed, thus solving the difficulty in autonomous grasping control of the robot's off-line arm in an autonomous obstacle-crossing. The controller uses the depth-independent interaction matrix to map the errors on the image plane onto the joint space of the power transmission lines inspection robot. The unknown intrinsic and extrinsic parameters and the 3-D positions of the transmission power lines are estimated during the control process by using an adaptive algorithm. We have proved the asymptotic convergence of the errors to zero by the Lyapunov method. Finally, through experiment on a simulation, the good performance of the visual servoing grasping method has been proved.

Acknowledgments This work was supported in part by Shanghai Rising-Star Program under Grant 14QA1402500, in part by the Natural Science Foundation of China under Grant 61105095 and 61473191.

References

Andreff N, Espiau B, Horaud R (2002) Visual servoing from lines. Int J Robot Res 21(8):679–699
Chaumette F, Hutchinson S (2006) Visual servo control. I. basic approaches. IEEE Robot Autom Mag 13(4):82–90
Chaumette F, Hutchinson S (2007) Visual servo control, part II: advanced approaches. IEEE Robot Autom Mag 14(1):109–118

Chen Z, Xiao H, Wu G (2006) Electromagnetic sensor navigation system of robot for high-voltage transmission line inspection. Transducer Microsys Technol 9:30–39

Fonseca A, Abdo R, Alberto J (2012) Robot for inspection of transmission lines. In: 2nd international conference on applied robotics for the power industry (CARPI), pp 83–87

Han SS, Lee JM (2008) Path-selection control of a power line inspection robot using sensor fusion. In: IEEE international conference on multisensor fusion and integration for intelligent systems, MFI 2008, pp 8–13

Liu Y, Wang H, Lam K (2005) Dynamic visual servoing of robots in uncalibrated environments. In: Proceedings of the 2005 IEEE international conference on robotics and automation, ICRA 2005, pp 3131–3136

Liu Y, Wang H, Wang C, Lam K (2006) Uncalibrated visual servoing of robots using a depth-independent interaction matrix. IEEE Trans Robot 22(4):804–817

Liu Y, Wang H, Chen W, Zhou D (2013) Adaptive visual servoing using common image features with unknown geometric parameters. Automatica 49(8):2453–2460

Montambault S, Pouliot N (2006) Linescout technology: development of an inspection robot capable of clearing obstacles while operating on a live line. In: IEEE 11th international conference on transmission & distribution construction, operation and live-line maintenance, ESMO 2006

Pouliot N, Montambault S (2009) Linescout technology: from inspection to robotic maintenance on live transmission power lines. In: IEEE international conference on robotics and automation, ICRA'09, pp 1034–1040

Slotine JJE, Li W (1987) On the adaptive control of robot manipulators. Int J Robot Res 6 (3):49–59

Tamura K, Kimura A (2010) Toward a practical robot for inspection of high voltage lines. J Inst Electr Eng Jpn 130:28–31

Wang W, Sun C (2011) Line-grasping control for the deicing robot on high voltage transmission line. J Mech Eng 9:004

Wang L, Wang H, Fang L, Zhao M (2007a) Visual-servo-based line-grasping control for power transmission line inspection robot. Robot 29(5):451–455

Wang H, Liu YH, Zhou D (2007b) Dynamic visual tracking for manipulators using an uncalibrated fixed camera. IEEE Trans Robot 23(3):610–617

Wang H, Liu YH, Wang Z (2008a) Uncalibrated dynamic visual servoing using line features. In: IEEE/RSJ international conference on intelligent robots and systems, IROS 2008, pp 3046–3051

Wang H, Liu YH, Zhou D (2008b) Adaptive visual servoing using point and line features with an uncalibrated eye-in-hand camera. IEEE Trans Robot 24(4):843–857

Wang W, Wu G, Bai Y, Xiao H, Yang Z, Yan Y, He Y, Xu X, Su F (2014) Hand-eye-vision based control for an inspection robot's autonomous line grasping. J Central South Univ 21:2216–2227

Zhang Y, Liang Z, Tan M, Ye W, Lian B (2007) Visual servo control of obstacle negotiation for overhead power line inspection robot. Robot 29(2):111–116

A Novel Design of Laser In-Frame Robot for Electron Cyber Scalpel Therapy and Its Rotary Joint

Jian-jun Yuan, Xi Chen, Chang-guang Tang and Kazuhisa Nakajima

Abstract This paper presents a conceptual design of laser in-frame robot for electron cyber scalpel therapy, and a detailed design of its rotary joint, particularly. This manipulator system is expected to be a tool, which transmits strong pulse laser beams from its base to tip, where the electron beams are stimulated for therapy, also controls and delivers such electron beams for precise treatment. Among all the research issues of radiosurgery robotics the rotary joint, which acts as the driving/guiding part and supporting structure, comes out to be extremely significant. With the novel proposal, the mentioned joint is designed with a big diameter vacuum hollow structure inside, to serve for the transmission channel of laser. Thus, all the driven components are outward distributed. Relative analysis or discussion is given to illuminate design goal, high positional precision, high rigidity and motion stability, low manufacturing cost, simple structure, and easy maintainability, also welcome for laser transmission.

Keywords Radiation therapy · Robot manipulator · Pulse laser beam · Rotary joint design

1 Introduction

The medical robots have been acting as very important equipment for many clinical applications. In addition, they have extended human being's capabilities and limitations in tremor reduction, repeatability, accuracy, for instance. These decades, umbers of robotic devices have been successfully utilized to the medical areas. Such

J. Yuan (✉) · X. Chen · C. Tang
Advanced Robotics Lab, Robotics Institute, Shanghai Jiao Tong University,
Shanghai 200240, China
e-mail: yuanjj@sjtu.edu.cn

K. Nakajima
Center for Relativistic Laser Science, Institute for Basic Science,
Gwangiu 500-712, Korea

© Zhejiang University Press and Springer Science+Business Media Singapore 2017   571
C. Yang et al. (eds.), *Wearable Sensors and Robots*, Lecture Notes in Electrical
Engineering 399, DOI 10.1007/978-981-10-2404-7_43

as, the famous PUMA 560 robot, was first used to place a needle for a brain biopsy in 1985, and in 1988, the PROBOT was used to perform prostatic surgery by Dr. Senthil Nathan (Kwoh et al. 1988). Some surgery robots such as Robodoc (Integrated Surgical Systems, Inc.) have had varying degrees of commercial success in medical area. The tele-operated robot, the da Vinci (Intuitive Surgical, Inc., USA), has allowed soft tissue procedures and has started benefiting from robotic enhancement. These systems have successfully made their way into cardiac and abdominal applications as an enhancement to traditional minimally invasive surgery (Olender et al. 2005). The Cyber Knife radiosurgery system (Accuray TM, Inc, USA), has pioneered the field of robotic radiosurgery by introducing the advantages of medical robots to perform more precise and accurate delivery of ionizing radiation.

Cyber Knife is a frameless robotic radiosurgery system used for treating benign tumors, malignant tumors, and other medical conditions.[1] The Cyber Knife system deploys a linac mounted on an agile robot and directed under image guidance for stereotactic radiotherapy using nonisocentric beam delivery (Dieterich and Gibbs 2011). Multiple small and very precise radiation beams are delivered and, as with other stereotactic systems, a small number of high-dose fractions are delivered (Murray and Robinson 2010). The Cyber Knife was said the first robotic device that a human was permitted to be present within a robot workspace. After that, several generations of the Cyber Knife robotic system have been successfully developed and equipped, since its initial prototype in 1990.[2]

This project of developing a novel laser in-frame robotic device for electron cyber scalpel therapy is to provide a revolutionary one which is significant different both in robotics and physics fields from the commercial radiation therapy systems. The Cyber Knife, for instance, is a simple combination of a traditional industrial manipulator and a small-sized X-ray linac attached to end tip of such manipulator. And it is a combination of robotic technology, image process, and medical treatment in its techniques. However, because of the great damage to normal tissue and the limited curative effect of using X-ray, the research of using electron beam or other heavy ion beam as substitute has received more and more observation. In physics field, researchers have already found that stable and powerful electron beam for therapy can be generated by focusing a compressed strong pulse laser into a small-sized laser electron accelerator (Nakajima 2000; Hidding et al. 2011). Our project proposal becomes clear, as shown in Fig. 1, a laser system, a compressor, an accelerator and a multi-degree of freedom robot manipulator are necessary.

The strong pulse laser is generated and transmitted to the base of the manipulator; then through its frame and joints, is finally delivered to the accelerator, where the electron beam is generated. The laser beams with a diameter no small than Ø50 mm has to be transmitted in vacuum before it is delivered to the accelerator. Therefore, in our robot proposal, all the frames and joints are designed in big-sized

---

[1]http://med.stanford.edu/neurosurgery/patient_care/radiosurgery.html.

[2]http://en.wikipedia.org/wiki/CyberKnife.

Fig. 1 Schematic view of the laser in-frame robot by SJTU

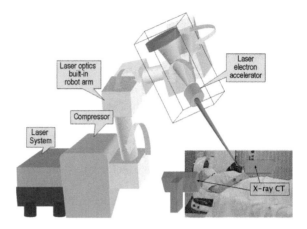

vacuum hollow structure; then, all of the driven components have to be outward distributed. Among them, the rotary joint that acts as the driving\guiding part and supporting structure comes out to be extremely significant.

The first step of this research is to design a new type of high-accuracy high-rigidity rotary joint to fulfill the above mentioned demand, and it must be low cost at the same time. At the beginning of the treatment, the joints shall find the location of the lesion determined by the computer. During the treatment, the irradiated point should be adjusted immediately once the data of the location of the lesion changes. This system must be real time and accurate positioning.

After detailed research and analysis on manipulator, outward load, the radiation oncology, and the motion scope of the electron beam, it is able to get the design parameters, such as the size, the rotational speed of the rotary joint. This rotary joint needs to have a vacuum channel inside, the diameter of which should be at least 80 mm to serve for the laser transmission. There is no such economical commercial mechanism that meets these requirements among the present industrial manipulators. Though the RV reducer and others may have both hollow inside structure and high accuracy, to reach all above-mentioned requirements, they must be custom made, which turns out to be really expensive.

Following of the paper is organized as five sections. Section 1 introduces the background of medical robotic and our research proposal. Section 2 introduces some overall structural design issues of the robotic system. Section 3 analyzes of the conventional design and present the design features. Section 4 proposes the design concept of the rotary joint. Finally, Sect. 5 concludes the paper.

2   Overall Structural Design Issues

2.1   *Robotic Design Analysis*

As an initial motion input, the robotic joint is directly related to the performance of the entire medical robot. The mechanical configurations of the manipulator are in many forms. The most common forms of the structure are described by their coordinate properties, as shown in Table 1. The choice of using what must base on operational requirements, such as working space, specified location and positioning accuracy, and so on. Therefore, selection of the best structure form should follow the analysis to the specific task.

The main features of typical structures for selection (Yongtao 2012)

2.1.1   Cylindrical Coordinate Robot (RPP)

It is mainly a combination of two translation joints and one rotation joint. In this way, the robot's working space forms a cylindrical surface.

2.1.2   Spherical Coordinate Robot (RRP)

The robot is like a turret of the tank. The manipulator can telescopically inside and outside, swing in the vertical plane and rotate around the base in the horizontal plane. This robotâ€™s working space forms part of a sphere.

2.1.3   Articulated Spherical Coordinate Robot (RRR)

The robot consists of the base, and several serial linked arms, whose joints are all parallel mounted. Basically, the upper arms are working in a vertical plane. The rotary motion in the horizontal plane can be either provided by the shoulder joint or the rotation of the base. The robot's working space occupies most of the sphere, and so it is called the articulated spherical robot.

According to the work environment and the tasks required by this laser in-frame robot, the robot arm design requirements can be summarized as follows:

1. Able to achieve high endpoint positioning accuracy;
2. In the same structure size, the working space should be as large as possible;
3. To achieve the same working space, the arm occupies as least space as possible;
4. The structure, especially the joints and frames, must be easy to transmit big diameter laser beams.

Table 1 Typical overall manipulator structure

Manipulator structure	Cylindrical coordinate	Spherical coordinate	Articulate spherical
Configuration	RPP	RRP	RRR
Working space	Medium	Big	Big
Occupied space positioning	Medium	Small	Small
Accuracy	High	Low	Medium
Motion intuitive	Good	Bad	Good

2.2 Robotic Overall Structure Design

Comparing to the operating characteristics of the articulated spherical coordinate robotic structure (RRR), which is able to obtains flexibility, large working space, and be as compact as possible to take up less space and all the driven components can be outward distributed, this kind of configuration is most suitable and recommended as a whole manipulator structure.

This type of robot arm motion is similar to the human's, the designed robot have three arms. After detail analysis, for well controlling of the laser beam, totally at least five degrees of freedom (DOFs) are needed. The movements of the proposed manipulator include waist rotation, the elbow pitch, the shoulder pitch, as well as the tilt of the wrist with two directions y and z, as Fig. 2 shown.

3 The Design Principle

3.1 Features of the Rotary Joint Design Task

The main objective of this project is to develop a novel laser in-frame robotic manipulator with five DOFs in which the laser beam is guided to negotiate inside. To fulfill the medical application, such manipulator requires to be small and compact in structure, to ensure the quality and positional accuracy of the laser beam target and to be as low cost as possible. In addition, the manipulator must ensure an internal hollow and vacuum structure for implantation of optical path. This demands all the mechanical and electrical devices including servo motors, gearboxes, and sensors are all external distributed. The design becomes to be a very challenging task by comparing to traditional manipulators in order to realize the above objective and techniques.

Rotary joints are a common solution for multiple DOFs manipulator, and consequently multiple rotation speeds and payloads should be able to be provided. According to the treatment plan for this project, the accelerator at the end of the robotic manipulator is going to be positioned at any location within a sphere that keeps a certain distance from the lesion area. The multiple rotation speeds are

Fig. 2 A recommended
structure for laser in-frame
robot system

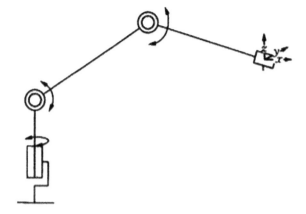

scheduled as following procedure: First, during the initialization mode of a surgery, the manipulator is expected a fast positioning speed to ensure the electron accelerators at the end quickly move to the patient's lesions nearby and ready for subsequent treatment. Secondly, a relatively low speed, under the servo control algorithm, is serving as the precise positioning of patients' lesions during the radiosurgery process mode.

During the design process of the rotary joint, we simplify the robot manipulator as a steel rigid beam with 1.5 m in length and 40 kg in mass, after the analysis of the actual operation condition is given. Also, the laser electron accelerator is simplified as a point with 80 kg in mass. Right below the steel beam is the cylindrical rotational waist with a height of 0.3 m and a diameter of 200 mm. With the above simplification, it becomes easy and convenient to calculate the approximately force onto the rotary joint and the velocity of both working condition and positioning condition. After detailed calculation, it is optimized that the reduction ratio of this rotary joint is 125.7. At the working condition, the velocity of the nut of the power screw is 50 mm/s, while at the positioning condition the velocity is 1000 mm/s.

3.2   Analysis of Conventional Design

For every robotic joint, one must consider how to load external drive to it, making its rotation a controllable and stable motion, and thus accuracy positioning of the electron accelerator at end of the manipulator. Usually there are two kinds of drive modes, the direct drive and indirect drive. Direct drive can achieve good static and dynamic stiffness, and significantly improve the bandwidth and the response speed of the system. While for indirect drive, a small driving torque can drive a relatively large inertia, easy to implement high-precision motion requirements (Li et al. 1995). For most devices, drivers often drive large inertia components. Usual the speed difference between the drive and the driven components is too large to apply direct drive. So in the actual situation, the use of indirect drive components is more common. However, the structure size of the indirect drives is larger than that of the direct ones. The structure is more complex and there is gap due to the reduction gears. Also there is a need to consider how to eliminate the impact of clearance to improve the control performance. Speed reducers are commonly used for indirect drive components. Reducer is mainly the mutual transmission of gears at all levels to achieve the indirect drive torque amplification and speed reduction by the meshing of the pinion gears. Traditional reducers are mainly the following:

1. Fixed axis gear drive reducer structure is simple with stable transmission ratio, high transmission efficiency, reliable operation, and long service life. However, its transmission ratio depends on the number of teeth of two gears. In order to obtain a larger transmission ratio, the diameter difference of the two gears will be relatively large, making the reducer bulky with cumbersome structure, while the excessive difference of teeth number could easily lead to gear failure.
2. Worm drive reducer is compact, smooth, and at the same time the transmission ratio is relatively large, and reverse self-locking can be achieved to protect the whole structure. However, sliding velocity between helical teeth on the worm and worm gear meshing tooth surface is so large, resulting in severe tooth wear, large heat release and prone to glue pitting. And the efficiency of the worm drive is relatively low.
3. The planetary drive speed reducer is using the composite of rotation and revolution of one or several transmission components to reduce speed. Its structure is more compact because the gear diameter difference of planetary drive speed reducer is small. At the same time it can ensure a large transmission ratio. But its main problem is prone to internal tooth meshing interference. In addition, when planetary gears are working, the centrifugal force generated by centrifugal torque is difficult to strike a balance. Combining with its eccentric output structure, the transmission power and transmission efficiency of this reducer is not ideal.
4. The harmonic gear drive reducer is made of elastic material of the flexible wheel with the wave generator to produce a controlled elastic deformation and the meshing of the drum to transmit motion. Harmonic gear drive reducer does not need the eccentric output structure and therefore the structure is simple. Its transmission efficiency is relatively high. But the requirements for material of the flexible wheel for meshing are relatively high, which make the processing more difficult (Seneczko 1984).

G. Hirzinger describes the development of light-weight manipulator, the use of special light weight reducer to replace the planetary gear reducer with the large reduction ratio. Through the high stiffness and torque transmission scheme high-precision end positioning and high dynamic performance can be achieved (Hirzinger 2000). However, similar to the above techniques, using the harmonic reducer or RV reducer in the field of robotics, the production costs are very high. In an environment with more emphasis on cost-effective, it is difficult to obtain a wide range of applications.

The proposed design in this paper, from the essence, is one kind of novel indirect drive of the manipulator. Because the followup manipulator device size is relatively large, the cumulative effect results in large errors from the rotary joint to the electron accelerator device at the manipulator end. So as a starting of the whole mechanical structure, the rotary joint should have a good speed reducing function, while maintaining high stiffness, high repeat accuracy, and positioning accuracy. Precision reducers on the market today do meet the accuracy requirements. However, the high prices will ultimately affect the future clinical application of this medical robotic system.

4 Proposed Rotary Joint Design

4.1 The Proposed Design Concept

Figure 3 gives the illustration of our proposed novel design concept of the rotary joint. As the Figure shown, the screw is installed in the fixation box, in which there are two guide rods passing through the slider. One guide rod is on the upper side while the other on the other side. The screw nut is fixed in the slider. The rotary platform is installed in the base. The distance between the rotary platform and the fixation box is the radius of the rotary platform. The sidewall of the rotary platform is fixed with a steel wire, which is winding to the back side of the slider and fixed there. The steel wires coming from two other sides constitute the steel wire dual. When fixing the steel wires, preload should be provided to make them in the tension state. A manipulator is installed in the rotary platform. Through the rotational movements offered by the steel wire dual, the whole manipulator structure can rotate clockwise and counterclockwise.

In this design, the crucial structural components are shown in Fig. 4: Fixation Box-1, Screw-2, and Screw nut-3, Guide rod-41, Guide rod-42, The Slider-5, The Base-6, Rotary Waist-7. There is one DOF for the screw installed in the fixation box. The screw is driven by a motor that is mounted outside from the junction-21.

Fig. 3 The draft of the design concept

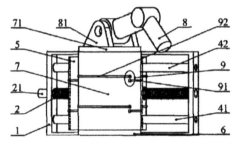

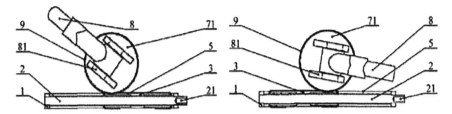

Fig. 4 Top view of the design, the *left* is running counterclockwise while the *right* is clockwise direction

Therefore, when the screw rotates along the axial direction, the slider with a screw nut-3 installed moving from left to right and from right to left along the guide rods. The guide rods not only make sure the slider is moving in a linear motion along the rail but also enhance the stiffness of the whole system. The rotary platform-71 above which is a manipulator-81 is installed in the base-6 by bearings. It rotates along its axis while the base keeps still. There is a fixed distance between the base-6 and the fixation box-1. The distance is the same as the radius of rotary waist-7. From the above figures, it is easy to understand how the steel wire dual (91 and 92) is installed and where to be preloaded.

Here is the principle: There is a source of power, usually provided by a motor, coming from outside. The motor is connected to the screw. Through a screw–nut combination, the screw nut installed in the slider is in a linear motion. So the slider is in the linear motion. The slider and the rotary platform are wrapped by two steel wires. When the slider is going in one direction, one steel wire is in a tension state while the other is in an elastic state. When the slider is going in the opposite direction, the first steel wire is in an elastic state and the other in a tension state. The two moving direction of the slider are corresponding to the clockwise and counterclockwise movements of the rotary waist of the manipulator.

The proposed design has following merits:

1. The structure consists of common components, which is easy to purchase in low price. The structure is compact, convenience to install and disassembly, and manufacturing cost is relatively low.
2. The transmission efficiency is high. It can transfer large transmission torque, with large reduction ratio, and meanwhile guarantee a large hollow structure.
3. With the screw drives, it has a high mechanical gain and transmission uniformity, easy to self-lock, accurate and stable. The screw–nut drive structure ensures a high repeat positioning accuracy and trajectory accuracy, and achieving seamless transmission.
4. Between the reducer and rotary waist, the steel wire ensure the connection a high transmission stiffness in the horizontal direction while a good flexibility in other supple direction, which reduces vibration and noise thus extending the life of bearings in the waist.

In all, this paper proposes a novel rotary joint with low cost but high positioning accuracy, thereby reducing the whole cost of the package of this medical robotic system. The rotary joint is designed capable of withstanding the whole weight of the manipulator as well as the external load. Fast and accurate positioning to the pathological location determined by the treatment plan is promised. During the course of treatment, the end of the robot manipulator is therefore capable of keeping/changing its orientation at any time according to the positioning data, and adjusting the irradiation site to achieve the accuracy of the real-time treatment.

4.2   The Prototype and Some Details

The rotary joint of this research is fastening on the back of an outside box, as Fig. 5 shown. Through the steel belt, the rotary waist is connected with the slider so that it will have a specific related motion. Through the waist rotation driven by the steel belt, the robot arm installed above will follow the movement, and thus accomplish the initial positioning. It can be seen from the figure, by controlling each rotational degree of freedom of the robot arm, it can achieve precise positioning of the terminal electron accelerators, which prepares for the subsequent laser treatment. The design of the robotic arm part structure will be in the future research. Herein shown is only the schematic.

Fig. 5 A whole model of the manipulator system with the design joint

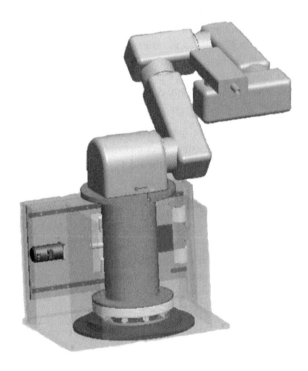

Fig. 6 Front view of the design joint

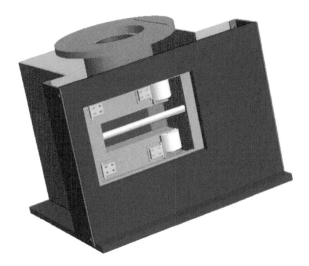

In this design, the steel wires are originally scheduled for motion transmission between the slider and the rotary waist. However, after the actual depth understanding, a steel wire is essentially many small steel wires spiral wound together. When under large tension, it is prone to suffer deformation, more specifically, the tensile deformation, which affects the accuracy of the system. So the steel belt is used as the motion transmission device to avoid this problem.

Illustrated from Figs. 6 and 7, the original guide rod is replaced by linear guide. This adjustment is based on the following considerations. First, the guide rod mounting accuracy is relatively poor. In addition, if using the rod as a slider guide, installation guide rod and slider become a problem. Due to the gap between the guide rod and the slider, it will not only cause tiny vibration of the slider during the movement, and thus exacerbate the loss of the device, but also increase the system noise. Moreover, due to sliding friction between the slider and the guide bar, both the guide and the slide wear quickly, and then cause energy dissipation. Therefore, the design of linear slide is to avoid or mitigate these adverse effects.

Shown as Fig. 8, the rotary joint structure in this design is divided into two structural units, the base part and the slider part. The slider part is driven by the ball screw, which is connected to the AC servo motor. Two sets of linear guides are utilized with up and down structure, and four small steel drums are installed into four corners respectively, through which, the steel belts are wounded and driven. This structure ensures the driven force distributed evenly and stable.

The rotary joint is driven to rotate without any backlash by considering the steel belts are preloaded with certain tension forces. Therefore, the whole rotary joint generates an extremely precise mechanism, but with very economical feature.

Fig. 7 Rear view of the
design joint

Fig. 8 The details of the
slider

4.3   Simulation Analysis

In this section, we resort to animation to the performance of the proposed design and demonstrate that it can indeed realize its function.

No mutual interference between components is noticed during the animation process, as shown in Fig. 9. The total structure can be operated normally. So the accelerator at the end of the manipulator can be well controlled.

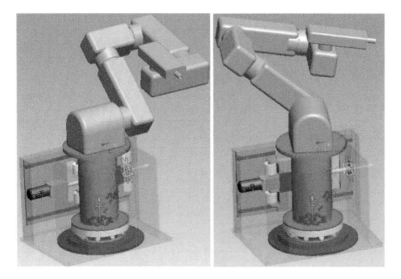

Fig. 9 Two simulation positions in the animation

5 Conclusion

The paper proposes a conceptual design of rotary joints for laser in-frame robot for electron cyber scalpel therapy. Many features and details have been discussed. The further work will be designing a manipulator with a large hollow structure. Experiments will be carried out after the whole system is set up and the structures are manufactured. More researches will be done in the very near future, such as how to control the laser beam inside the vacuum of the manipulator. The results will be updated and more paper will come up about this issue.

Acknowledgments This work is supported by National Natural Science Foundation of China (NSFC) No. 51175324.

References

Dieterich S, Gibbs IC (2011) The cyber knife in clinical use: current roles, future expectations. In: Meyer JL (ed) IMRT, IGRT, SBRT, vol. 43, pp 181â€"194
Hirzinger G (2000) A mechatronics approach to the design of light-weight arms and multifingered hands. In: IEEE international conference on robotics and automation, ICRA
Hidding B, Konigstein T, Willi O, Rosenzweig JB, Nakajima K, Pretzler G (2011) Laser-plasma-accelerators-a novel, versatile tool for space radiation studies. Nucl Instrum Methods Phys Res A
Kwoh YS, Hou J, Jonckheere EA, Hayall S (1988) A robot with improved absolute positioning accuracy for CT guided stereotactic brain surgery. IEEE Trans Biomed Eng 35(2):153–161

Li S, Zao K, Wu S, Zhang F, Liu Q (1995) The general design and the technique three-axis simulator. J Astronaut 16(2)

Murray LJ, Robinson MH (2010) Radiotherapy: technical aspects. Medicine 39(12):698–704

Nakajima K (2000) Recent progress on laser acceleration. Nucl Instrum Methods Phys Res A 455:140–147

Olender D, Kilby W, Schulz RA (2005) Robotic whole body stereotactic radiosurgery: clinical advantages of the CyberKnife integrated system. Int J Med Robot Comput Assist Surg 28–39

Seneczko M (1984) Machine design, gearless speed reducers

Yongtao S (2012) Urban-oriented EOD's complex wheeled mobile carrier and robot arm mechanical design and optimazition. Shanghai Jiao Tong University Master of Science thesis, pp 18–19

Index

© Zhejiang University Press and Springer Science+Business Media Singapore 2017
C. Yang et al. (eds.), *Wearable Sensors and Robots*, Lecture Notes in Electrical
Engineering 399, DOI 10.1007/978-981-10-2404-7

Printed by Printforce, the Netherlands